Lecture Notes in Physics

Lecture Notes in Physics

Edited by H. Araki, Kyoto, J. Ehlers, München, K. Hepp, Zürich
R. Kippenhahn, München, H. A. Weidenmüller, Heidelberg
and J. Zittartz, Köln

191

Photon Photon Collisions

Proceedings of the Fifth International Workshop
on Photon Photon Collisions
held at the Rheinisch-Westfälische Technische
Hochschule Aachen, April 13–16, 1983

Edited by Ch. Berger

Springer-Verlag
Berlin Heidelberg GmbH 1983

Editor

Ch. Berger
I. Physikalisches Institut der
Rheinisch-Westfälischen Technischen Hochschule Aachen
Sommerfeldstraße, D-5100 Aachen 1

ISBN 978-3-540-12691-1   ISBN 978-3-540-38683-4 (eBook)
DOI 10.1007/978-3-540-38683-4

2153/3140-543210

Preface

The fifth International Workshop on Photon Photon Collisions was held at Aachen from 13 to 16 April 1983. Continuing a tradition started only two years ago at the Paris Conference we had, besides the plenary talks, experimental and theoretical parallel sessions. These discussion sessions proved to be a remarkable success.

'Two Photon Physics' plays an important role in testing quantum chromodynamics, our candidate theory of strong interactions. A large amount of new data has been presented at the workshop and many bothersome theoretical problems are now more thoroughly understood. I hope that this continuous experimental and theoretical effort leads to a deeper understanding of the underlying strong interaction dynamics.

The Aachen Conference was financially supported by the German Federal Government and by the State of Nordrhein Westfalen. The administrational, technical, and financial support by the Technische Hochschule Aachen and especially the 1. Physikalisches Institut is gratefully acknowledged. I have also to thank the members of the International Advisory Committee for their work in preparing the conference. Last but not least, I want to express my gratitude to the speakers for their excellent talks.

Ch. Berger
July 30, 1983

International Advisory Committee

CONTENTS

Photon–Photon Production in Hadron–Hadron Collisions[*][†]

R. D. Field
Particle Theory Group, Department of Physics,
University of Florida, Gainesville, FL 32611

May 1983

ABSTRACT

Quantum Chromodynamic (QCD) estimates are made for the large transverse momentum production of single and double photons in pp, $\bar{p}p$, and $\pi^{\pm}p$ collisions. In addition to the pure QED annihilation term $q\bar{q} \to \gamma\gamma$, it is found that the QCD induced gluon-gluon subprocess $gg \to \gamma\gamma$, is an important source of double photons. Photon Bremsstrahlung contributions are also examined.

[*]Invited talk presented at the 5th International Workshop on Photon–Photon Collisions, 13–16 April 1983, Aachen, Germany.

[†]Work supported by the U.S. Department of Energy under contract No. DE–AS–05–81–ER40008.

Usually at a topical conference on a specific subject at least one speaker is asked to discuss related topics. Here I will examine the subject of photon-photon production at this workshop on photon-photon collisions. Fig. l(a) shows the dominate Born constituent subprocess, $\gamma\gamma \to q\bar{q}$, for the reaction $e^+e^- \to e^+e^-$+hadrons. The inverse process, $q\bar{q} \to \gamma\gamma$, shown in Fig. l(b) results in photon-photon production in hadron-hadron collisions. The QED annihilation term, $q\bar{q} \to \gamma\gamma$, together with the QCD induced gluon-gluon subprocess, $gg \to \gamma\gamma$, are the major sources of double photons at large p_T and produce quite spectacular events in which a large p_T proton on one side is balanced by another photon on the "away-side" with roughly the same transverse momentum. I present here estimates made by E. Berger, E. Braaten, and myself [1] on single and double photons in $\pi^\pm p$, pp, and $\bar{p}p$ collisions [2]. Although the rates for the production of two photons is small, experimental study of the systematics of these processes will provide valuable information on the size of the strong interaction coupling constant, $\alpha_s(Q)$, and on the charge of the quark [3]. Also, knowledge of the gluon distributions within hadrons and of the effective transverse momentum of partons in the hadrons can be gained.

Let us begin our examination of photon-photon production by considering the production of single and double photons in $\pi^\pm p$ collisions. Except for couplings and color factors the differential cross section for $q\bar{q} \to \gamma\gamma$ is identical to that for $q\bar{q} \to \gamma g$. The ratio is given by

$$\frac{d\hat{\sigma}/d\hat{t}(q\bar{q} \to \gamma\gamma)}{d\hat{\sigma}/d\hat{t}(q\bar{q} \to \gamma g)} = \frac{3}{4} \left(\frac{\alpha_{em}}{\alpha_s}\right) e_q^2, \tag{1}$$

where e_q is the quark charge. If we are naive and assume that single and double photons are produced solely by these annihilation subprocesses and further consider only valance quarks then we arrive at several interesting

(a) Photon-Photon Collisions:

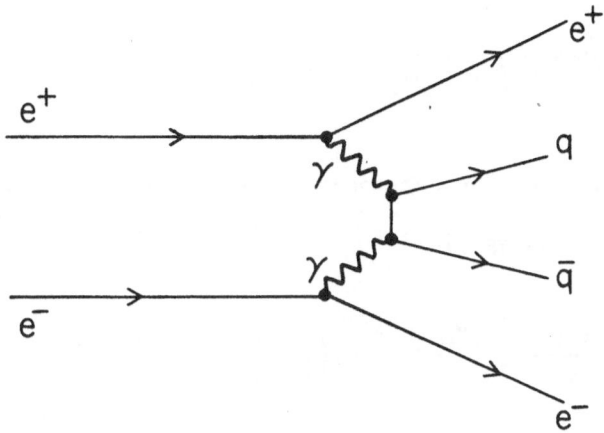

(b) Photon-Photon Production:

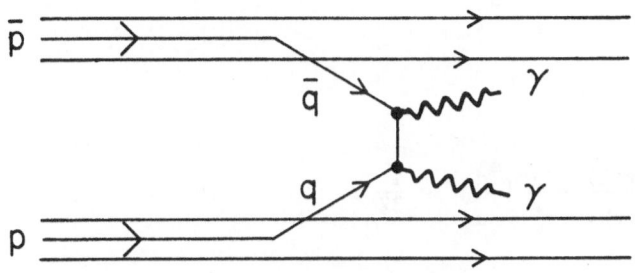

Fig. 1. (a) Born term for the production of quark jets from the collision of two photons in the reaction $e^+e^- \to e^+e^- q\bar{q}$. (b) Born term for the production of two photons by quark-antiquark annihilation in hadron-hadron collisions.

predictions. For $\pi^- p$ collisions we arrive at

$$\frac{Ed\sigma/d^3p(\pi^- p \to \gamma\gamma + X)}{Ed\sigma/d^3p(\pi^- p \to \gamma + X)} = \frac{3}{4} \left(\frac{\alpha_{em}}{\alpha_s}\right) e_u^2 = \frac{1}{3} \left(\frac{\alpha_{em}}{\alpha_s}\right), \tag{2a}$$

where I have only included the subprocesses $\bar{u}u \to \gamma\gamma$ and $\bar{u}u \to \gamma g$. Similarly the subprocesses $\bar{d}d \to \gamma\gamma$ and $\bar{d}d \to \gamma g$ yield

$$\frac{Ed\sigma/d^3p(\pi^+ p \to \gamma\gamma + X)}{Ed\sigma/d^3p(\pi^+ p \to \gamma + X)} = \frac{3}{4} \left(\frac{\alpha_{em}}{\alpha_s}\right) e_d^2 = \frac{1}{12} \left(\frac{\alpha_{em}}{\alpha_s}\right) \tag{2b}$$

for $\pi^+ p$ collisions. The added approximation that there are twice as many u quarks than d quarks within the proton gives

$$\frac{Ed\sigma/d^3p(\pi^+ p \to \gamma + X)}{Ed\sigma/d^3p(\pi^- p \to \gamma + X)} \approx \frac{1}{2} \left(\frac{e_d}{e_u}\right)^2 = \frac{1}{8}, \tag{3a}$$

and

$$\frac{Ed\sigma/d^3p(\pi^+ p \to \gamma\gamma + X)}{Ed\sigma/d^3p(\pi^- p \to \gamma\gamma + X)} \approx \frac{1}{2} \left(\frac{e_d}{e_u}\right)^2 = \frac{1}{32}. \tag{3b}$$

Our naive reasoning leads to a π^- beam that is 32 times more efficient at producing double photons and 8 times more efficient at producing single photons than a π^+ beam. Furthermore, we see from (2) that, in this approximation, the double to single photon ratio provides a direct measure of the strong interaction coupling constant, α_s, and that one expects this ratio to be about 0.01 for $\pi^- p$ collisions and 4 times smaller for $\pi^+ p$ collisions (with $\alpha_s \approx 0.25$). This would mean that for $\pi^- p$ collisions one would expect that, on the average, one out of every 100 large p_T photon triggers are balanced on the "away-side" by a photon of roughly the same p_T. If there were

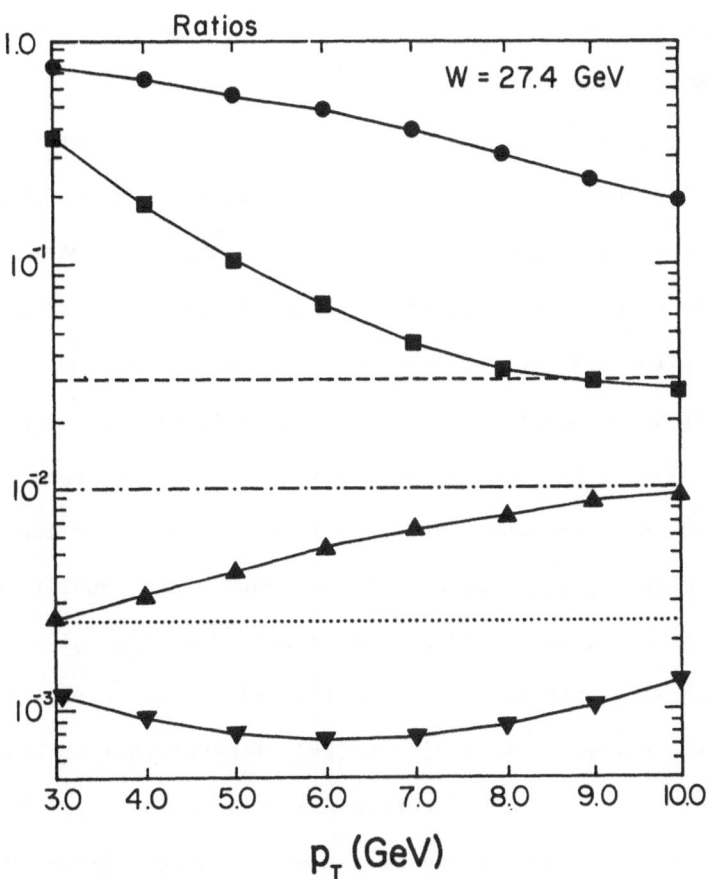

Fig. 2. Beam ratios and double to single photon ratios in πp collisions at $W=27.4$ GeV and $\theta_1=90^0$: $(\pi^+/\pi^-)p\rightarrow\gamma_1+X$ (solid dots); $(\pi^+/\pi^-)p\rightarrow\gamma_1\gamma_2+X$ (solid squares); $\pi^-p\rightarrow(\gamma_1\gamma_2/\gamma_1)+X$ (up pointing triangles); $\pi^+p\rightarrow(\gamma_1\gamma_2/\gamma_1)+X$ (down pointing triangles). Also shown are the naive estimates of (2) and (3): $\frac{1}{2}(e_d/e_u)^2$ (dashed line); $\frac{3}{4}(\alpha_{em}/\alpha_s)e_u^2$ (dot-dashed line); $\frac{3}{4}(\alpha_{em}/\alpha_s)e_d^2$ (dotted line).

no internal transverse momentum of the quarks within the initial hadrons the p_T of the two photons would exactly match.

For the region of x_T ($x_T = 2P_T/W$) accessible to experiment, the situation is more complicated than the above naive estimates. There are many other sources of single and double photons. Fig. 2 shows the beam ratios (3) and the double to single photon ratios (2) for πp collisions at W=27.4 GeV and $\theta = 90^0$ after all the contributions in Table 1 have been included. The naive estimates are approached only at extremely high p_T values. At low p_T values π^+/π^- beam ratio for the production of double photons (3b) is much closer to one and approaches 1/32 only at $p_T > 9$ GeV. At low p_T the gluon-gluon subprocess $gg \rightarrow \gamma\gamma$ is important and contributes equally to both π^+ and π^- collisions. At low and intermediate p_T values the double to single photon ratio for both $\pi^- p$ and $\pi^+ p$ collisions is significantly smaller than our naive estimates in (2a) and (2b), respectively. This is due to additional single photon contributions in this region from the Compton term, $gq \rightarrow \gamma q$, and from photon Bremsstrahlung. Table 2 lists the contributions to single and double photon production in $\pi^- p$ and $\pi^+ p$ collisions at W=27.4 GeV, $\theta = 90^0$, and $p_T = 4$ GeV. At this p_T the double to single photon ratio for $\pi^- p$ collisions is only about 0.003 considerably less than the value of 0.01 estimated from (2a). For $\pi^+ p$ collisions the $\gamma\gamma/\gamma$ ratio is only 0.0009.

The small $\gamma\gamma/\gamma$ ratio predicted in Fig. 2 means that one must collect thousands of single γ triggers in order to observe a few back-to-back double photons. This may be possible at the high intensity pion beam scheduled at Fermilab (Tevatron II).

Cahn and Gunion [4] showed that the subprocess $\gamma\gamma \rightarrow gg$ makes only a small contribution in photon-photon collisions. However, it is interesting to see

TABLE 1

The order in the strong, α_s, and electromagnetic, α_{em}, coupling constants of various subprocesses contributing to the production of single and double photons in hadron-hadron collisions.

Description	Order	Subprocess
Annihilation	$\alpha_{em}\alpha_s$	$q\bar{q} \to \gamma g$
Compton	$\alpha_{em}\alpha_s$	$qg \to \gamma q$
Examples of Single Bremsstrahlung	$\alpha_{em}\alpha_s^2$	$qq \to q(q \to \gamma)$
		$gq \to g(q \to \gamma)$
		$qg \to q(g \to \gamma)$
		$gg \to g(g \to \gamma)$
QCD Induced Gluon-Photon Coupling	$\alpha_{em}\alpha_s^3$	$gg \to \gamma g$
Pure QED Annihilation	α_{em}^2	$q\bar{q} \to \gamma\gamma$
Single Bremsstrahlung Contribution to Two Photon	$\alpha_{em}^2\alpha_s$	$qg \to \gamma(q \to \gamma)$
Double Bremsstrahlung	$\alpha_{em}^2\alpha_s^2$	$qq \to (q \to \gamma)(q \to \gamma)$
		$gq \to (g \to \gamma)(q \to \gamma)$
		$gg \to (g \to \gamma)(g \to \gamma)$
QCD Induced Gluon-Photon Coupling	$\alpha_{em}^2\alpha_s^2$	$gg \to \gamma\gamma$

TABLE 2

Contributions to the pion-proton single photon invariant cross section, $E d\sigma/d^3 p$ ($\mu b/GeV^2$) at CM energy $W=27.4$ GeV, $\theta = 90^0$, and $p_T = 4.0$ GeV from various constituent subprocesses. (Note that $A(-n)=A \times 10^{-n}$)

Subprocess	$\pi^+ p \to \gamma + X$		$\pi^- p \to \gamma + X$	
	$E d\sigma/d^3 p$	Fraction	$E d\sigma/d^3 p$	Fraction
1. $gq \to \gamma_1 q$	6.7(-5)	0.71	6.7(-5)	0.48
2. $q\bar{q} \to \gamma_1 g$	9.6(-6)	0.10	5.3(-5)	0.38
3. $gg \to \gamma_1 g$	2.3(-7)	2.4(-3)	2.3(-7)	1.7(-3)
4. $qq \to q(q \to \gamma_1)$	7.8(-6)	8.3(-2)	9.0(-6)	6.5(-2)
5. $gq \to g(q \to \gamma_1)$	8.6(-6)	9.1(-2)	8.6(-6)	6.2(-2)
6. $gq \to (g \to \gamma_1)q$	4.1(-7)	4.3(-3)	4.1(-7)	3.0(-3)
7. $gg \to \bar{q}(q \to \gamma_1)$	2.6(-7)	2.7(-3)	2.6(-7)	1.8(-3)
8. $gg \to g(g \to \gamma_1)$	3.0(-7)	3.1(-3)	3.0(-7)	2.1(-3)
9. $gg \to \gamma_1 \gamma_2$	3.6(-8)	3.8(-4)	3.6(-8)	2.6(-4)
10. $q\bar{q} \to \gamma_1 \gamma_2$	3.8(-8)	4.1(-4)	4.1(-7)	3.0(-3)
11. $gq \to \gamma_2(q \to \gamma_1)$	1.4(-8)	1.5(-4)	1.4(-8)	9.9(-5)
12. $q\bar{q} \to \gamma_2(g \to \gamma_1)$	4.7(-11)	5.0(-7)	3.5(-10)	2.5(-6)
13. $gg \to \gamma_2(g \to \gamma_1)$	4.6(-13)	4.9(-9)	4.6(-13)	3.3(-9)
Compton + Annihilation (1 + 2)	7.7(-5)	0.81	1.2(-4)	0.87
Bremsstrahlung (4 + 5 + 6 + 7 + 8)	1.4(-5)	0.19	1.9(-5)	0.13
Double Photon (9 + 10 + 11 + 12 + 13)	8.8(-8)	9.4(-4)	4.6(-7)	3.3(-3)
Total	9.4(-5)	1.00	1.4(-4)	1.00

that due to the large numbers of gluons within hadrons at low x_T, the inverse process $gg \to \gamma\gamma$ shown in Fig. 3(b) does make a significant contribution to the double photon rate in this region [5]. On the other hand, the QCD induced single photon subprocess $gg \to \gamma g$ in Fig. 3(a) makes a negligible contribution to the single photon rate. The ratio of these two subprocesses is given by

$$\frac{d\hat{\sigma}/d\hat{t}(gg \to \gamma\gamma)}{d\hat{\sigma}/d\hat{t}(gg \to \gamma g)} = \frac{12}{5} \frac{(\sum_{i=1}^{n_f} e_{q_i}^2)^2}{(\sum_{i=1}^{n_f} e_{q_i})^2} (\frac{\alpha_{em}}{\alpha_s}), \tag{4a}$$

where n_f is the number of flavors included in the quark loop [6]. For $n_f = 4$ (i.e. u, d, s, c) this ratio becomes

$$\frac{d\hat{\sigma}/d\hat{t}(gg \to \gamma\gamma)}{d\hat{\sigma}/d\hat{t}(gg \to \gamma g)} = (\frac{20}{3})(\frac{\alpha_{em}}{\alpha_s}) \approx \frac{1}{5} \tag{4b}$$

with $\alpha_s = 0.25$. The $gg \to \gamma\gamma$ contribution is of order $\alpha^2 \alpha_s^2$ (see Table 1) and thus down by two powers of α_s from the pure QED quark-antiquark annihilation diagram $q\bar{q} \to \gamma\gamma$. Nevertheless, the large numbers of gluons within hadrons at small x makes this subprocess an important source of double photons. Table 2 and Figs. 4 and 5 show that the gluon-gluon annihilation term represents 41% of the double photon rate for $\pi^+ p$ and 8% for $\pi^- p$ collisions at W=27.4 GeV and $p_T = 4.0$ GeV. In pp collisions at W=63 GeV and $p_T = 6$ GeV Table 3 and Fig. 6 show that the $gg \to \gamma\gamma$ term makes up about 30% of the double photon rate.

Figs. 4 and 5 and Table 2 give the single and double photon rates for $\pi^- p$ collisions at W=27.4 GeV. The results for pp and $\bar{p}p$ collisions at W=63 GeV are presented in Fig. 6-8 and Table 3. One expects a larger single and double photon yield in $\bar{p}p$ collisions over pp collisions due to the

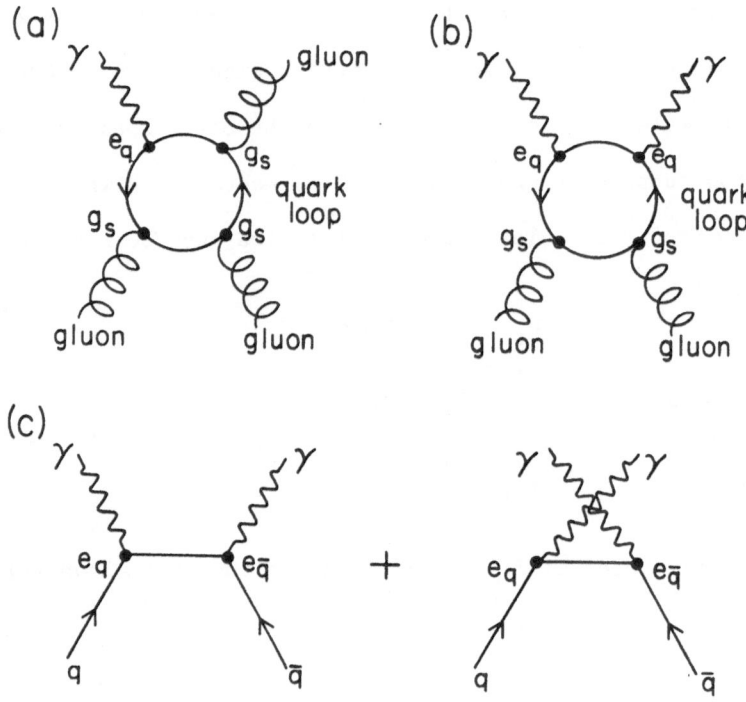

Fig. 3. (a) QCD induced single photon subprocess gg→γg. (b) QCD induced double photon subprocess gg→γγ. (c) Pure QED quark-antiquark annihilation q̄q→γγ.

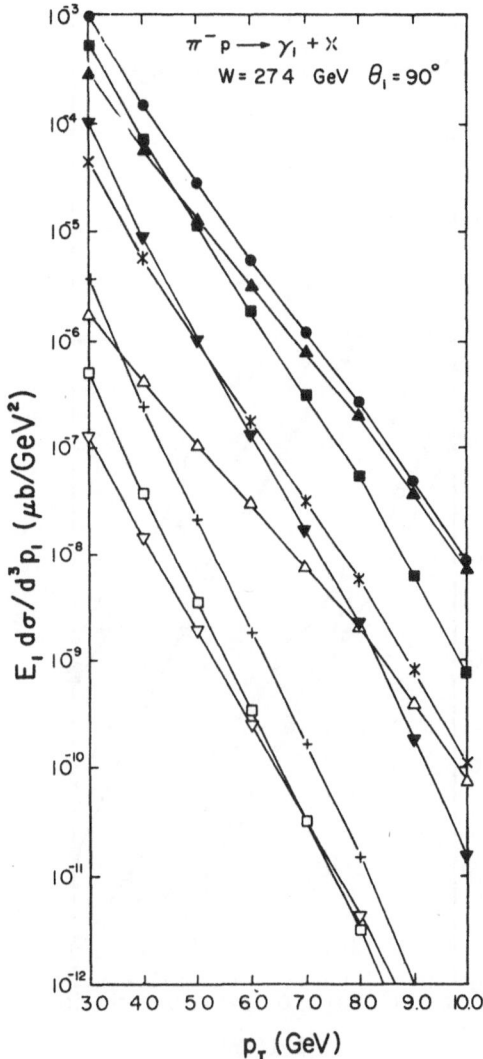

Fig. 4. Invariant cross section, $E_1 d\sigma/d^3 p_1$, for the production of photons in $\pi^- p$ collisions at W=27.4 GeV and $\theta_1 = 90^0$: total (solid dots); $gq \rightarrow \gamma_1 q$ (solid squares); $q\bar{q} \rightarrow \gamma_1 g$ (up pointng solid triangles); $qq \rightarrow (q \rightarrow \gamma_1)q$ (asterisk); $gq \rightarrow g(q \rightarrow \gamma_1)$ (down pointing solid triangles); $gg \rightarrow \gamma_1 g$ (plus sign); $q\bar{q} \rightarrow \gamma_1 \gamma_2$ (up pointing open triangles); $gg \rightarrow \gamma_1 \gamma_2$ (open squares); $gq \rightarrow \gamma_2 (q \rightarrow \gamma_1)$ (down pointing open triangles).

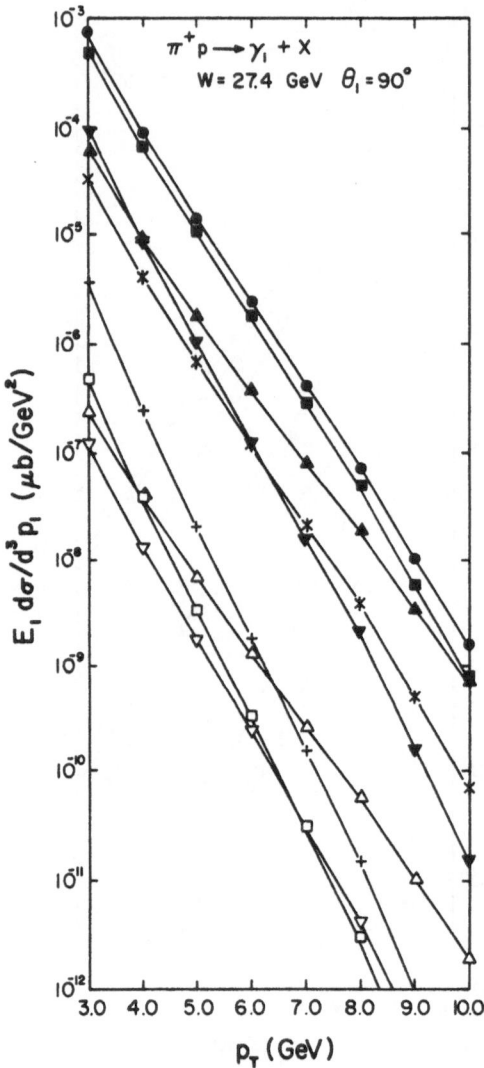

Fig. 5. Invariant cross section, $E_1 d\sigma/d^3 p_1$, for the production of photons in $\pi^+ p$ collisions at $W=27.4$ GeV and $\theta_1 = 90^0$. The individual subprocesses are labeled as in Fig. 4.

annihilation contributions to the former reaction. However, as can be seen in Fig. 8 where I plot the ratio of photons produced in the two reactions, this increase is not as large as one might naively expect. At W=63 GeV and say, p_T = 6.0 GeV, one is at the rather small x_T value of only about 0.2. At these x_T values there are lots of gluons within the proton and also a non-negligible amount of antiquarks. Table 3 shows that at this x_T the Compton term, gq→γq, gives the dominate single photon contribution to both reactions and the annihilation component to $\overline{p}$p is only a factor 6.7 times that for pp collisions. This together with other contributions results in the rather modest increase of the high p_T photon yield in $\overline{p}$p over pp collisions seen in Fig. 8.

At present one cannot make reliable predictions for the large p_T photon rate below p_T of about 4.0 GeV and below x_T of about 0.1. Large p_T is necessary to insure a large Q and thus justify the use of QCD perturbation theory. In addition, at large p_T one can avoid the uncertainties of smearing and of higher twist contributions [7]. One must also avoid the small x_T region where the structure functions are changing rapidly. In this region, the leading order QCD evolution formula is not accurate. One is particularly skeptical of the gluon distribution $G_{A \to g}(x,Q)$ which is increasing quite rapidly with increasing Q at low x. On the other hand, one would like to estimate photon yields at collider energies which unless one is at extremely high p_T values involves the low x_T region. I include here estimates of the photon rates at W=540 and 1600 GeV, however below x_T values of about 0.1 these results should be considered only as rough guesses.

Figs. 9, 10, and 11 show the results for pp and $\overline{p}$p collisions at W=540 GeV while Figs. 12, 13, and 14 display the predictions at W=1600 GeV. Table 4 lists contributions to the single photon rate for pp and $\overline{p}$p collisions at

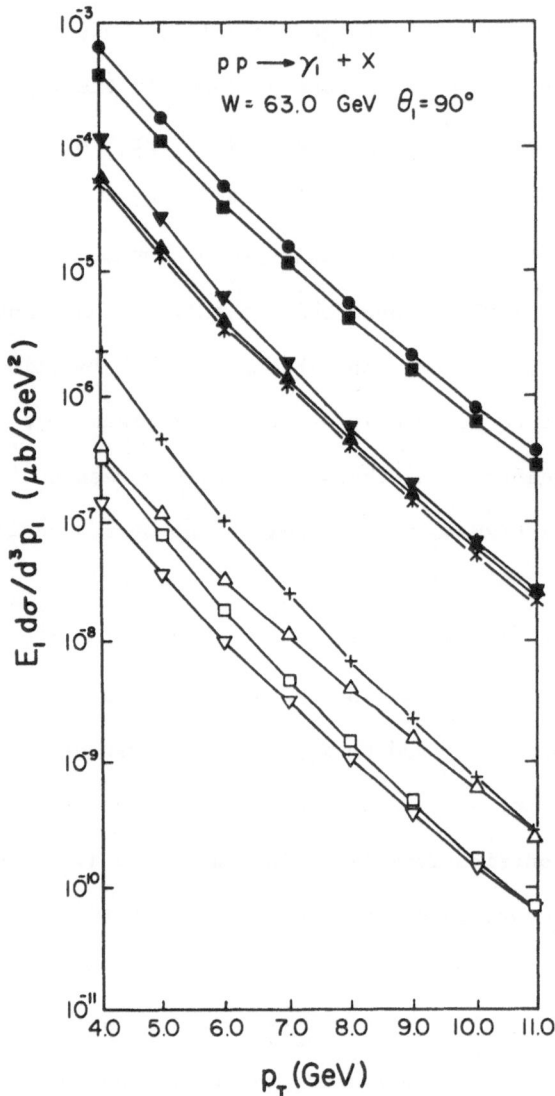

Fig. 6. Invariant cross section, $E_1 d\sigma/d^3 p_1$, for the production of photons in pp collisions at W=63 GeV and $\theta_1 = 90^0$. The individual subprocesses are labeled as in Fig. 4.

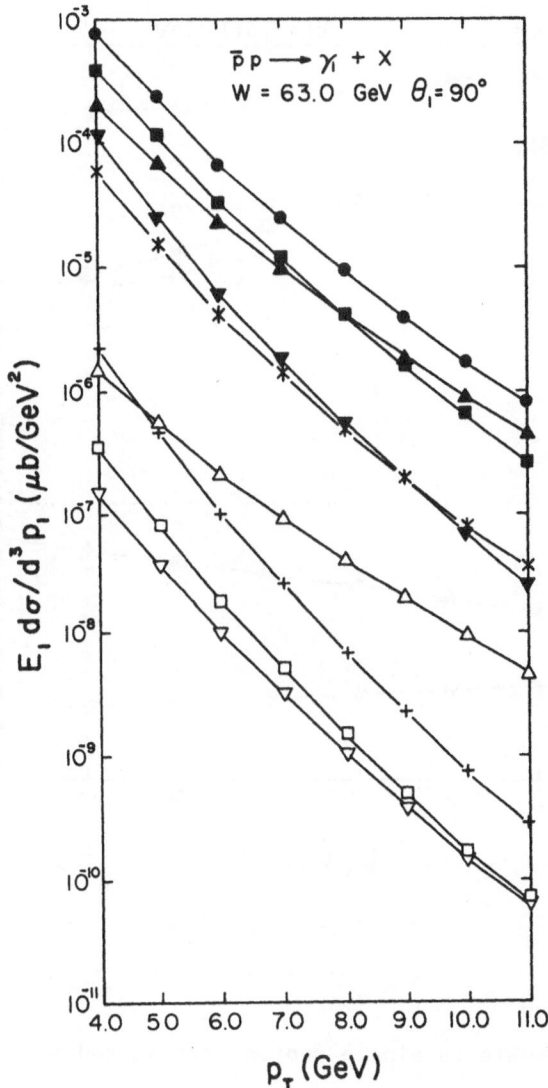

Fig. 7. Invariant cross section, $E_1 d\sigma/d^3 p_1$, for the production of photons in $\bar{p}p$ collisions at W=63 GeV and $\theta_1=90^0$. The individual subprocesses are labeled as in Fig. 4.

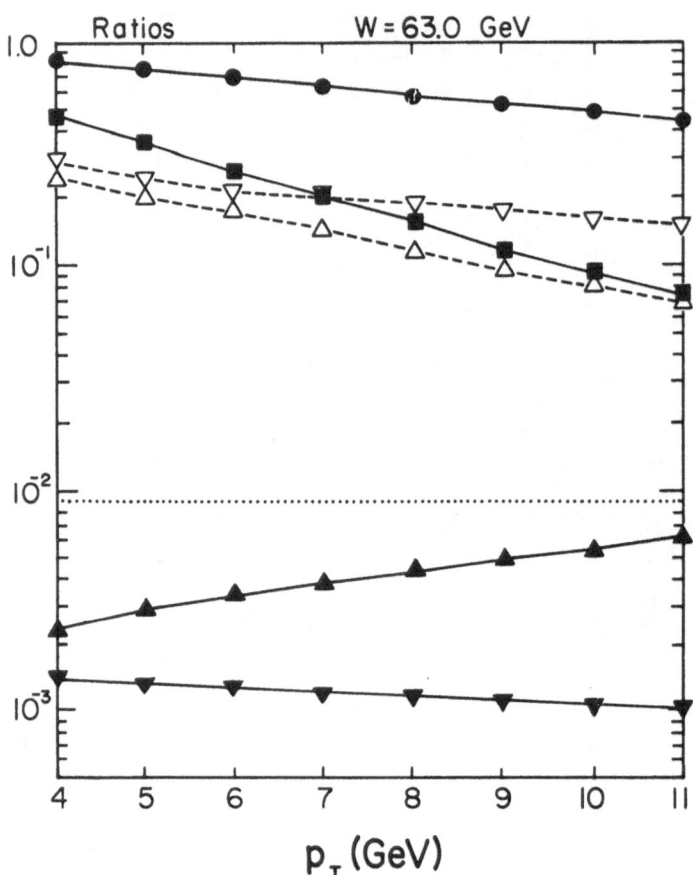

Fig. 8. Beam ratios, double to single photon ratios, and Bremsstrahlung fractions for pp and $\bar{p}p$ collisions at W=63 GeV and $\theta_1 = 90^0$: $(p/\bar{p})p \rightarrow \gamma_1 + X$ (solid dots); $(p/\bar{p})p \rightarrow \gamma_1 \gamma_2 + X$ (solid squares); $\bar{p}p \rightarrow (\gamma_1 \gamma_2 / \gamma_1) + X$ (up pointing solid triangles); $pp \rightarrow (\gamma_1 \gamma_2 / \gamma_1) + X$ (down pointing solid triangles); $\bar{p}p \rightarrow (\gamma_{BREM} / \gamma_{tot}) + X$ (up pointing open triangles); $pp \rightarrow (\gamma_{BREM} / \gamma_{tot}) + X$ (down pointing open triangles). Also shown is the naive estimate from (2), $\frac{3}{4} (\alpha_{em} / \alpha_s)(\langle e_q^4 \rangle / \langle e_q^2 \rangle)$ (dotted line).

TABLE 3

Contributions to the pp and $\bar{p}p$ single photon invariant cross section, $Ed\sigma/d^3p$ ($\mu b/GeV^2$) at CM energy W=63.0 GeV, $\theta=90^0$, and p_T=6.0 GeV from various constituent subprocesses. (Note that $A(-n)=A \times 10^{-n}$)

Subprocess	$pp \rightarrow \gamma+X$		$\bar{p}p \rightarrow \gamma+X$	
	$Ed\sigma/d^3p$	Fraction	$Ed\sigma/d^3p$	Fraction
1. $gq \rightarrow \gamma_1 q$	3.3(-5)	0.68	3.3(-5)	0.48
2. $q\bar{q} \rightarrow \gamma_1 g$	4.2(-6)	8.7(-2)	2.4(-5)	0.35
3. $gg \rightarrow \gamma_1 g$	1.0(-7)	2.1(-3)	1.0(-7)	1.4(-3)
4. $qq \rightarrow q(q \rightarrow \gamma_1)$	3.7(-6)	7.6(-2)	4.3(-6)	6.2(-2)
5. $gq \rightarrow g(q \rightarrow \gamma_1)$	6.3(-6)	0.13	6.3(-6)	9.1(-2)
6. $gq \rightarrow (g \rightarrow \gamma_1)q$	2.7(-7)	5.6(-3)	2.7(-7)	3.9(-3)
7. $gg \rightarrow \bar{q}(q \rightarrow \gamma_1)$	2.3(-7)	4.8(-3)	2.3(-7)	3.3(-3)
8. $gg \rightarrow g(g \rightarrow \gamma_1)$	3.2(-7)	6.6(-3)	3.2(-7)	4.6(-3)
9. $gg \rightarrow \gamma_1 \gamma_2$	1.8(-8)	3.7(-4)	1.8(-8)	2.6(-4)
10. $q\bar{q} \rightarrow \gamma_1 \gamma_2$	3.5(-8)	7.2(-4)	2.1(-7)	3.0(-3)
11. $gq \rightarrow \gamma_2(q \rightarrow \gamma_1)$	1.0(-8)	2.1(-4)	1.0(-8)	1.4(-4)
12. $q\bar{q} \rightarrow \gamma_2(g \rightarrow \gamma_1)$	2.4(-11)	5.0(-7)	2.5(-10)	3.6(-6)
13. $gg \rightarrow \gamma_2(g \rightarrow \gamma_1)$	4.0(-13)	8.3(-9)	4.0(-13)	5.8(-9)
Compton + Annihilation (1 + 2)	3.7(-5)	0.77	5.7(-5)	0.83
Bremsstrahlung (4 + 5 + 6 + 7 + 8)	1.1(-5)	0.23	1.1(-5)	0.17
Double Photon (9 + 10 + 11 + 12 + 13)	6.3(-8)	1.3(-3)	2.4(-7)	3.5(-3)
Total	4.8(-5)	1.00	6.9(-5)	1.00

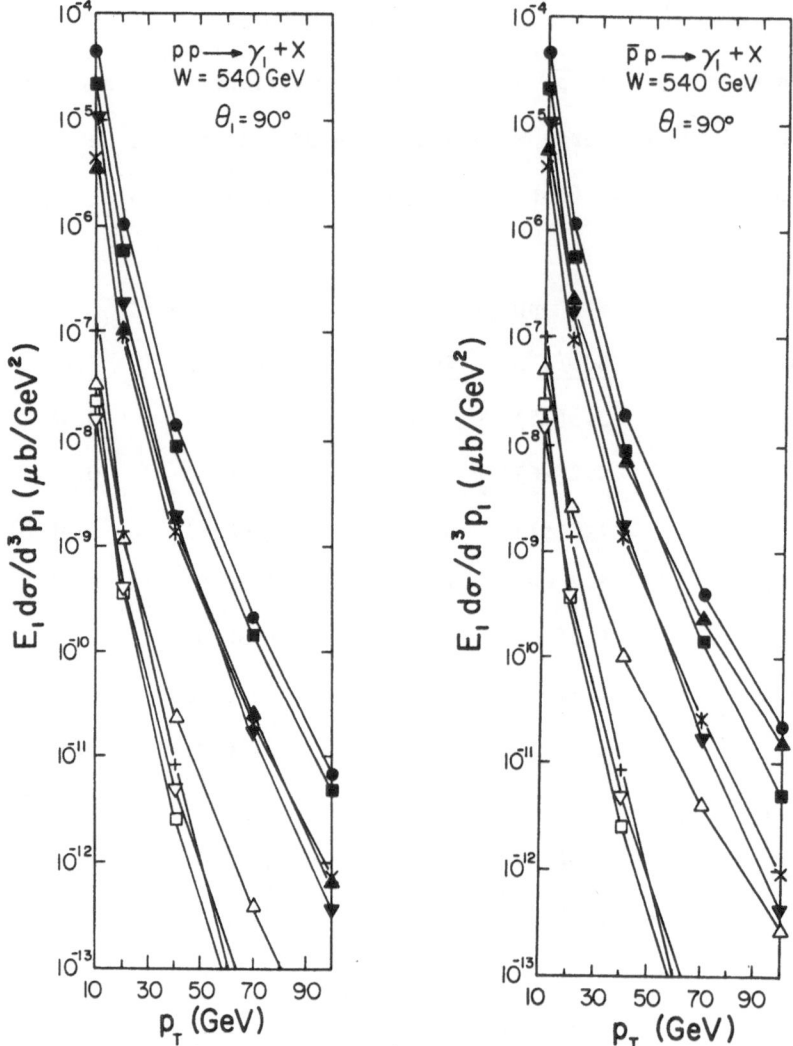

Fig. 9.   Invariant cross section, $E_1 d\sigma/d^3 p_1$, for the production of photons in pp collisions at W=540 GeV and $\theta_1=90^0$. The individual subprocesses are labeled as in Fig. 4.

Fig. 10.   Invariant cross section, $E_1 d\sigma/d^3 p_1$, for the production of photons in $\bar{p}$p collisions at W=540 GeV and $\theta_1=90^0$. The individual subprocesses are labeled as in Fig. 4.

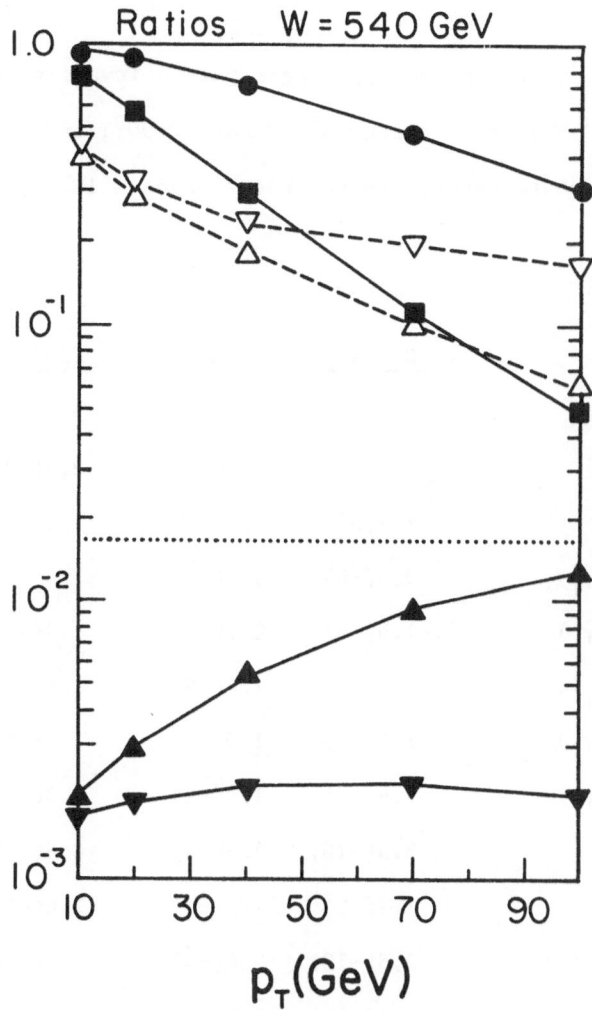

Fig. 11. Beam ratios, double to single photon ratios, and Bremsstrahlung fractions for pp and $\overline{p}p$ collisions at W=540 GeV and $\theta_1=90^0$. The curves are labeled as in Fig. 8. Also shown is the naive estimate from (2), $\frac{3}{4}$ (α_{em}/α_s) ($\langle e_q^4\rangle/\langle e_q^2\rangle$) (dotted line).

TABLE 4

Contributions to the pp and $\overline{p}p$ single photon invariant cross section, $E d\sigma/d^3p$ ($\mu b/GeV^2$) at CM energy W=540 GeV, $\theta=90^0$, and p_T=20 GeV from various constituent subprocesses. (Note that A(-n)=A x 10^{-n})

Subprocess	pp→γ+X		$\overline{p}p$→γ+X	
	$E d\sigma/d^3p$	Fraction	$E d\sigma/d^3p$	Fraction
1. $gq \to \gamma_1 q$	5.6(-7)	0.56	5.6(-7)	0.50
2. $q\bar{q} \to \gamma_1 g$	1.0(-7)	0.10	2.2(-7)	0.20
3. $gg \to \gamma_1 g$	1.3(-9)	1.3(-3)	1.3(-9)	1.2(-3)
4. $qq \to q(q \to \gamma_1)$	9.5(-8)	9.4(-2)	1.0(-7)	9.1(-2)
5. $gq \to g(q \to \gamma_1)$	1.9(-7)	0.18	1.9(-7)	0.17
6. $gq \to (g \to \gamma_1)q$	1.4(-8)	1.4(-2)	1.4(-8)	1.2(-2)
7. $gg \to \bar{q}(q \to \gamma_1)$	1.0(-8)	1.0(-2)	1.0(-8)	9.3(-3)
8. $gg \to g(g \to \gamma_1)$	1.9(-8)	1.9(-2)	1.9(-8)	1.7(-2)
9. $gg \to \gamma_1 \gamma_2$	3.5(-10)	3.5(-4)	3.5(-10)	3.1(-4)
10. $q\bar{q} \to \gamma_1 \gamma_2$	1.2(-9)	1.2(-3)	2.6(-9)	2.3(-3)
11. $gq \to \gamma_2(q \to \gamma_1)$	3.8(-10)	3.8(-4)	3.8(-10)	3.4(-4)
12. $q\bar{q} \to \gamma_2(g \to \gamma_1)$	2.0(-12)	2.0(-6)	7.5(-12)	6.7(-6)
13. $gg \to \gamma_2(g \to \gamma_1)$	1.5(-14)	1.5(-8)	1.5(-14)	1.3(-8)
Compton + Annihilation (1 + 2)	6.7(-7)	0.66	7.9(-7)	0.70
Bremsstrahlung (4 + 5 + 6 + 7 + 8)	3.3(-7)	0.32	3.3(-7)	0.29
Double Photon (9 + 10 + 11 + 12 + 13)	1.9(-9)	1.9(-3)	3.4(-9)	3.0(-3)
Total	1.0(-6)	1.00	1.1(-6)	1.00

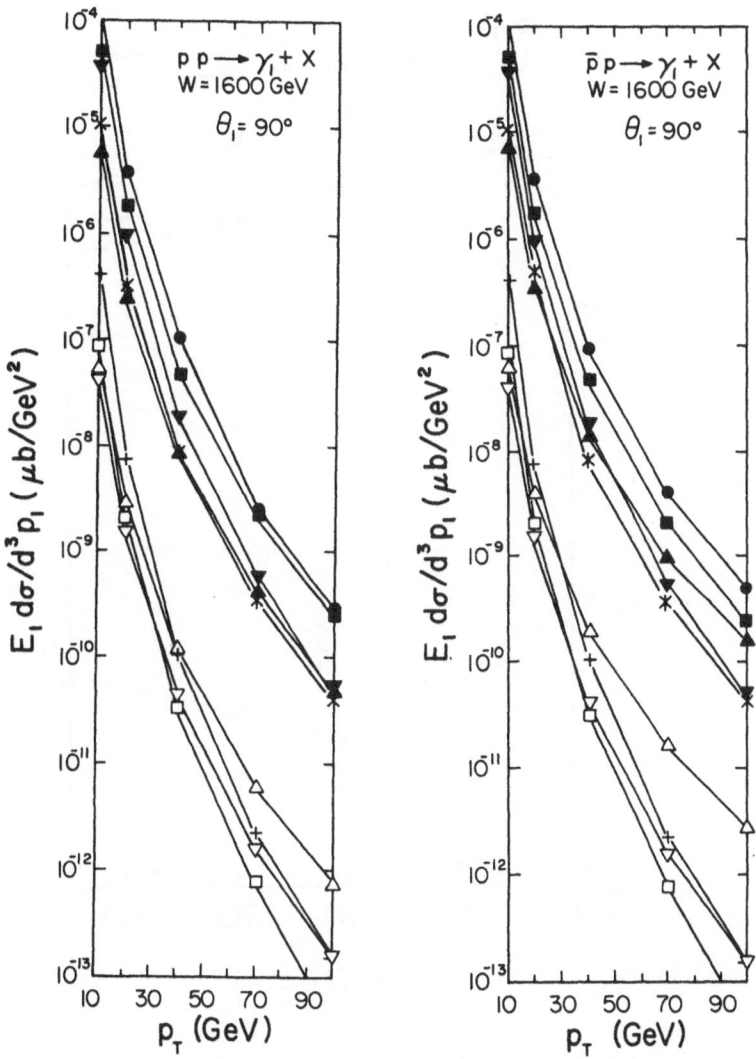

Fig. 12. Invariant cross section, $E_1 d\sigma/d^3 p_1$, for the production of photons in pp collisions at W=1600 GeV and $\theta_1=90^0$. The individual subprocesses are labeled as in Fig. 4.

Fig. 13. Invariant cross section, $E_1 d\sigma/d^3 p_1$, for the production of photons in $\bar{p}p$ collisions at W=1600 GeV and $\theta_1=90^0$. The individual subprocesses are labeled as in Fig. 4.

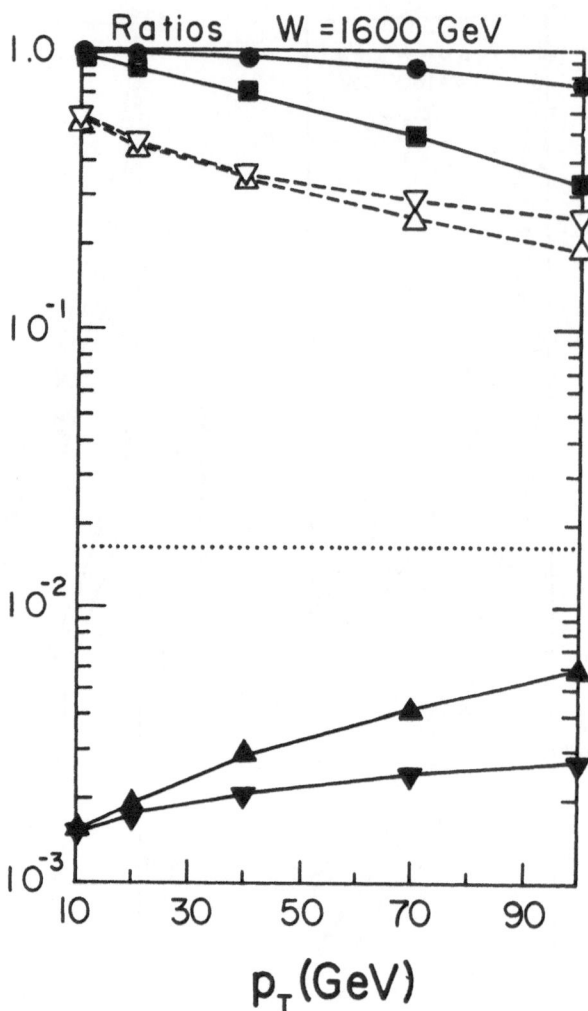

Fig. 14. Beam ratios, double to single photon ratios, and Bremsstrahlung fractions for pp and $\overline{p}p$ collisions at W=1600 GeV and θ_1=90^0. The curves are labeled as in Fig. 8. Also shown is the naive estimate from (2), $\frac{3}{4}$ (α_{em}/α_s) ($<e_q^4>/<e_q^2>$) (dotted line).

TABLE 5

Contributions to the pp and $\overline{p}p$ single photon invariant cross section, $Ed\sigma/d^3p$ ($\mu b/GeV^2$) at CM energy W=1600 GeV, $\theta=90^0$, and p_T=40 GeV from various constituent subprocesses. (Note that $A(-n)=A \times 10^{-n}$)

Subprocess	$pp \rightarrow \gamma + X$		$\overline{p}p \rightarrow \gamma + X$	
	$Ed\sigma/d^3p$	Fraction	$Ed\sigma/d^3p$	Fraction
1. $gq \rightarrow \gamma_1 q$	4.7(-8)	0.54	4.7(-8)	0.50
2. $q\overline{q} \rightarrow \gamma_1 g$	8.3(-9)	9.5(-2)	1.4(-8)	0.15
3. $gg \rightarrow \gamma_1 g$	1.0(-10)	1.2(-3)	1.0(-10)	1.1(-3)
4. $qq \rightarrow q(q \rightarrow \gamma_1)$	8.3(-9)	9.5(-2)	8.7(-9)	9.3(-2)
5. $gq \rightarrow g(q \rightarrow \gamma_1)$	1.8(-8)	0.21	1.8(-8)	0.19
6. $gq \rightarrow (g \rightarrow \gamma_1)q$	1.6(-9)	1.8(-2)	1.6(-9)	1.7(-2)
7. $gg \rightarrow \overline{q}(q \rightarrow \gamma_1)$	1.2(-9)	1.4(-2)	1.2(-9)	1.3(-2)
8. $gg \rightarrow g(g \rightarrow \gamma_1)$	2.5(-9)	2.8(-2)	2.5(-9)	2.7(-2)
9. $gg \rightarrow \gamma_1 \gamma_2$	3.1(-11)	3.6(-4)	3.1(-11)	3.3(-4)
10. $q\overline{q} \rightarrow \gamma_1 \gamma_2$	1.1(-10)	1.3(-3)	1.9(-10)	2.0(-3)
11. $gq \rightarrow \gamma_2(q \rightarrow \gamma_1)$	4.1(-11)	4.7(-4)	4.1(-11)	4.3(-4)
12. $q\overline{q} \rightarrow \gamma_2(g \rightarrow \gamma_1)$	2.4(-13)	2.8(-6)	6.6(-13)	7.0(-6)
13. $gg \rightarrow \gamma_2(g \rightarrow \gamma_1)$	1.5(-15)	1.8(-8)	1.5(-15)	1.6(-8)
Compton + Annihilation (1 + 2)	5.6(-8)	0.63	6.1(-8)	0.65
Bremsstrahlung (4 + 5 + 6 + 7 + 8)	3.2(-8)	0.36	3.2(-8)	0.34
Double Photon (9 + 10 + 11 + 12 + 13)	1.8(-10)	2.1(-3)	2.6(-10)	2.8(-3)
Total	8.7(-8)	1.00	9.4(-8)	1.00

W=540 GeV, θ_{cm} = 90^0 and p_T = 20 GeV, while the W=1600 GeV predictions at p_T = 40 GeV are tabulated in Table 7. One must reach p_T values of about 20 GeV at W=540 and 50 GeV at W=1600 GeV before one can see a significant difference between pp and $\overline{p}p$ collisions. At lower p_T values everything is dominated by subprocesses involving gluons.

All the QCD matrix elements for subprocesses in Table 1 of order $\alpha_{em}\alpha_s^2$ in which two partons scatter and produce two partons plus a photon have not been calculated and so it is not possible to include them exactly. However, the dominant contribution from these diagrams arises when the outgoing large p_T photon is parallel to one of the outgoing partons. This allows us to approximate all such photon Bremsstrahlung diagrams by introducing the parton to photon fragmentation functions, $D_{i \to \gamma}(z,Q)$, as illustrated in Fig. 15. One takes the photon to be parallel to the outgoing parton with $D_{i \to \gamma}(z,Q)$ giving the number of photons carrying fractional longitudinal momentum between z and z+dz. The invariant cross section for the reaction A+B→γ+X arising from photon Bremsstrahlung can be estimated using

$$E \frac{d\sigma}{d^3p} (A+B \to \gamma+X; s, p_T, \theta_{cm}) = \int_{x_a^{min}}^{1.0} dx_a \int_{x_b^{min}}^{1.0} dx_b \ G_{A \to a}(x_a, Q) G_{B \to b}(x_b, Q)$$

$$D_{c \to \gamma}(z_c, Q) \ \left(\frac{1}{\pi z_c}\right) \frac{d\hat{\sigma}}{d\hat{t}} (ab \to cd; \hat{s}, \hat{t}, \hat{u}), \tag{5}$$

where $d\hat{\sigma}/d\hat{t}(ab \to cd; \hat{s}, \hat{t}, \hat{u})$ is the hard scattering parton differential cross section, a+b→c+d, calculated to order α_s^2 from perturbation theory and where one must sum over all eight two-to-two parton subprocesses qq→qq, $q\overline{q} \to q\overline{q}$, $\overline{qq} \to \overline{qq}$, gq→gq, $\overline{gq} \to \overline{gq}$, gg→$q\overline{q}$, $q\overline{q} \to$ gg, and gg→gg. The invariants $\hat{s}$, $\hat{t}$ and $\hat{u}$ are given by

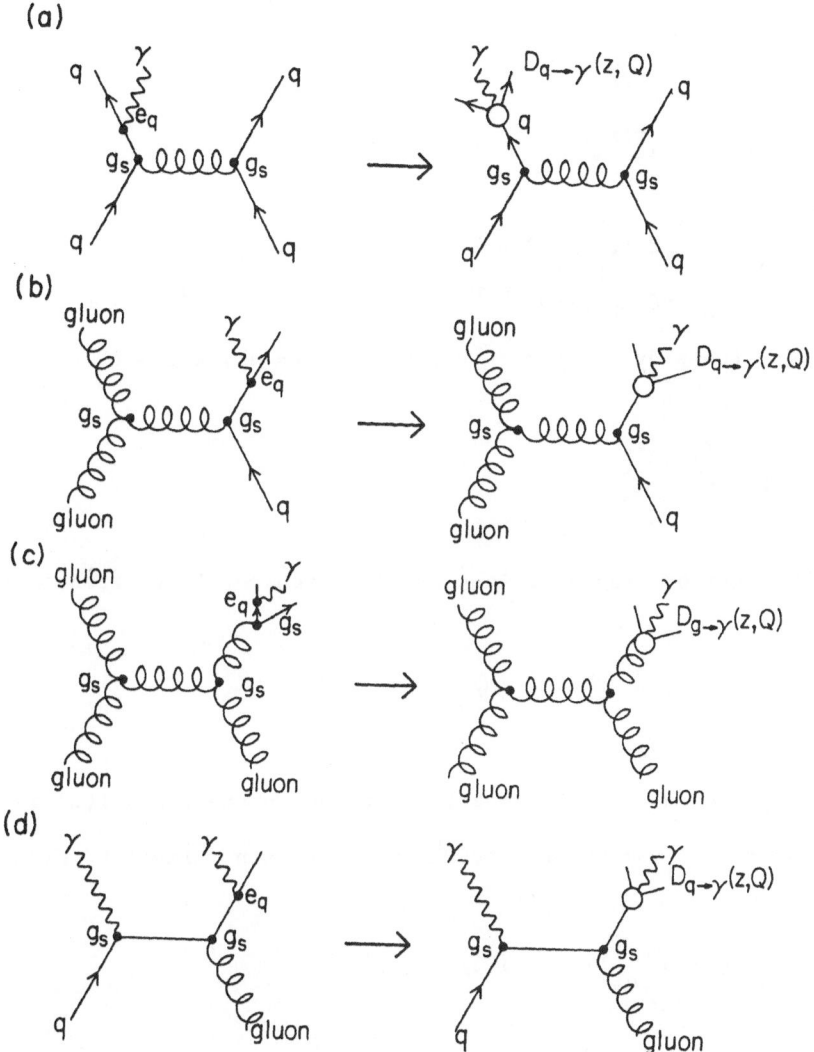

Fig. 15. (a)-(c). Illustration of how constituent subprocesses of the form parton+parton→parton+parton+photon (order $\alpha_s^2 \alpha_{em}$) can be estimated by introducing parton→γ fragmentation functions, $D_{i \to \gamma}(z, Q)$. (d) Illustration of how the constituent subprocess qg→γγq (order $\alpha_s \alpha_{em}^2$) can be estimated by the use of the q→γ fragmentation function.

$$\hat{s} = x_a x_b s \tag{6a}$$

$$\hat{t} = - x_a x_2 s/z_c \tag{6b}$$

$$\hat{u} = - x_b x_1 s/z_c \tag{6c}$$

where $x_1 = \frac{1}{2} x_T \cot(\frac{1}{2} \theta_{cm})$ and $x_2 = \frac{1}{2} x_T \tan(\frac{1}{2} \theta_{cm})$ and where one must integrate over both x_a and x_b with $x_a^{min} = x_1/(1-x_2)$ and $x_b^{min} = x_a x_2/(x_a-x_1)$. The massless two-body scattering constraint, $\hat{s} + \hat{t} + \hat{u} = 0$, gives

$$z_c = \frac{x_1}{x_a} + \frac{x_2}{x_b} . \tag{6d}$$

The parton fragmentation into a photon can be written in the form

$$D_{i \to \gamma}(z,Q) = \frac{\alpha_{em}}{\pi} \log (Q/\Lambda) f_{i \to \gamma}(z), \tag{7}$$

where α_{em} is the electromagnetic coupling constant and $f(z)$ is independent of Q. The simple photon Bremsstrahlung "Born approximation" gives

$$f_{q \to \gamma}(z) = e_q^2 (1+(1-z)^2)/z \tag{8a}$$

$$f_{g \to \gamma}(z) = 0, \tag{8b}$$

where e_q is the electric charge of the quark. In leading order QCD modifies the shape of the function $f(z)$ and induces an indirect coupling of the gluon to the photon as shown in Fig. 16 [8]. The fragmentation of a quark into a photon is softened because the quark may emit a gluon (or several gluons) before radiating the photon. An incident gluon may convert to a $q\bar{q}$ pair which subsequently radiates a photon. To leading log order all these effects are

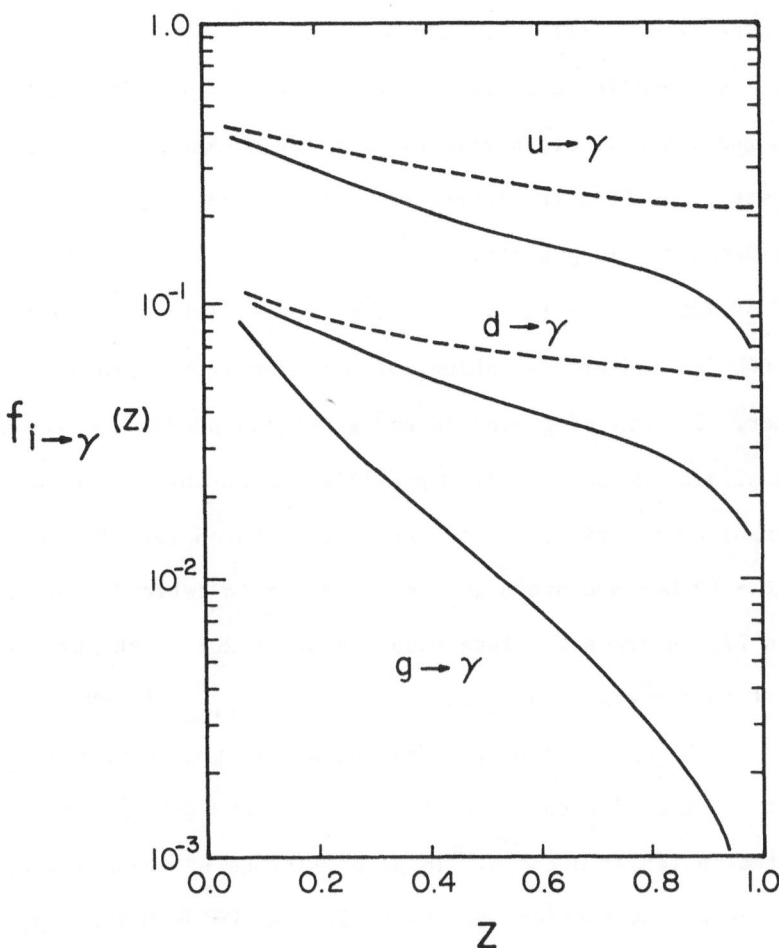

Fig. 16. Leading order QCD parton→γ fragmentations, $D_{i \to \gamma}(z,Q)$ from [8]. These functions can be written in the form $D_{i \to \gamma}(z,Q) = (\alpha_{em}/\pi)\log(Q/\Lambda)$ $f_{i \to \gamma}(z)$, with the functions $f_{i \to \gamma}(z)$ displayed in this figure.

contained in the fragmentation functions.

Figs. 4 and 5 and Table 2 give the contributions to the single photon rate for $\pi^- p$ and $\pi^+ p$ collisions arising from the Bremsstrahlung diagrams. Photon Bremsstrahlung from the subprocesses $qq \to qq$ and $gq \to gq$ are comparable, with a total Bremsstrahlung contribution at W=27.4 GeV, $\theta_{cm} = 90^0$, and p_T = 4.0 GeV of 13% and 19% for $\pi^- p$ and $\pi^+ p$, respectively. As one would expect, the QCD induced Bremsstrahlung of a photon from a gluon does not have a large effect. The sum of $gq \to (g \to \gamma)q$ and $gg \to g(g \to \gamma)$ produces only about 0.7% of the total single photon rate in $\pi^- p$ collisions at this energy and p_T value.

Bremsstrahlung contributions to the single photon rate for pp and $\overline{p}p$ collisions at W=63 GeV are shown in Figs. 6-8 and in Table 3. As can be seen explicitly in Fig. 8 Bremsstrahlung produces about 20% of the predicted single photon rate at this energy and p_T in the range 4 < p_T < 10 GeV.

The Bremsstrahlung contributions to the single photon rate for pp and $\overline{p}p$ collisions at the collider energies of W=540 and 1600 GeV are shown in Figs. 9-14 and listed in Tables 3 and 4. In pp collisions at these low x_T values the Bremsstrahlung contributions are sizeable. At W=540 GeV and p_T = 20 GeV the Bremsstrahlung contribution is 32% while at W=1600 GeV, p_T = 40 GeV it is 36% of the total photon rate. Because of the abundance of gluons at low x_T the QCD induced gluon-photon contributions are not negligible at collider energies. In pp collisions at W=1600 GeV and p_T = 40 GeV the $gg \to g(g \to \gamma)$ Bremsstrahlung contribution is almost 3% of the total single photon yield.

The single photon Bremsstrahlung off the outgoing quark in the Compton subprocess, $gg \to \gamma q$, also produces a non-negligible contribution to the double photon rate. This subprocess is of order $\alpha_{em}^2 \alpha_s$ and contributes about 16% to the double photon rate in $\pi^+ p$ collisions at W=27.4 GeV and p_T = 4.0 GeV and makes a similar contribution to pp collisions at W=63 GeV and p_T = 6 GeV.

Double Bremsstrahlung subprocesses (order α_{em}^2) also make sizeable contributions to the double photon rate. On the other hand, the single and double Bremsstrahlung subprocesses do not exhibit a back-to-back two photon signal characteristic of the direct $q\bar{q} \rightarrow \gamma\gamma$ and $gg \rightarrow \gamma\gamma$ subprocesses. One can select the latter two subprocesses by examining the momentum distribution of the photons on the away-side of the large p_T photon trigger.

In the absence of internal parton transverse momentum within the initial hadrons, measurement of the two photons arising from either the subprocess $q\bar{q} \rightarrow \gamma\gamma$ or $gg \rightarrow \gamma\gamma$ completely specifies the momentum fractions, x_a and x_b, of the incoming partons. The double differential cross section for the reaction shown in Fig. 17a is given by [9]

$$E_1 \frac{d\sigma}{d^3 p_1 dy_2 dz_e} (s, p_{T_1}, p_{T_2}, \theta_1, \theta_2) = x_a x_b G_{A \rightarrow a}(x_a, Q) G_{B \rightarrow b}(x_b, Q)$$

$$(\frac{1}{\pi}) \frac{d\hat{\sigma}}{dt} (\hat{s}, \hat{t}, \hat{u}) \delta(1-z_e), \qquad (9)$$

where x_a and x_b are given in terms of the angles of the two photons by

$$x_a = \frac{1}{2} x_T (\frac{1}{T_1} + \frac{1}{T_2}) \qquad (10a)$$

$$x_b = \frac{1}{2} x_T (T_1 + T_2) \qquad (10b)$$

with

$$x_{T_1} = x_{T_2} = x_T = 2 p_T/W. \qquad (10c)$$

The invariants $\hat{s}$, $\hat{t}$, and $\hat{u}$ are given by

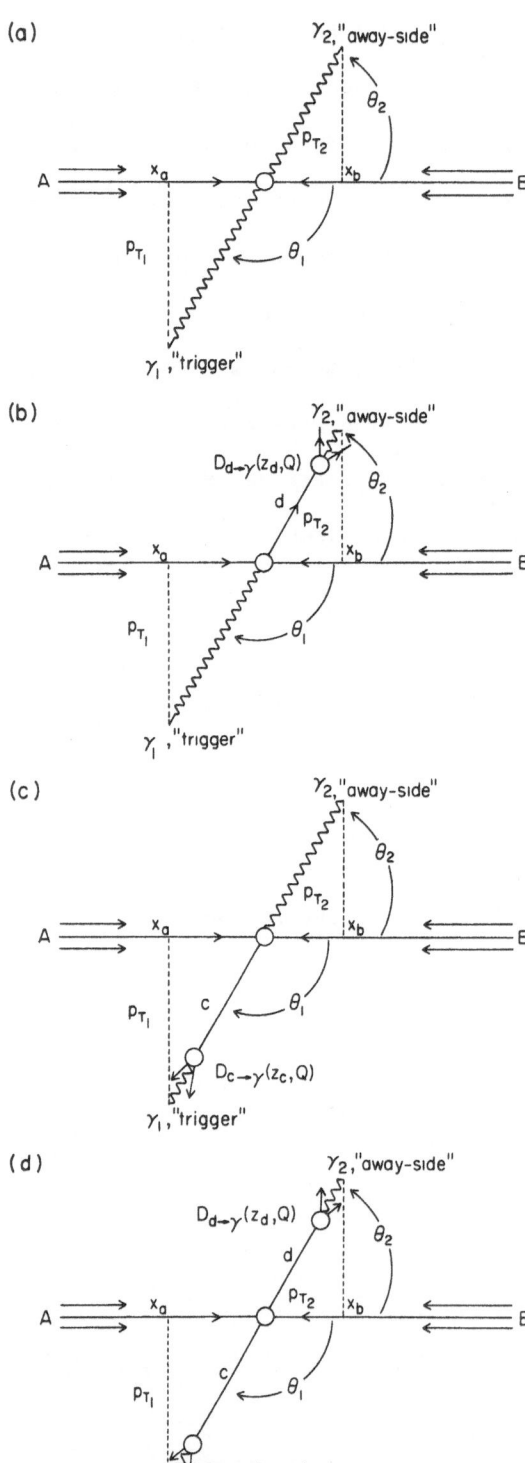

Fig. 17. Diagrams for the process $A+B \to \gamma_1 \gamma_2 + X$, where γ_1 is defined as the "trigger" and γ_2 as the "away-side" photon: (a) Both photons are direct; (b) trigger photon is direct and away photon results from Bremsstrahlung; (c) away photon is direct and trigger photon arises from Bremsstrahlung; (d) both photons come from Bremsstrahlung.

$$\hat{s} = x_a x_b s, \tag{11a}$$

$$\hat{t} = -\frac{1}{2} x_a s x_T T_1, \tag{11b}$$

$$\hat{u} = -\frac{1}{2} x_b s x_T / T_1, \tag{11c}$$

with

$$T_i = \tan\left(\frac{1}{2}\theta_i\right) \tag{12}$$

and where y_2 is the rapidity of photon number 2,

$$y_2 = -\log(T_2). \tag{13}$$

The variable z_e is defined as the ratio of the transverse momentum of the two photons

$$z_e = p_{T_2}/p_{T_1} = x_{T_2}/x_{T_1}, \tag{14}$$

or defining photon 1 as the trigger (see Fig. 17), z_e is the fraction of the trigger p_T carried by the away-side photon (number 2). The delta function in (3.5) ensures that the two photons have equal transverse momentum.

Integrating over the away-side photon angle θ_2 yields

$$E_1 \frac{d\sigma}{d^3 p_1 dz_e}(s, p_{T_1}, z_e, \theta_1) = \delta(1-z_e) \frac{2}{T_1 x_T} \int_{x_a^{min}}^{1.0} dx_a x_b^2 G_{A \to a}(x_a, Q) G_{B \to b}(x_b, Q)$$

$$\left(\frac{1}{\pi}\right) \frac{d\hat{\sigma}}{d\hat{t}}(\hat{s}, \hat{t}, \hat{u}), \tag{15}$$

where s is the cm energy square ($s = W^2$) and

$$x_a^{min} = x_T/(2T_1 - x_T T_1^2), \tag{16a}$$

and

$$x_b = x_a x_T T_1^2/(2x_a T_1 - x_T). \tag{16b}$$

The integration over x_a spans a range in away-side angle given by

$$(T_2)_{min} = x_T T_1/(2T_1 - x_T), \tag{17a}$$

$$(T_2)_{max} = (2 - x_T T_1)/x_T. \tag{17b}$$

Eq. (15) gives the away-side photon z_e distribution for a photon trigger. If we normalize by dividing by the single photon trigger rate we arrive at

$$\frac{1}{\sigma}\frac{d\sigma}{dz_e}(s,p_{T_1},z_e,\theta_1) \equiv \frac{E_1 d\sigma/d^3 p_1 dz_e(s,p_{T_1},z_e,\theta_1)}{E_1 d\sigma/d^3 p_1(s,p_{T_1},\theta_1)} \tag{18}$$

which measures the number of away-side photons between z_e and $z_e + dz_e$ for a given photon trigger.

For the case shown in Fig. 17b where the away-side photon arises from Bremsstrahlung (i.e. $gq \to \gamma_1(g \to \gamma_2)$) we have $z_e < 1$ and a double differential cross section of the form

$$E_1 \frac{d\sigma}{d^3p_1 dy_2 dz_e} (s, p_{T_1}, z_e, \theta_1, \theta_2) = x_a G_{A \to a}(x_a, Q) x_b G_{B \to b}(x_b, Q)$$

$$D_{d \to \gamma}(z_d, Q) \frac{1}{\pi} \frac{d\hat{\sigma}}{dt} (\hat{s}, \hat{t}, \hat{u}), \tag{19}$$

where $\hat{s}$, $\hat{t}$, and $\hat{u}$ are given by (3.7) and the momentum fractions x_a, x_b, and z_d are again specified by the photon angles and p_T values,

$$x_a = \frac{1}{2} x_{T_1} \left(\frac{1}{T_1} + \frac{1}{T_2} \right), \tag{20a}$$

$$x_b = x_a T_1 T_2, \tag{20b}$$

$$z_d = z_e. \tag{20c}$$

Integrating over θ_2 yields

$$E_1 \frac{d\sigma}{d^3p_1 dz_e} (s, p_{T_1}, z_e, \theta_1) = \frac{2}{T_1 x_{T_1}} \int_{x_a^{min}}^{1.0} dx_a \, x_b^2 G_{A \to a}(x_a, Q) G_{B \to b}(x_b, Q)$$

$$D_{d \to \gamma}(z_d, Q) \left(\frac{1}{\pi} \right) \frac{d\hat{\sigma}}{dt} (\hat{s}, \hat{t}, \hat{u}) \tag{21}$$

where

$$x_a^{min} = x_{T_1} / (2T_1 - x_{T_1} T_1^2) \tag{22a}$$

and

$$x_b = x_a x_{T_1} T_1^2 / (2x_a T_1 - x_{T_1}) \tag{22b}$$

The alternative case, $gq \to \gamma_2 (q \to \gamma_1)$, shown in Fig. 17c where the trigger photon arises from Bremsstrahlung and the away-photon is direct is similar with the double differential cross section given by

$$E_1 \frac{d\sigma}{d^3p_1 dy_2 dz_e}(s,p_{T_1},z_e,\theta_1,\theta_2) = x_a G_{A \to a}(x_a,Q) x_b G_{B \to b}(x_b,Q) D_{c \to \gamma}(z_c,Q)$$

$$(\frac{1}{\pi}) \frac{d\hat{\sigma}}{d\hat{t}}(\hat{s},\hat{t},\hat{u}), \tag{23}$$

with

$$x_a = x_{T_2} \frac{1}{2}(\frac{1}{T_1} + \frac{1}{T_2}), \tag{24a}$$

$$x_b = x_a T_1 T_2, \tag{24b}$$

$$z_c = \frac{1}{z_e} . \tag{24c}$$

The away-side z_e-distribution is given by

$$E_1 \frac{d\sigma}{d^3p_1 dz_e}(s,p_{T_1},z_e,\theta_1) = \frac{2}{T_1 x_{T_2}} \int_{x_a^{min}}^{1.0} dx_a x_b^2 G_{A \to a}(x_a,Q) G_{B \to b}(x_b,Q) D_{c \to \gamma}(z_c,Q)$$

$$(\frac{1}{\pi}) \frac{d\hat{\sigma}}{d\hat{t}}(\hat{s},\hat{t},\hat{u}), \tag{25}$$

with

$$x_a^{min} = x_{T_2}/(2T_1 - x_{T_2} T_1^2), \tag{26a}$$

and

$$x_b = x_a x_{T_2} T_1^2 / (2 x_a T_1 - x_{T_2}). \tag{26b}$$

For the final case of double Bremsstrahlung shown in Fig. 17d, measurement of the two outgoing photons does not completely specify the incoming parton momenta. The double differential cross section involves an integration over parton momenta and is given by

$$E_1 \frac{d\sigma}{d^3 P_1 dy_2 dz_e} (s, P_{T_1}, z_e, \theta_1, \theta_2) = \int_{x_a^{min}}^{1.0} dx_a x_b G_{A \to a}(x_a, Q) G_{B \to b}(x_b, Q)$$

$$D_{c \to \gamma}(z_c, Q) D_{d \to \gamma}(z_d, Q) \left(\frac{1}{\pi}\right) \frac{d\hat{\sigma}}{d\hat{t}} (\hat{s}, \hat{t}, \hat{u}), \tag{27}$$

where the invariants $\hat{s}$, $\hat{t}$, and $\hat{u}$ are given by

$$\hat{s} = x_a x_b s \tag{28a}$$

$$\hat{t} = -\frac{1}{2} x_a s \, x_{T_1} T_1 / z_c \tag{28b}$$

$$\hat{u} = -\frac{1}{2} x_b s \, x_{T_1} / (T_1 z_c) \tag{28c}$$

with

$$x_b = x_a T_1 T_2, \tag{29a}$$

$$z_c = \frac{1}{2} \left(\frac{x_{T_1}}{x_a}\right) \left(\frac{1}{T_1} + \frac{1}{T_2}\right), \tag{29b}$$

$$z_d = \frac{1}{2} \left(\frac{x_{T_2}}{x_a}\right) \left(\frac{1}{T_1} + \frac{1}{T_2}\right). \tag{29c}$$

The integration covers the range

$$x_a^{min} = \frac{1}{2} x_T^{max} \left(\frac{1}{T_1} + \frac{1}{T_2}\right), \tag{30a}$$

$$x_b^{min} = x_a^{min} T_1 T_2, \tag{30b}$$

$$z_c^{min} = \frac{1}{2} x_{T_1} \left(\frac{1}{T_1} + \frac{1}{T_2}\right), \tag{30c}$$

$$z_d^{min} = \frac{1}{2} x_{T_2} \left(\frac{1}{T_1} + \frac{1}{T_2}\right), \tag{30d}$$

where x_T^{max} is the maximum of x_{T_1} and x_{T_2}. Integrating over the away-side angle θ_2 yields the double integral

$$E_1 \frac{d\sigma}{d^3 p dz_e} (s, z_e, \theta) = \int_{x_a^{min}}^{1.0} dx_a \int_{x_b^{min}}^{1.0} dx_b \ G_{A \to a}(x_a, Q) G_{B \to b}(x_b, Q)$$

$$D_{c \to \gamma}(z_c, Q) D_{d \to \gamma}(z_d, Q) \left(\frac{1}{\pi}\right) \frac{d\hat{\sigma}}{d\hat{t}} (\hat{s}, \hat{t}, \hat{u}), \tag{31}$$

with

$$x_a^{min} = x_T^{max} / (2T_1 - x_T^{max} T_1^2), \tag{32a}$$

and

$$x_b^{min} = x_a x_T^{max} T_1^2 / (2x_a T_1 - x_T^{max}). \tag{32b}$$

The away-side z_e-distributions for a photon trigger at W=27.4 GeV, p_{T_1} = 4.0 GeV, and $\theta_1 = 90^0$ for $\pi^- p$ and $\pi^+ p$ collisions are shown in Figs. 18 and

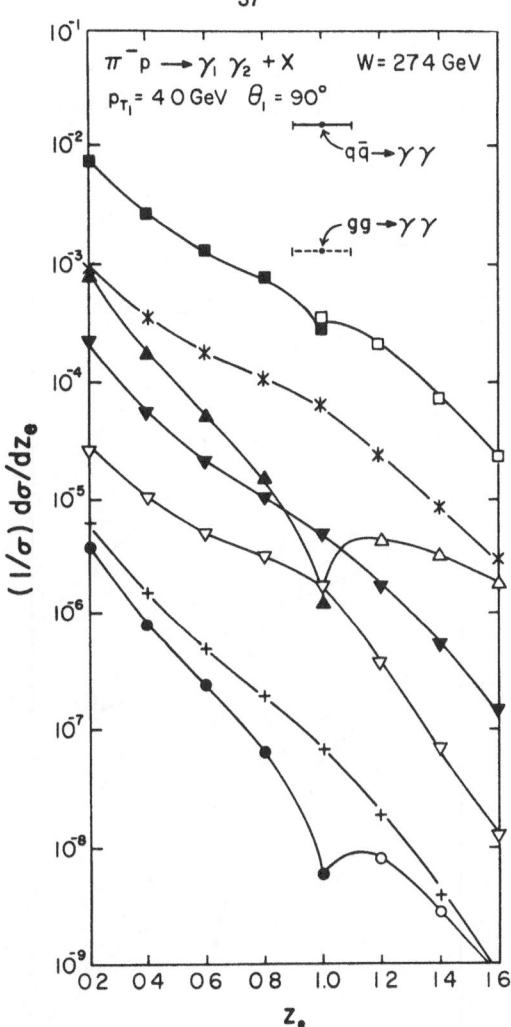

Fig. 18. The z_e-distribution of the "away-side" photon, γ_2, for a photon

trigger, γ_1, in $\pi^- p$ collisions at W=27.4 GeV, θ_1=90⁰, and p_{T_1} = 4.0 GeV, where

$z_e = p_{T_2}/p_{T_1}$. The quantity $(1/\sigma)d\sigma/dz_e$ measures the number of away-side

photons between z_e and $z_e + dz_e$ for the given photon trigger. The individual

subprocesses are labeled as follows: $gq \rightarrow \gamma_1(q \rightarrow \gamma_2)$ (solid squares);

$gq \rightarrow \gamma_2(q \rightarrow \gamma_1)$ (open squares); $qq \rightarrow (q \rightarrow \gamma)(q \rightarrow \gamma)$ (asterisk); $q\bar{q} \rightarrow \gamma_1(g \rightarrow \gamma_2)$ (up

pointing solid triangles); $q\bar{q} \rightarrow \gamma_2(g \rightarrow \gamma_1)$ (up pointing open triangles);

$gq \rightarrow (g \rightarrow \gamma)(q \rightarrow \gamma)$ (down pointing solid triangles); $gg \rightarrow (q \rightarrow \gamma)(\bar{q} \rightarrow \gamma)$ (down pointing

open triangles); $gg \rightarrow (g \rightarrow \gamma)(g \rightarrow \gamma)$ (plus signs); $gg \rightarrow \gamma_1(g \rightarrow \gamma_2)$ (solid dots);

$gg \rightarrow \gamma_2(g \rightarrow \gamma_1)$ (open dots). The z_e = 1 delta function contributions from the

direct subprocesses $q\bar{q} \rightarrow \gamma\gamma$ and $gg \rightarrow \gamma\gamma$ have arbitrarily been spread over 0.2

units of z_e.

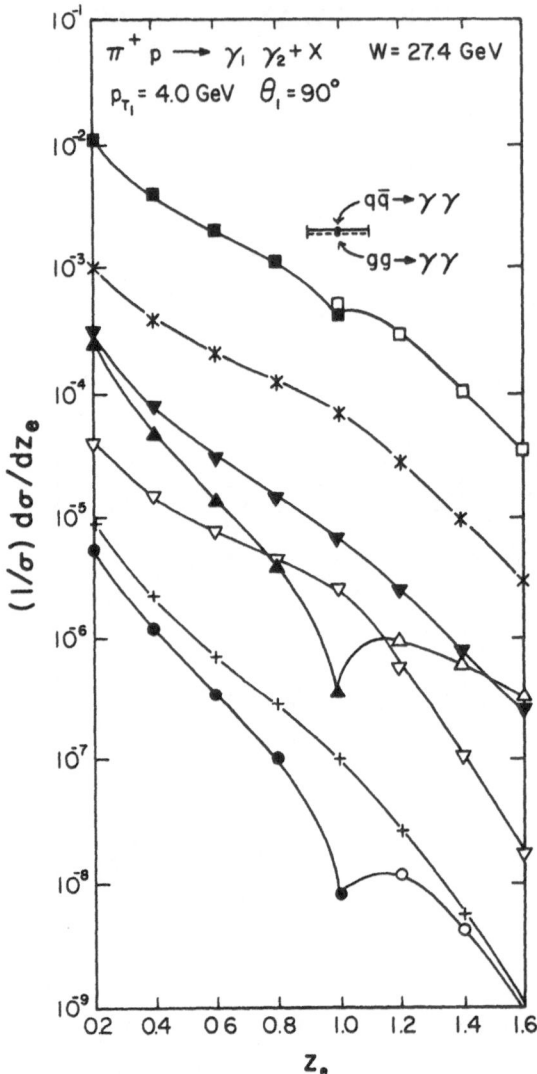

Fig. 19. The z_e-distribution of the "away-side" photon, γ_2, for a photon trigger, γ_1, in $\pi^+ p$ collisions at W=27.4 GeV, $\theta_1 = 90^0$, and $p_{T_1} = 4.0$ GeV, where $z_e = p_{T_2}/p_{T_1}$. The quantity $(1/\sigma)d\sigma/dz_e$ measures the number of away-side photons between z_e and $z_e + dz_e$ for the given photon trigger. The individual subprocesses are labeled as in Fig. 18. The $z_e = 1$ delta function contributions from the direct subprocesses $q\bar{q} \to \gamma\gamma$ and $gg \to \gamma\gamma$ have arbitrarily been spread over 0.2 units of z_e.

19, respectively. I have not smeared the distributions over the internal parton transverse momentum or experimental resolution. However, the delta function contributions (i.e. $\delta(1-z_e)$) arising from the direct two photon subprocesses $\bar{q}q \to \gamma\gamma$ and $gg \to \gamma\gamma$ have been spread over a z_e bin from 0.9 to 1.1 for display purposes. Even for the π^+p case, the direct contributions stand out clearly near $z_e = 1$.

The away-side z_e-distributions in (21) and (25) (Fig. 17b and c, respectively) are directly proportional to the Bremsstrahlung fragmentation functions, $D_{i \to \gamma}(z,Q)$. Experimental measurements would provide a determination of these interesting functions. In addition, the width of the $z_e = 1$ peak arising form the direct two photon subprocess $\bar{q}q \to \gamma\gamma$ and $gg \to \gamma\gamma$ provides information on the intrinsic transverse momentum of the partons within the initial hadrons.

ACKNOWLEDGEMENTS

This work was done in collaboration with Ed Berger and Eric Braaten. I would like to congratulate and thank Chris Berger for a very fruitful and enjoyable workshop.

REFERENCES

1. E. Berger, E. Braaten, and R. D. Field, "The Large p_T Production of Single and Double Photons in Proton-Proton and Pion-Proton Collisions", University of Florida preprint UFTP-83-10.

2. There are many estimates of single photon production available in the literature. See, for example, H. Fritzsch and P. Minkowski, Phys. Lett. 69B, 316 (1977); R. Ruckl, S. Brodsky, and J. Gunion, Phys. Rev. D18, 2469 (1978); F. Halzen and D. Scott, Phys. Rev. Lett. 40, 1117 (1978); Phys. Rev. D18, 3378 (1978); R. D. Field, "Dynamics of High Energy Reactions", plemary session talk at the XIX International Conference on High Energy Physics, Tokyo, 1978; L. Cormell and J. F. Owens, Florida State University preprint FSU-HEP-030780 (1980); R. Horgan and P. Scharbach, Nucl. Phys. B181, 421 (1981); A. P. Contogouris, S. Papadopoulos, and C. Papavassiliou, Nucl. Phys. B179, 461 (1981); F. Paige and I. Stumer, Proc. ISABELLE Summer Study, Brookhaven, 1981.

3. Direct two-photon production in hadronic interactions by the QED annihilation subprocess, $q\bar{q} \to \gamma\gamma$, was discussed as early as 1971 by S. Berman, J. D. Bjorken, and J. Kogut, Phys. Rev. D4, 3388 (1971).

4. R. N. Cahn and J. F. Gunion, Phys. Rev. D20, 2253 (1979).

5. The importance of the $gg \to \gamma\gamma$ was first pointed out by B. L. Combridge, Nucl. Phys. B174, 243 (1980). See also, C. Carimalo, M. Crozon, P. Kessler, and J. Parisi, Phys. Lett. 98B, 105 (1981): M. Krawsgyk and W. Ochs, Phys. Lett. 79B, 119 (1978).

6. The $gg \to \gamma g$ and $gg \to \gamma\gamma$ differential cross sections are arrived at by inserting the appropriate color factors into the $\gamma\gamma \to \gamma\gamma$ amplitudes calculated by B. De Tollis, Nuovo Cimento 35, 1182 (1965).

7. I have not included the effects of the parton intrinsic transverse momentum within the initial hadrons in the analysis presented here. The effect of "smearing" over the parton intrinsic transverse momentum is, however, important at low p_T and in regions where the cross section is steeply falling. For example, smearing over the parton intrinsic transverse momentum can raise the W=27.4 GeV single photon rates by as much as a factor of two. On the other hand, the ratios of the double to single photon production presented in Figs. 2, 8, 11, and 14 are not affected greatly by smearing since I have not required the two photons to balance transverse momentum. The ratio simply measures the fraction of photon triggers that contain a photon on the away-side.

8. A. Nicolaidis, Nucl. Phys. 163, 156 (1980).

9. I adopt here the notation of S. D. Ellis and M. B. Kislinger, Phys. Rev. D9, 2027 (1974).

DISCUSSION

<u>J. Badier (Ecole Polytechnique)</u>: What is the consequence of the quark
intrinsic transverse momentum for your calculations?

<u>R. D. Field (University of Florida)</u>: I have not included effects of the quark
intrinsic transverse momentum in the analysis presented here. These effects
are, however, important at low. p_T and in regions where the cross section is
steeply falling. For example, smearing over the parton intrinsic transverse
momentum can raise the W=27.4 GeV single photon rates by as much as a factor
of two. On the other hand, the ratios of the double to single photon
production presented in Figs. 2, 8, 11, and 14 are not affected greatly by
smearing since I have not required the two photons to balance transverse
momentum. The ratio measures the fraction of photon triggers that contain a
photon on the away-side.

<u>Vermaseren (NIKHEF-Amsterdam)</u>: Considering the "K-factors" in the Drell-Yan
process, wouldn't you expect a difference the data and the theoretical
estimates.

<u>R. D. Field (University of Florida)</u>: Your question is a good one. As is well
known, there are large order α_s corrections to the Drell-Yan production of
muon pairs in hadron-hadron collisions. This large correction to the
annihilation subprocess $q\bar{q} \rightarrow \gamma^* \rightarrow \mu^+\mu^-$ is predominately multiplicative and arises
from the continuation from the space-like deep inelastic region scattering
where the structure functions are defined to the time-like Drell-Yan region.
One picks up π^2 pieces from the double logarithms (i.e. $\log(-q^2)=\log(q^2)+\pi^2$).
It is likely that the two-photon subprocess $q\bar{q} \rightarrow \gamma\gamma$ will also have large

order α_s corrections. However, these corrections are very process dependent
and one must actually do the calculation before one can know for sure. I
might add that the pion structure functions have been determined from the
Drell-Yan process, $\pi p \rightarrow \mu^+ \mu^- + X$, using the leading order formulas and not from
deep inelastic scattering. For these structure functions one does not need to
continue from the space-like to the time-like region and hence one will not
pick up the same type of π^2 terms. It is likely that in this case the
order α_s corrections will not be large, but one really needs to do the
calculations.

P. Kessler (College de France, Paris): There is another interesting aspect
of $\gamma\gamma$ production (I shall discuss it tomorrow in the parallel session), namely
the search for massive structures coupled to two photons: massive quarkonium
states, massive gluonia and perhaps other exotic structures.

R. D. Field (University of Florida): You are right. I have calculated the
"background" to such processes. There may indeed be additional two-photon
contributions arising from the production and decay of massive quarkonium and
gluonium states.

D. Scott (Cambridge University): What can you learn about parton intrinsic
transverse momentum in $\gamma\gamma$ production that you can't learn in Drell-Yan?

R. D. Field (University of Florida): Nothing.

S. Brodsky (SLAC): I recall that R. Cahn and J. Gunion computed $\gamma\gamma \rightarrow gg$ and
found it a small (5%) contribution to $\gamma\gamma \rightarrow jet+jet$.

R. D. Field (University of Florida): Right. See Ref. [4]. The inverse process gg→γγ is relatively more important in γγ production because of the abundance of (low x) gluons within hadrons.

RESONANCE PRODUCTION IN γγ REACTIONS

J.E. Olsson

D E S Y, Hamburg, West Germany

Abstract

Experimental results on the exclusive production of resonances in γγ collisions are reviewed. These include new measurements of the radiative widths of the pseudoscalar (η, η') and the tensor mesons (f, A_2, f'). A comparison of these results with SU(3) is made. Upper limits for other states than f in $\gamma\gamma \rightarrow \pi\pi$ are given. The searches for γγ production of the states ι and Θ as well as η_c are presented and upper limits are given. Finally a limit is given for the rare decay $f \rightarrow \pi^+\pi^-2\pi^0$.

This review covers the recent results in single resonance production in $\gamma\gamma$ collisions. Reactions which involve double resonance production are dealt with in the talk by H. Kolanoski in these proceedings.

Resonance production in $\gamma\gamma$ collisions was first studied by using the Primakoff effect[1,2]; with a photon beam, resonances (π^o, η) were produced in the Coulomb field of nuclei. Today however, the main research is carried out at e^+e^- storage rings. The basic diagram is shown below :

$$e^+e^- \to e^+e^-R \tag{1}$$

A characteristic feature of this diagram is the dominating low Q^2 of the emitted photons and the scattering of the electrons at small angles. The produced system R is boosted along the beam direction and its decay products are the only particles detected in the event, since the outgoing electrons are normally not seen (notag). The overall transverse momentum p_t of the system R is balanced with respect to the beam and this also constitutes an important criterium for the selection of exclusive final states in the experimental analysis. Furthermore, the C-parity of the resonances is C = +1 and the spin is different from 1, since the photons are almost real[3].

The cross section for reaction (1) is given by

$$\sigma(e^+e^- \to e^+e^-R) = \int \sigma_{\gamma\gamma\to R}(s)\,L_{\gamma\gamma}(z)\,dz \tag{2}$$

Here s is the squared CM energy of the two photons and $\sigma_{\gamma\gamma}$ is the cross section for $\gamma\gamma \to R$. The density $L_{\gamma\gamma}$ is the luminosity function[4] for the two photons and the variable $z = \sqrt{s}/2E_b$, with E_b the electron beam energy. The cross section $\sigma_{\gamma\gamma}$ is given by the Breit-Wigner formula

$$\sigma_{\gamma\gamma\to R} = 8\pi(2J+1)\Gamma_{R\gamma\gamma}\frac{\Gamma_R}{(m_R^2 - s)^2 + m_R^2\Gamma_R^2} \tag{3}$$

where 2J+1 is a spin factor, $\Gamma_{R\gamma\gamma}$ is the decay width into two photons and Γ_R is the full width of R. For a narrow resonance, $\Gamma_R/m_R \ll 1$ and the Breit-Wigner can be replaced by a δ-function, so that

$$\sigma(e^+e^- \to e^+e^-R) = \frac{8\pi^2\Gamma_{R\gamma\gamma}}{m_R^2\,4E_b}\,L_{\gamma\gamma}\left(\frac{m_R}{2E_b}\right)(2J+1) \tag{4}$$

In this expression the luminosity function $L_{\gamma\gamma}$ is still a complicated function. By introducing the equivalent photon approximation[5], (4) simplifies to[6] :

$$\sigma\left(e^+e^- \rightarrow e^+e^-R\right) = \frac{16\alpha^2 \Gamma_{R\gamma\gamma}}{m_R^3} \left(\ln \frac{E_b}{m_e}\right)^2 f\left(\frac{m_R}{2E_b}\right) (2J+1) \qquad (5)$$

The Low function $f(z)$ is given by $f(z) = (2+z^2)^2 \ln(1/z) - (1-z^2)(3+z^2)$ and is a slow-ly varying function of resonance mass m_R and beam energy E_b. In (5) the dependence of the cross section $\sigma(e^+e^- \rightarrow e^+e^-R)$ on $\ln(E_b)$ as well as on the third power of the produced mass is clearly demonstrated.

This is all straightforward and a measurement of $\sigma(e^+e^- \rightarrow e^+e^-R)$ will give the corres-ponding value of the decay width of R into two photons, from (2), (4) or (5). However, there are several complications: firstly, the width $\Gamma_{R\gamma\gamma}$ refers to real photons and although the photons in reaction (1) have low Q^2, they are virtual and the cross section has a Q^2 dependence which is not explicit in (3). This Q^2 dependence can be taken into account by introducing form factors for the photons, either with the standard ρ -pole, $1/(1+Q^2/m_\rho^2)^2$ or with a GVDM Ansatz[7]. A difference in the measured $\Gamma_{R\gamma\gamma}$ of 5-10% may result. Experiments rarely take this into account. Secondly, for decaying resonances care has to be taken to use correct matrix elements to describe the decays. Angular correlations among decay products and other dynamic effects may influence the overall detection efficiency and thereby the measurement of $\Gamma_{R\gamma\gamma}$. An example discussed at this conference is the decay $\eta' \rightarrow \gamma\rho^0$, which is a dipole transition[8]; $M^2 \sim q^2 k^2 m_{\pi\pi}^2 \sin^2\theta |\text{Breit-Wigner}|^2$, where q and k are the momenta of the pions and the photon and θ is the decay angle of the pions, all in the ρ rest system. In the recent measurements of $\Gamma_{\eta'\gamma\gamma}$ in the reaction $e^+e^- \rightarrow e^+e^-\eta'$ [9-11], this matrix element seems not to be fully taken into account. A third complication is the small overall detection efficiency, typically a few % to a few ‰. This is caused by the boost along the beam axis, which together with the geometry of the detector tends to bring the final state particles out of the detector acceptance. The difficulties of triggering efficiently on an overall low energy final state at high energy e^+e^- storage rings also contributes, as well as the fact that the thresholds for detection of low energy particles, in particular photons, are in most detectors close to the typical energies of the involved particles.

All these considerations as well as other uncertainties combine to give relatively large systematic errors, typically 20% or more. Nevertheless the measurements are important, for several reasons. The main interest lies in the measurements of $\Gamma_{R\gamma\gamma}$. The absolute sizes of these widths are predicted in many models; in simple flavour SU(3) the quark content and mixing angle of the neutral mesons can be related to the relative sizes of the involved radiative widths. In some models glueballs mix with the standard $q\bar{q}$ mesons and again predictions are given about the radiative widths. Thus the measurements are needed to test such models. But there is also other interest in reaction (1). With increasing integrated luminosities at the storage rings, the possibility to discover new C = +1 mesons should be kept in mind. γγ collisions form a clean source of meson production and bump hunting is a vivid part of two photon physics. And last but not least, the large integrated luminosities offer possibilities to study rare decay modes of well known resonances. As an example, ~ 200 000 f mesons have so far been produced in each of the interaction regions at PETRA. Even with detection efficiencies in the 1% range, signals of rare decay modes may be detected and measured.

In the following, the present experimental data for the radiative widths of the pseudoscalars and the tensor mesons will be compared to the expectations from SU(3). It is therefore good to summarize the relevant formulae. In the meson nonet, the three neutral members are described by the following quark wave functions[12] :

$$|8,3\rangle = 1/\sqrt{2} \ (d\bar{d}-u\bar{u}) \qquad \pi^{o}, \ A_{2}^{o} \qquad \text{isovector}$$
$$|8,1\rangle = 1/\sqrt{6} \ (u\bar{u}+d\bar{d}-2s\bar{s}) \qquad \eta_{8}, \ f_{8} \qquad \text{isoscalar octet}$$
$$|1,1\rangle = 1/\sqrt{3} \ (u\bar{u}+d\bar{d}+s\bar{s}) \qquad \eta_{1}, \ f_{1} \qquad \text{isoscalar singlet}$$

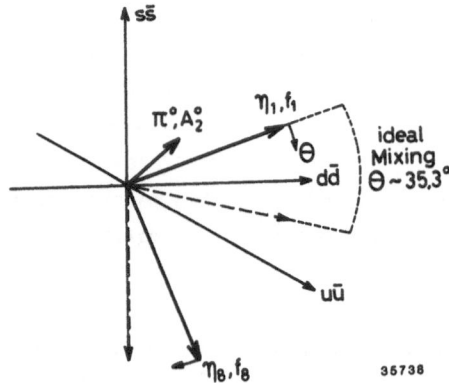

<u>Fig. 1</u>

Quark wave functions for the neutral members of the pseudoscalar and tensor meson nonets.

Fig. 1 shows these states in $u\bar{u}$, $d\bar{d}$ and $s\bar{s}$ space . The physical isoscalar particles are mixtures of the pure octet and singlet states[1]:

$$\eta = \cos\theta \, \eta_{8} - \sin\theta \, \eta_{1} \qquad f' = \cos\theta \, f_{8} - \sin\theta \, f_{1}$$
$$\eta' = \sin\theta \, \eta_{8} + \cos\theta \, \eta_{1} \qquad f = \sin\theta \, f_{8} + \cos\theta \, f_{1} \tag{6}$$

The mixing angle thus represents a rotation about the isovector axis. For a value of ~ 35.3°, the so called ideal mixing is obtained, with pure $s\bar{s}$ and pure $u\bar{u}$, $d\bar{d}$ states. With some further assumptions (e.g. equality of the quark magnetic moments), the couplings of the states to two photons can be related[12],

---

[1] In some textbooks, the states $|8,3\rangle=1/\sqrt{2}$ ($u\bar{u}-d\bar{d}$) and $|8,1\rangle=1/\sqrt{6}$ ($2s\bar{s}-u\bar{u}-d\bar{d}$) are used.

$$M_{\eta_0\gamma\gamma} = -\sqrt{\frac{1}{3}}\,M_{\pi^0\gamma\gamma} \qquad\qquad M_{\eta_1\gamma\gamma} = -\sqrt{\frac{8}{3}}\,M_{\pi^0\gamma\gamma} \qquad (7)$$

and one obtains the following relations between the corresponding radiative widths:

$$\Gamma_{\eta\gamma\gamma} = \Gamma_{\pi^0\gamma\gamma}\left(\frac{m_\eta}{m_{\pi^0}}\right)^3 \frac{1}{3}(\sqrt{8}\sin\theta - \cos\theta)^2$$

$$\Gamma_{\eta'\gamma\gamma} = \Gamma_{\pi^0\gamma\gamma}\left(\frac{m_{\eta'}}{m_{\pi^0}}\right)^3 \frac{1}{3}(\sin\theta + \sqrt{8}\cos\theta)^2 \qquad (8)$$

$$\Gamma_{\eta'\gamma\gamma} = \Gamma_{\eta\gamma\gamma}\left(\frac{m_{\eta'}}{m_\eta}\right)^3 \left(\frac{\sin\theta + \sqrt{8}\cos\theta}{\sqrt{8}\sin\theta - \cos\theta}\right)^2$$

$$\Gamma_{f'\gamma\gamma} = \Gamma_{A_2\gamma\gamma}\left(\frac{m_{f'}}{m_{A_2}}\right)^3 \frac{1}{3}(\sqrt{8}\sin\theta - \cos\theta)^2$$

$$\Gamma_{f\gamma\gamma} = \Gamma_{A_2\gamma\gamma}\left(\frac{m_f}{m_{A_2}}\right)^3 \frac{1}{3}(\sin\theta + \sqrt{8}\cos\theta)^2 \qquad (9)$$

$$\Gamma_{f\gamma\gamma} = \Gamma_{f'\gamma\gamma}\left(\frac{m_f}{m_{f'}}\right)^3 \left(\frac{\sin\theta + \sqrt{8}\cos\theta}{\sqrt{8}\sin\theta - \cos\theta}\right)^2$$

$$\frac{\Gamma_{A_2\gamma\gamma}}{\Gamma_{f\gamma\gamma}} = \left(\frac{m_{A_2}}{m_f}\right)^3 \frac{3}{(\sin\theta + \sqrt{8}\cos\theta)^2} \qquad (10)$$

Here the mass powers are phase space factors. Thus the measured radiative widths give values for the mixing angles which can be compared with the mixing angle values obtained in studies of other reactions involving these particles.

When presenting the experimental situation of today, the results divide naturally into three parts. After starting with the pseudoscalars, the tensor mesons are presented and discussed and finally the remaining results can conveniently be grouped under the heading "upper limits". But let us begin with the neutral members of the pseudo-scalar nonet. These are all well established mesons and the branching ratios into $\gamma\gamma$ are well known. The width of π^0 is known from measurements of the lifetime[13] as well as from photoproduction experiments[1]. It is here worthwhile to mention that a new experiment (NA30)[14] is under way at the CERN SPS, with the goal to measure τ_{π^0} to 1% precision (present accuracy 7%). The method consists essentially of producing π^0's in a gold foil via the proton beam and measuring the rate of conversion electrons behind a second foil; this rate depends on whether the pions decay before or after passing the second foil. Thus the rate of electrons as function of foil distance will allow a measurement of τ_{π^0} and thereby $\Gamma_{\pi^0\gamma\gamma}$, since the branching ratio $B(\pi^0 \to \gamma\gamma)$ is very well known, 0.98787 ± 0.00030[16]. The main run for the experiment is scheduled for the autumn 1983.

Measurements of $\Gamma_{\pi^0\gamma\gamma}$ at e^+e^- storage rings have not yet been performed, mainly because of the problems of triggering and detecting such a low energy final state. For the same reasons, the measurement of $\Gamma_{\eta\gamma\gamma}$ is a difficult task. However, at this conference the Crystal Ball group now reports the first measurement of the process $e^+e^- \to e^+e^-\eta$, $\eta \to \gamma\gamma$, at SPEAR. For this purpose, a special "topology" trigger was installed for low energy neutral final states. Essentially, the ball is divided into two opposite hemispheres in several different ways and the trigger condition is given by a certain minimum energy in each of the two hemispheres. The main background in such a trigger mode comes from cosmic radiation, which however can be well rejected with a series of cuts, using chamber information, energy patterns in the crystals and the timing of the energy measurement. The $\gamma\gamma$ mass spectrum for exclusively two photons in the final state, after all cuts, is shown in Fig. 2a. A clear peak at the η mass is seen. However, η's may also come from electroproduction off the residual gas in the vacuum pipe. To obtain information on this, data were also taken with separated beams and the corresponding mass spectrum is shown in Fig. 2b. Also here a peaking at the η mass is seen. Normalizing this spectrum to the colliding beam data in Fig. 2a and subtracting, the mass distribution in Fig. 3 is obtained. It is well fitted by a Gaussian close to the η mass and the width agrees with the values from other η spectra in the Crystal Ball. The total integrated luminosity is only 2.7 pb^{-1} and this accounts for the low statistics; nevertheless, one cannot enough stress the beauty of this spectrum.

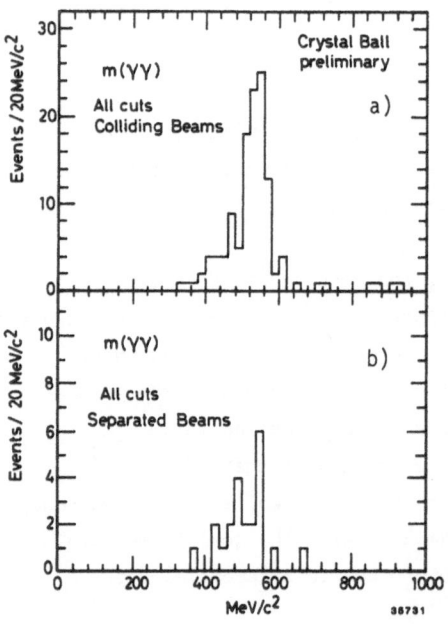

Fig. 2

a) Mass spectrum of 2γ in events with exclusively 2γ in the final state. Data with colliding beams.

b) Same, but data with separated beams. The normalization corresponds to about 37% of the data in a).

From these data, the group obtains the radiative width value,

$$\Gamma_{\eta\gamma\gamma} = 0.56 \pm 0.12 \text{ (stat.)} \pm 0.09 \text{ (syst.) keV} \qquad \text{(preliminary)}$$

which is larger than the previous measurement of Browman et al.[2] at Cornell in 1974. This experiment used the Primakoff effect and obtained the value $\Gamma_{\eta\gamma\gamma}$ = 324 ± 46 keV. These are the only measurements available so far. They represent two different experimental techniques, each with its own kind of systematic error sources. Comparing them, one should remember that the first one still has a large statistical error, which hopefully can be improved in the future; and the Cornell group also measured $\Gamma_{\pi^0\gamma\gamma}$[1], in good agreement with measurements of τ_{π^0}[13].

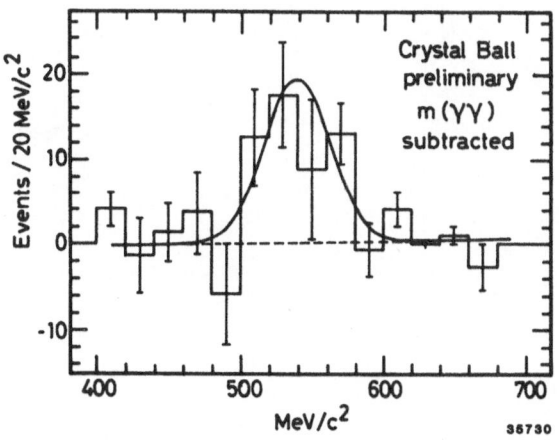

Fig. 3

Mass spectrum of 2γ in exclusive 2 γ events. The beam gas background in Fig. 2b has been normalized and subtracted from the data in Fig. 2a. The curve is a gaussian fit.

The γγ width of the third pseudoscalar, η', has now been measured by several groups in e^+e^- collisions. After the first measurement by the MARK II collaboration[9], also the CELLO[10] and JADE[11] collaborations published measurements last year. Like the MARK II collaboration, both groups used the decay mode η' → γρ⁰ and the final state is $\pi^+\pi^-\gamma$ exclusive. The two $\pi^+\pi^-\gamma$ mass spectra from such events are shown in Figs. 4a and b, respectively. In both figures, events with the $\pi^+\pi^-$ mass in a ρ band are shown as hatched distributions. The second peak in Fig. 4a is due to the incompletely reconstructed decay $A_2 \to \rho^\pm \pi^\mp \to \pi^+\pi^-\pi^0$, where a photon from an asymmetric π⁰-decay is lost. In Fig. 4b, events with $\pi^+\pi^-$ masses above 1 GeV/c² are excluded, which reduces the background in the A_2 region. For the width $\Gamma_{\eta'\gamma\gamma}$, the values $\Gamma_{\eta'\gamma\gamma}$ = 6.2 ± 1.1 (stat.) ± 0.8 (syst.) keV and $\Gamma_{\eta'\gamma\gamma}$ = 5.0 ± 0.5 ±0.9 keV were obtained, respectively.

At this conference a new, preliminary measurement by the TASSO collaboration is presented. Again, the decay mode η' → γρ⁰ is used; the mass spectrum of $\pi^+\pi^-\gamma$ is shown in Fig. 5 and exhibits a nice η' signal. The γρ⁰ decay is shown by the shaded distribution with the $\pi^+\pi^-$ mass limited to the ρ⁰ interval. The spectrum is in fact the sum of the two spectra obtained from the two different sets of shower counters used in TASSO to detect photons, liquid argon (LA) in top and bottom and lead-scintillator sandwich counters (SC) at the end of the lateral spectrometer arms. The data comprise 75 pb^{-1}. The value of the radiative width,

$$\Gamma_{\eta'\gamma\gamma} = 4.1 \pm 0.4 \text{ (stat.)} \pm 1.5 \text{ (syst.) keV} \qquad \text{(preliminary)}$$

is the mean of the two values obtained from the LA and SC parts, both of which however are close to 4 keV. The large systematic error is mainly connected with the Monte Carlo simulation, which is still under study. A similar measurement is under way in the PLUTO collaboration[15], but no value of $\Gamma_{\eta'\gamma\gamma}$ is available yet.

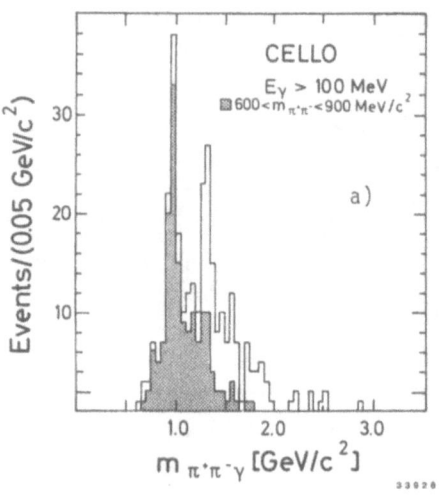
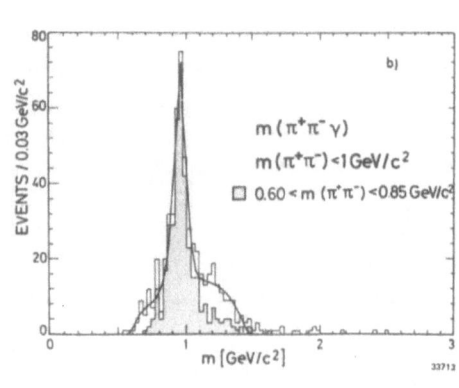

Fig. 4

Mass spectrum of the $\pi^+\pi^-\gamma$ system in exclusive $\pi^+\pi^-\gamma$ events. The shaded distributions correspond to events with the $\pi^+\pi^-$ mass in a ρ-band.
a) CELLO, data from 11 pb^{-1}. A cut, $p_t(\pi^+\pi^-\gamma) < 0.2$ GeV/c, has been applied.
b) JADE, data from 36 pb^{-1}. A cut, $p_t(\pi^+\pi^-\gamma) < 0.6$ GeV/c, has been applied.

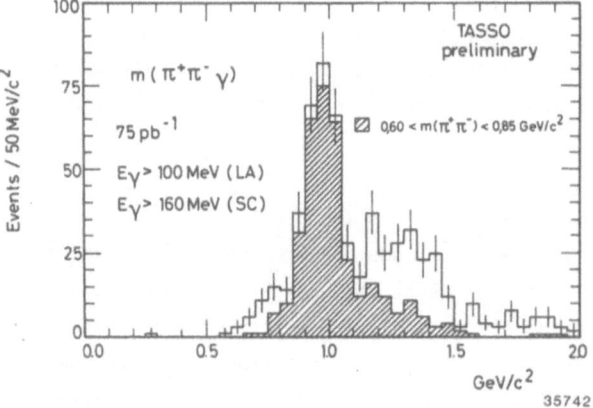

Fig. 5

Mass spectrum of $\pi^+\pi^-\gamma$, from events with exclusively two charged pions and one photon. The shaded distribution corresponds to events with the $\pi^+\pi^-$ mass in a ρ-band. A cut, $p_t(\pi^+\pi^-\gamma) < 0.1$ GeV/c, has been applied.

The present situation of $\Gamma_{\eta\gamma\gamma}$ and $\Gamma_{\eta'\gamma\gamma}$ measurements is summarized in Table 1. The $\Gamma_{\eta'\gamma\gamma}$ values agree well with the various predictions from fractionally charged quark models, 4-7 keV, while the large values, ~26 keV, from integrally charged quark models are clearly ruled out[16,29]. The mean value is 5.3 ± 0.6 keV. Taking the mean of only the e^+e^- measurements and using $B(\eta' \to \gamma\gamma) = 1.9 \pm 0.2\%$, the total width of η' is obtained, $\Gamma_{\eta'} = 276 \pm 45$ keV, which agrees very nicely with the missing mass measurement of Binnie et al.[17], $\Gamma_{\eta'} = 280 \pm 100$ keV. It is also interesting to note (Budney and Kaloshin[16]) that the measurement of the total width, which gives $\Gamma_{\eta'\gamma\gamma}$ by using $B(\eta' \to \gamma\gamma)$, together with the e^+e^- values give a measurement of the spin $J_{\eta'}$ of η', since $\sigma(e^+e^- \to e^+e^-\eta')$ is proportional to $(2J_{\eta'} + 1)\Gamma_{\eta'\gamma\gamma}$.

Quark model predictions for $\Gamma_{\eta\gamma\gamma}$ range from 300 - 800 eV[16]. Here the distinction between integral and fractional charge quarks is not so marked, although the data in Table 1 favours the latter. The measurement of $\Gamma_{\eta'\gamma\gamma}$ is in fact one of the strongest experimental evidences for fractionally charged quarks.

We now compare the measured values of $\Gamma_{\eta\gamma\gamma}$ and $\Gamma_{\eta'\gamma\gamma}$ with the SU(3) relations (8). This is done in Figs. 6a and b. Fig. 6a shows the widths $\Gamma_{\eta'\gamma\gamma}$ and $\Gamma_{\eta\gamma\gamma}$ as functions of the mixing angle Θ, with the width $\Gamma_{\pi^0\gamma\gamma}$ fixed at the Particle Data Group (PDG) value[18], 7.85 eV. The mean value of $\Gamma_{\eta'\gamma\gamma}$ and the two measurements of $\Gamma_{\eta\gamma\gamma}$ from Crystal Ball and Cornell are also shown, together with the corresponding ranges in the mixing angle. One sees that the $\Gamma_{\eta'\gamma\gamma}$ value and the Crystal Ball value of $\Gamma_{\eta\gamma\gamma}$ cover the same range, $\Theta \sim -18 \pm 5°$, while the Cornell measurement gives $\Theta \sim -7.5 \pm 2.5°$.

In Fig. 6b $\Gamma_{\eta'\gamma\gamma}$ is shown as function of Θ and of $\Gamma_{\eta\gamma\gamma}$; the bands from the two measurements of $\Gamma_{\eta\gamma\gamma}$ are indicated, together with the Θ ranges. Here a small region of overlap in Θ is present, which however would correspond to rather small values of $\Gamma_{\pi^0\gamma\gamma}$ (~7 eV). As comparison, one may note that the quadratic mass formula[18] gives the value $\Theta = -11.1 \pm 0.2°$. Larger negative values, $\Theta = -17$ to $-20°$ were found in a QCD calculation[19]; $\Theta = -16 \pm 2°$ was found in a study of the reaction $\pi^-p \to \eta'n \to \gamma\gamma n$[20].

When making comparisons like this, one should remember that the relations (8) are based on naive and simplifying assumptions. The mixing situation in the pseudoscalar nonet is certainly much more complex and to clarify the large deviation from ideal mixing encountered in the pseudoscalar nonet is a longstanding and famous theoretical problem[21]. It is however clear that the measurement of the $\gamma\gamma$ widths is an important input to its eventual solution and that in particular better measurements of $\Gamma_{\eta\gamma\gamma}$ in e^+e^- storage rings are needed.

* * *

Table 1

Radiative Widths of Pseudoscalar Mesons

Experiment	$\Gamma_{\eta\gamma\gamma}$ ± (stat.) ± (syst.) keV
Browman et al. Crystal Ball	0.324 ± 0.046 0.56 ± 0.12 ± 0.09 (prel.)
Weighted Mean :	0.344 ± 0.044 keV

Experiment	$\Gamma_{\eta'\gamma\gamma}$ ± (stat.) ± (syst.) keV
Binnie et al. MARK II CELLO JADE TASSO	5.4 ± 2.1 5.8 ± 1.1 ± 1.2 6.2 ± 1.1 ± 0.8 5.0 ± 0.5 ± 0.9 4.1 ± 0.4 ± 1.5 (prel.)
Weighted Mean :	5.3 ± 0.6 keV

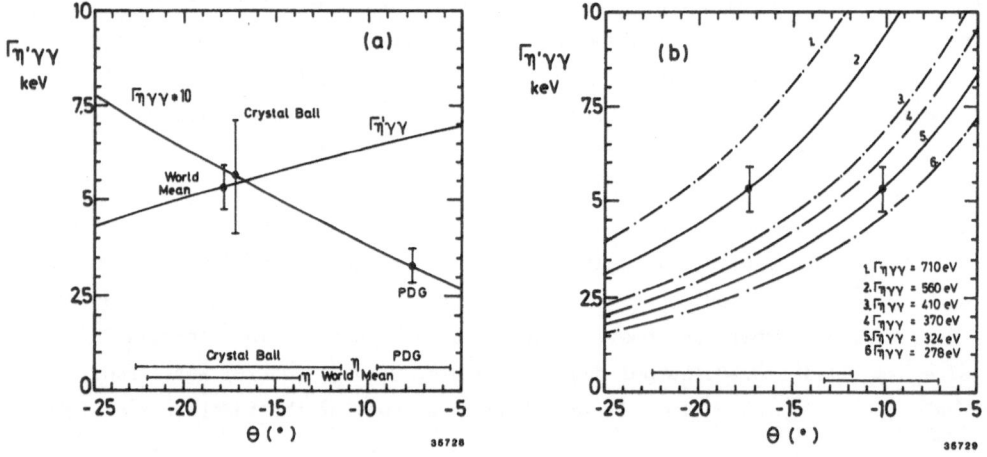

Fig. 6 a) $\Gamma_{\eta'\gamma\gamma}$ and $\Gamma_{\eta\gamma\gamma}$ as function of the mixing angle Θ, with $\Gamma_{\pi^{0}\gamma\gamma} = 7.85$ eV. $\Gamma_{\eta\gamma\gamma}$ is multiplied by 10. The measurements of $\Gamma_{\eta'\gamma\gamma}$ and $\Gamma_{\eta\gamma\gamma}$ are shown as data points. PDG stands for the measurement of $\Gamma_{\eta\gamma\gamma}$ from Cornell. The corresponding ranges in Θ are given by the errors.

b) $\Gamma_{\eta'\gamma\gamma}$ as function of mixing angle Θ and $\Gamma_{\eta\gamma\gamma}$. The curves correspond to the two measurements of $\Gamma_{\eta\gamma\gamma}$. The data point in each band is the mean value of $\Gamma_{\eta'\gamma\gamma}$ from Table 1.

The two photon production of the tensor mesons has also received a lot of work in the last years. The f(1270) meson has long posed a problem of measurement[22-24] in the reaction $e^+e^- \rightarrow e^+e^-f$, $f \rightarrow \pi^+\pi^-$, since the resonance appears together with a sizeable continuum from the reaction $\gamma\gamma \rightarrow \pi^+\pi^-$ and shows a distortion in mass and width. An early explanation for this was given by the MARK II group[23], who showed that the data could be explained with an interference between a continuum Born amplitude and a d-wave Breit-Wigner. At this conference, the CELLO collaboration presents

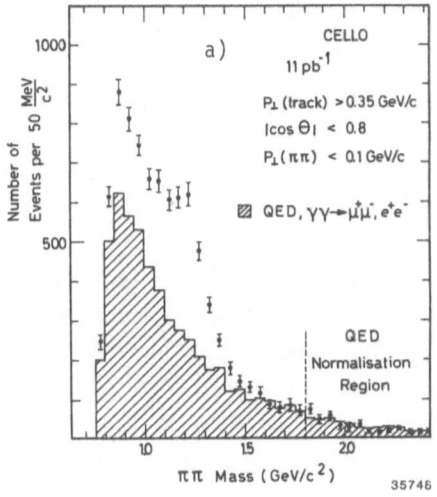
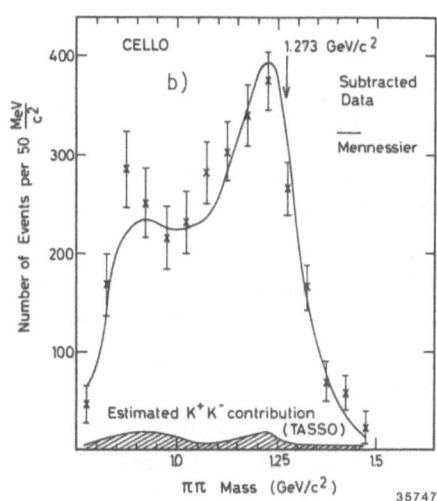

Fig. 7

a) $\pi^+\pi^-$ mass spectrum in exclusive two prong events. The hatched distribution shows the QED contribution.

b) The same data with the QED contribution subtracted. The hatched distribution shows the K^+K^- part. The curve is explained in the text.

a new analysis of this reaction. The mass spectrum of $\pi^+\pi^-$ from exclusive two prong events is shown in Fig. 7a. The data also contain events from the QED reactions $e^+e^- \rightarrow e^+e^-e^+e^-$ and $e^+e^- \rightarrow e^+e^-\mu^+\mu^-$. This contribution is calculated using the exact cross section[4] and is shown in the hatched distribution in Fig. 7a. It has been normalized to the data for $\pi\pi$ masses above 1.8 GeV/c^2, since other studies[25] show that the hadronic contribution is small at high masses. The calculated absolute rate however agrees with the event rate in this mass region and also other distributions are in excellent agreement with the QED expectations.

After subtraction of the QED contribution the mass spectrum in Fig. 7b remains. It shows clearly the f resonance situated on a large $\pi\pi$ continuum and with a visibly displaced mass. To analyze this spectrum, the model of G. Mennessier[26] has been used. In this model, the Born amplitude is unitarized, i.e. strong interactions in the final state are included, with help of measured $\pi\pi$ and KK phase shifts and inelasticities. The size of the K^+K^- contribution (misinterpreted as $\pi^+\pi^-$) is taken from the TASSO measurement[27]. In the version of the model used here, there is only one free parameter, namely $\Gamma_{f\gamma\gamma}$. The standard values[18] for f mass and full width are used as input in the calculations and pure helicity 2 is assumed for the f amplitude, in accordance with the result in Ref. 24 (dominance of helicity 2 was found in a study of the decay angular distribution of the reaction $\gamma\gamma \to f \to \pi^0\pi^0$). With this application of the Mennessier model the observed shift of the f mass of ~50 MeV/c^2 is again explained as due to interference between the Born term and the f resonance, constructive below and destructive above f. In the fit of the model to the data, shown in Fig. 7b, the following value for the radiative width is found,

$$\Gamma_{f\gamma\gamma} = 2.7 \pm 0.2 \text{ (stat.) } \pm 0.2 \text{ /syst.) keV} \qquad \text{(preliminary)}.$$

It is in good agreement with previous measurements[22-24]. It is however preliminary; the absolute normalization of the QED contribution (Fig. 7a) is still being studied.

The decay mode $f \to \pi^0\pi^0$ was first measured by the Crystal Ball group[24] at SPEAR and recently also by the JADE collaboration at PETRA. The JADE data are shown in Figs. 7 and 8. The final state is now 4 γ exclusive. The scatter plot in Fig. 8a of $\gamma\gamma$ mass vs. $\gamma\gamma$ mass (3 combinations per event) shows a strong signal of associated $\pi^0\pi^0$ production; the sum of the projections is shown in Fig. 8b. For the events in the $2\pi^0$ interval, the photon energies are now adjusted so that the final state is $2\pi^0$. The corresponding mass of $2\pi^0$ is shown in Fig. 9 and exhibits a clear f peak. The integrated luminosity is 32 pb^{-1} at 17.3 GeV beam energy. The radiative width of f, assuming pure helicity 2 in the production, is determined to be

$$\Gamma_{f\gamma\gamma} = 2.3 \pm 0.2 \text{ (stat.)} \pm 0.5 \text{ (syst.) keV} \qquad \text{(preliminary)}.$$

Apart from the f signal, only few events are present in the spectrum in Fig. 9. This was also found in the Crystal Ball analysis. The CELLO group has applied the Mennessier model also to this decay mode and finds that the model explains the Crystal Ball data well if the ω exchange term is suppressed, since its inclusion would lead to too high a level of $\pi^0\pi^0$ continuum. The level predicted by the final state rescattering process $\gamma\gamma \to \pi^+\pi^- \to \pi^0\pi^0$ agrees however. A small mass shift of f is also predicted, in agreement with the observed shift in the Crystal Ball analysis; the model does not, however, explain the large total f width, $\Gamma_f = 248 \pm 38$ MeV, observed in the same experiment. For the JADE data no values for the f mass and width in the $\pi^0\pi^0$ mass spectrum have been given yet.

It thus seems that interference between the $\gamma\gamma \to \pi^+\pi^-$ Born amplitude and the f resonance gives a satisfactory description of the $\pi^+\pi^-$ spectrum produced in $\gamma\gamma$ reactions. It should here be mentioned that also other explanations could be advanced, notably

Fig. 8

a) Scatterplot of m(γγ) vs. m(γγ) in exclusive 4γ events. All photon energies are > 90 MeV.

b) Sum of the projections of the scatterplot in a).

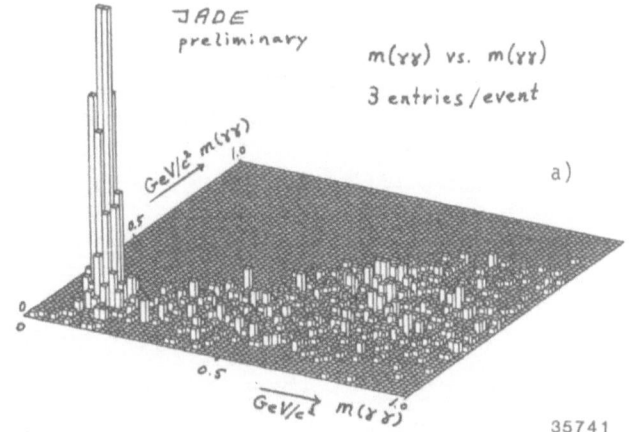

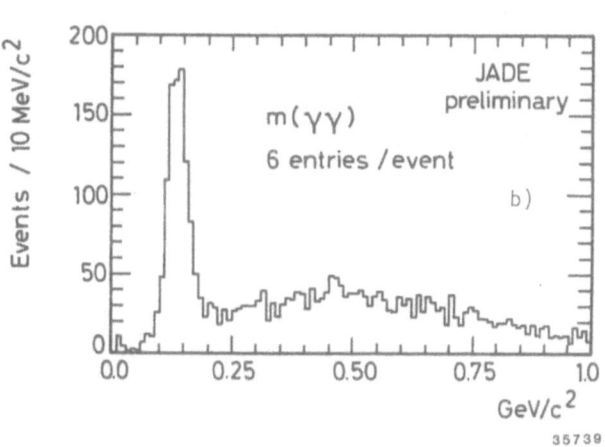

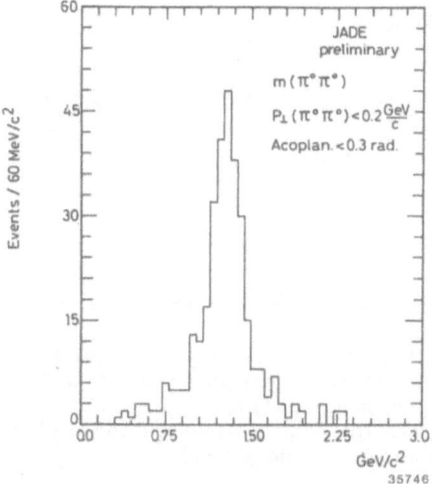

Fig. 9

Mass of 2π°, for events with two m(γγ) combinations in the π° band, 90-190 MeV/c². The photon energies have been adjusted so that m(γγ) = m_{π^0}.

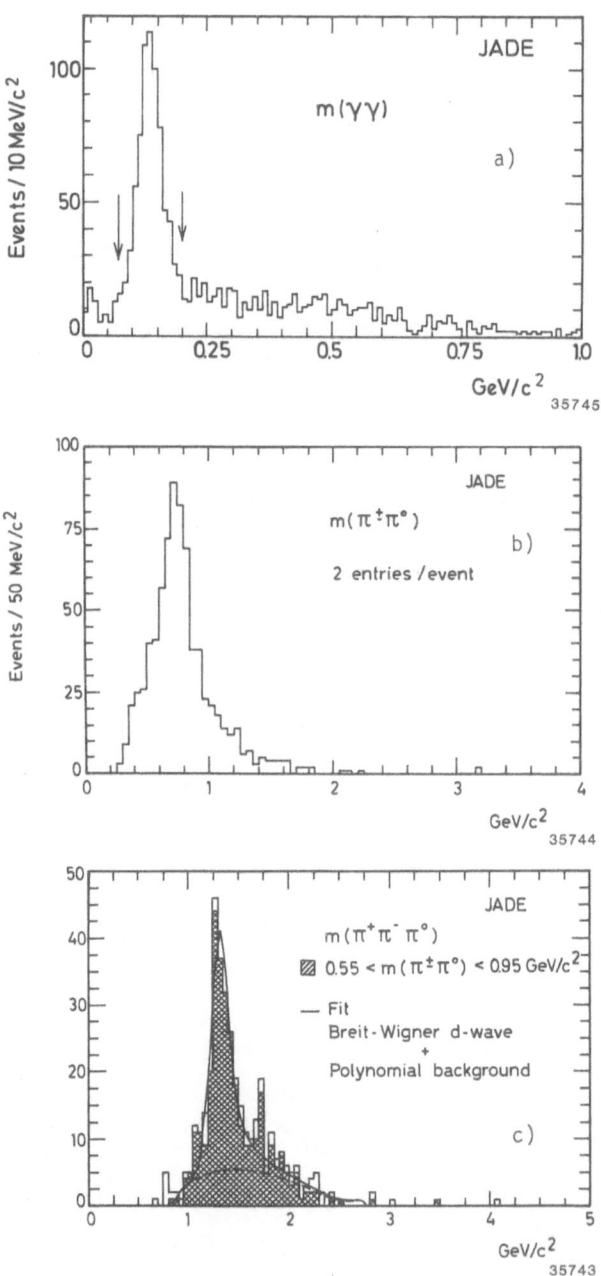

Fig. 10

a) Mass of 2γ in exclusive π⁺π⁻2γ events. The arrows indicate the π⁰ band, 70-200 MeV/c². Photon energies are > 90 MeV.

b) Mass of π⁺π⁰ with two entries per event. Only those events from a) with a m(γγ) combination in the π⁰ band are included and m(γγ) is adjusted to the π⁰ mass. A cut, $p_t(π⁺π⁻π⁰) < 0.2$ GeV/c, has been applied.

c) Mass of π⁺π⁻π⁰ for the same events as in b). The shaded distribution contains events with at least one π±π⁰ mass combination in the ρ band.

the presence of another resonance state close to the f, which could give rise to interference effects. Production of $\epsilon(1300)$ however would give a different angular distribution of the π's than the one observed and is probably small. A limit of $\Gamma_{\epsilon\gamma\gamma} \cdot B(\epsilon \rightarrow \gamma\gamma) < 1.5$ keV (95% C.L.) was set by the TASSO group[22]. The mixing of the f with a nearby 2^{++} gluonium state has also been suggested[28]; how to exclude or establish such a possibility is not clear.

The various measurements of $\Gamma_{f\gamma\gamma}$ are summarized in Table 2. The two values given for the Crystal Ball experiment correspond to the pure helicity 2 assumption and to the fit of different helicities to the $\pi^0\pi^0$ angular distribution. Only the former was used in calculating the weighted mean, $\Gamma_{f\gamma\gamma} = 2.8 \pm 0.2$ keV. The experimenal values all fall in the lower end of the large range of values which have been predicted by various models[29,34], ~ 1 keV to ~ 30 keV. Note here that many of the theoretical predictions are calculated with help of coupling constants determined from measurements of other decays; some of these measurements were grossly wrong, so that some predictions of $\Gamma_{f\gamma\gamma}$, if recalculated today, would change substantially.

$A_2(1320)$ production in $\gamma\gamma$ collisions was first seen by the Crystal Ball group, in the decay mode $A_2 \rightarrow \eta\pi^0 \rightarrow 4\gamma$. Last year the CELLO collaboration published[10] a measurement of $\Gamma_{A_2\gamma\gamma}$, using the decay mode $A_2 \rightarrow \rho^{\pm}\pi^{\mp} \rightarrow \pi^+\pi^-\pi^0$. Only one of the photons from the π^0 was observed and the A_2 peak therefore appears in the same mass spectrum as η', i.e. in the $\pi^+\pi^-\gamma$ spectrum (Fig. 4a). The JADE collaboration has made a complete reconstruction of the $2\pi^{\pm}2\gamma$ final state and now presents data corresponding to an integrated luminosity of 77 pb^{-1}. The mass spectrum of the two photons is shown in Fig. 10a; it is dominated by the π^0 peak. Again, for events with a $\gamma\gamma$ mass combination in the π^0 mass band, the photon energies are adjusted so that the final state is $\pi^+\pi^-\pi^0$. The $\pi^{\pm}\pi^0$ mass is shown in Fig. 10b with a clear $\rho^{\pm}$ signal. In fact, since there are two entries per event, almost all events are compatible with having a charged ρ. This is also obvious in Fig. 10c with the $\pi^+\pi^-\pi^0$ mass spectrum; the shaded distribution contains events with at least one $\pi^{\pm}\pi^0$ combination in the ρ band. The A_2 peak contains ~ 200 events. The value of the radiative width,

$$\Gamma_{A_2\gamma\gamma} = 0.84 \pm 0.07 \text{ (stat.)} \pm 0.15 \text{ (syst.)} \text{ keV} \qquad \text{(preliminary)},$$

is still preliminary, due to the not yet completed studies of the helicity structure of the reaction. The latter is studied with help of the angular distributions of the decay products. A preliminary result is 90% dominance of helicity 2, in good agreement with the findings for the f meson[24] and also with theoretical expectations for the tensor mesons[29]. The value of $\Gamma_{A_2\gamma\gamma}$ was obtained with the assumption of pure helicity 2, like the previous two measurements.

The three measurements of $\Gamma_{A_2\gamma\gamma}$ are summarized in Table 2. The mean value is $\Gamma_{A_2\gamma\gamma} = 0.82 \pm 0.13$ keV. Theoretical predictions again span a wide range of values [29], from 0.3 keV to 30 keV, and the same remarks must be made as above for $\Gamma_{f\gamma\gamma}$. In SU(3), the ratio of $\Gamma_{A_2\gamma\gamma}$ to $\Gamma_{f\gamma\gamma}$ is a function of the tensor nonet mixing angle. Table 2 lists the values of the ratio $\Gamma_{A_2\gamma\gamma}/\Gamma_{f\gamma\gamma}$ for those groups which have measured both $\Gamma_{f\gamma\gamma}$ and $\Gamma_{A_2\gamma\gamma}$. The weighted mean is 0.32 ± 0.05. Taking the ratio of the two mean values of $\Gamma_{A_2\gamma\gamma}$ and $\Gamma_{f\gamma\gamma}$ in Table 2, respectively, one obtains 0.30 ± 0.05.

TABLE 2

Radiative Widths of Tensor Mesons

Experiment	$\Gamma_{f\gamma\gamma} \pm$ (stat.) $\pm$ (syst.) keV	
PLUTO	$2.3 \pm 0.5 \pm 0.35$	
MARK II	$3.6 \pm 0.3 \pm 0.5$	
TASSO	$3.2 \pm 0.2 \pm 0.6$	
Crystal Ball	$2.7 \pm 0.2 \pm 0.6$	(helicity=2)
Crystal Ball	$2.9 \, {}^{+\,0.6}_{-\,0.4} \pm 0.6$	(helicity fit)
CELLO	$2.7 \pm 0.2 \pm 0.2$	(preliminary)
JADE	$2.3 \pm 0.2 \pm 0.5$	(preliminary)
Weighted Mean :	2.8 ± 0.2 keV	

Experiment	$\Gamma_{A_2\gamma\gamma} \pm$ (stat.) $\pm$ (syst.) keV	
Crystal Ball	$0.77 \pm 0.18 \pm 0.27$	
CELLO	$0.81 \pm 0.19 \pm 0.27$	
JADE	$0.84 \pm 0.07 \pm 0.15$	(preliminary)
Weighted Mean :	0.82 ± 0.13 keV	

Experiment	$\Gamma_{A_2\gamma\gamma}/\Gamma_{f\gamma\gamma}$	
Crystal Ball	$0.29 \pm 0.07 \pm 0.07$	
CELLO	0.30 ± 0.11	(preliminary)
JADE	0.36 ± 0.08	(preliminary)
Weighted Mean :	0.32 ± 0.05	

The SU(3) prediction, given by eq.(10) and including the phase space factor $(m_{A_2}/m_f)^3$, is shown in Fig. 13a as function of the mixing angle Θ. The two experimental ratios are shown as hatched bands. There is disagreement on the 1σ level with this simple SU(3) expectation and the experimental ratios do not limit the range of the mixing angle; however, one notes that the JADE value is in reasonable agreement with the ideal mixing expectation (~ 0.40).

It should be commented here that normally the phase space factor is neglected in this ratio and the value 0.36 (9/25) is quoted as the SU(3) ideal mixing expectation.

The two photon production of the third neutral tensor meson, f'(1515), was first observed by the TASSO collaboration[27] last year. It was seen in both decay modes, $f' \rightarrow K^+K^-$ and $f' \rightarrow K^o_s K^o_s$. In Fig. 11a the scatterplot of negative charge mass^2 vs. positive charge mass^2 is shown, as determined by TOF analysis in exclusive two prong events. The K^+K^- association is clearly seen. Fig. 11b shows the $\pi^+\pi^-$ mass (four combinations per event) from exclusive $4\pi^\pm$ events. A small K^o_s enhancement is visible. The shaded distribution shows the recoil $\pi^+\pi^-$ mass against those $\pi^+\pi^-$ combinations with a mass ($\pi^+\pi^-$) in a K^o band. The associated $K^o_s K^o_s$ contribution is now obvious. The corresponding mass distributions of K^+K^- and $K^o_s K^o_s$ are given in Figs. 11a and b, respectively. Both spectra show an f' signal. In Fig. 11b, the $\pi\pi$ combinations have been constrained in a fit to the K^o mass. In both samples the full statistics is used, 74 and 79 pb^{-1}, respectively. The analysis of these data samples is complicated by the fact that both f and A_2 decay into $K\bar{K}$ and they are close enough in mass to interfere with the f'. This interference is given by the following coherent sum(30),

$$\sigma_{\gamma\gamma\rightarrow K\bar{K}}(W_{\gamma\gamma}) = 40\pi/W^2_{\gamma\gamma} \cdot \left[\left| \Gamma_{f\gamma\gamma} \cdot B(f \rightarrow K\bar{K}) \right|^{1/2} \cdot BW(f) \right.$$

$$\pm \left| \Gamma_{A_2\gamma\gamma} \cdot B(A_2 \rightarrow K\bar{K}) \right|^{1/2} \cdot BW(A_2)$$

$$\left. + \left| \Gamma_{f'\gamma\gamma} \cdot B(f' \rightarrow K\bar{K}) \right|^{1/2} \cdot BW(f') \right]^2$$

where the different signs in the A_2 term refer to the different quark content in the two decay modes; the interference between A_2 and the isoscalars is destructive in the $K^o_s K^o_s$ decay.

The result of the analysis is shown in the fitted curves in Fig. 12a and b. The corresponding product of radiative width and branching ratio is

$$\Gamma_{f'\gamma\gamma} \cdot B(f' \rightarrow K\bar{K}) = 0.11 \pm 0.02 \text{ (stat.) } \pm 0.04 \text{ (syst.) keV } .$$

Pure helicity 2 was assumed for the production, in accordance with the experimental findings for f[24] and with theory[29]. Helicity 0 contributions would make the above value larger, due to the difference in angular distribution in the decay of f'. Part of the large systematic error is due to the uncertainties in the A_2 and f contributions, which come from the measured values of $\Gamma_{f\gamma\gamma}$ and $\Gamma_{A_2\gamma\gamma}$. This is also indicated in Fig. 12a.

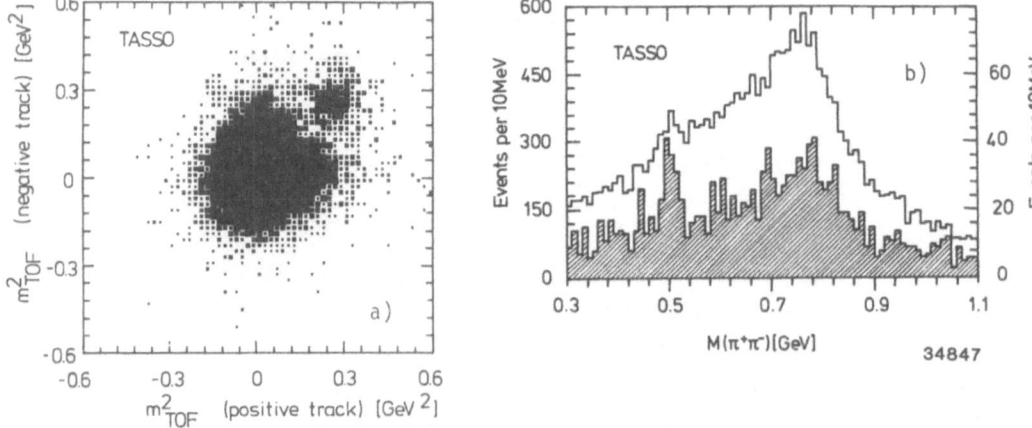

Fig. 11 a) Scatterplot of negative charge mass^2 vs. positive charge mass^2 for exclusive two prong events. The mass is determined by TOF analysis. b) Mass of $\pi^+\pi^-$, 4 combinations per event, in exclusive $4\pi^\pm$ events. The shaded distribution shows the $\pi^+\pi^-$ mass recoiling against a $\pi^+\pi^-$ combination within a K^0 band.

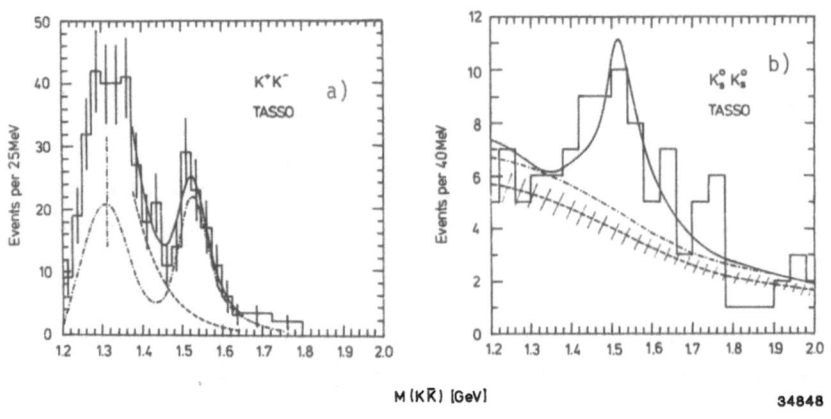

Fig. 12 a) Mass (K^+K^-) in exclusive K^+K^- events. A cut, $p_t(K^+K^-) < 0.15$ GeV/c, has been applied. The dashed curve shows the background in the fit and the dash-dotted curve the contribution from the interfering resonances f, A_2 and f'. The full curve shows the full fit. The error on the interference term is mainly due to uncertainties in $\Gamma_{A_2\gamma\gamma}$ and $\Gamma_{f\gamma\gamma}$. b) Mass ($K^0_S K^0_S$) in exclusive $4\pi^\pm$ events. A cut, $p_t(K^0_S K^0_S) < 0.15$ GeV/c, has been applied. The dashed curve shows the non $K^0_S K^0_S$ background and its uncertainty, the dash-dotted curve the full background, including the $K^0_S K^0_S$ part. The full curve shows the fit including the interference described in the text.

63

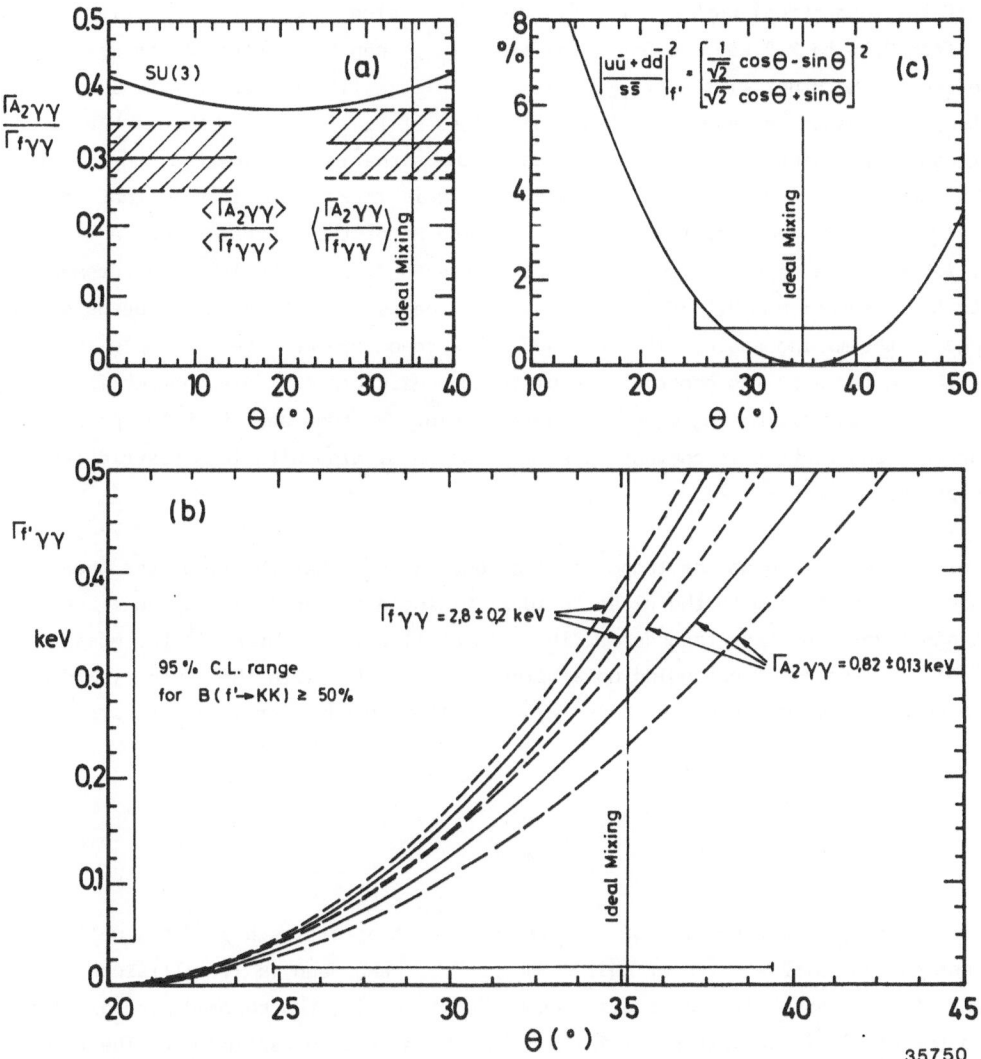

Fig. 13 a) The ratio $\Gamma_{A_2\gamma\gamma}/\Gamma_{f\gamma\gamma}$ as function of mixing angle Θ. The curve shows the SU(3) expectation, as given by eq.(10). The shaded bands are explained in the text. b) $\Gamma_{f'\gamma\gamma}$ as function of the mixing angle and the radiative widths $\Gamma_{A_2\gamma\gamma}$ and $\Gamma_{f\gamma\gamma}$. The dependence on the latter is given by eq.(9). The two bands correspond to the weighted means in Table 2. The range in Θ is given by the 95% C.L. range of $\Gamma_{f'\gamma\gamma}$ (the TASSO measurement and assuming $B(f'\to K\bar{K}) \geq 50\%$), through the intercepts with the f and A_2 bands. c) Percentage of non-strange quarks in f', as function of the mixing angle Θ. The intercept is the same range of Θ as determined in Fig. 13b.

Some of the theoretical work[29] for $\Gamma_{f\gamma\gamma}$ and $\Gamma_{A_2\gamma\gamma}$ also considers $\Gamma_{f'\gamma\gamma}$. Predictions range from 0.14 to 2.8 keV. In SU(3) the width $\Gamma_{f'\gamma\gamma}$ can be related to the mixing angle and to the other widths, $\Gamma_{A_2\gamma\gamma}$ and $\Gamma_{f\gamma\gamma}$. This is illustrated in Fig. 13b. Here the two curve bands correspond to the mean values of $\Gamma_{A_2\gamma\gamma}$ and $\Gamma_{f\gamma\gamma}$ from Table 2. Their non-overlapping is of course related to the situation in Fig. 13a; these values do not restrict the value of the mixing angle. It is interesting to note that for $\Theta \approx 19.5°$ the $\gamma\gamma$ width of f' vanishes, independent of the widths of f or A_2. In Fig. 13b also the range in $\Gamma_{f'\gamma\gamma}$ is shown, corresponding to the TASSO measurement and to the assumption of $B(f'\rightarrow K\bar{K}) \geq 50\%$. The intercepts with the A_2 and f bands define a range in the mixing angle, $\sim 25 < \Theta < 40°$. This range can be given a simple inter- pretation in terms of the content of non-strange quarks in f'. This content is clearly a function of the mixing angle and is shown in Fig. 13c. One sees that the present measurement restricts this content to at most a few %, also with less restrictive assumptions about $B(f'\rightarrow K\bar{K})$.

To finish this section on the tensor mesons, one may say that the $\gamma\gamma$ widths show reasonable agreement with the close to ideal mixing expected from studies of other reactions involving these resonances. There is still a large spread in the measure- ments of $\Gamma_{f\gamma\gamma}$ and a corresponding uncertainty in its ratio with the other $\gamma\gamma$ widths. More precise measurements of all these widths will be needed for more definite con- clusions.

* * *

We come now to the last section of this review. No other states beyond the well known low mass pseudoscalar and tensor mesons have so far been seen in $\gamma\gamma$ collisions. Searches for in particular the scalar mesons (ε, S* and δ), the charmed pseudoscalar η_c and the glueball candidates $\iota(1440)$ and $\Theta(1640)$ have been carried out. The results, in form of upper limits, are summarized below.

The $\pi\pi$ spectrum from $\gamma\gamma$ collisions has a particular interest in such searches, since the two scalar mesons ε and S* can be expected to show up here, both having major decay modes into $\pi\pi$. The $\pi°\pi°$ spectrum is the most sensitive place to look for such states, since it has no QED background to subtract, and a low level of $\pi°\pi°$ continuum. A limit for S* production was set by the Crystal Ball group[24]. At this conference the JADE group presents preliminary limits on the occurrence of narrow states between 0.55 and 1.0 GeV/c^2. These limits are derived from the mass spectrum in Fig. 9 and are shown as a curve in Fig. 14. Also shown is the S* limit from the Crystal Ball group. Since the branching ratio $B(S*\rightarrow\pi\pi)$ is now known, $78 \pm 3\%$[18], the following limits on $\Gamma_{S*\gamma\gamma}$ can be derived,

$\Gamma_{S*\gamma\gamma} < 1.0$ keV   95% C.L.   (Crystal Ball)

$\Gamma_{S*\gamma\gamma} < 0.8$ keV   95% C.L.   (JADE preliminary)

The limits in Fig. 14 can be compared with the absence of narrow states in the radiative decay $J/\psi \rightarrow \gamma\pi^0\pi^0$[31]. In the latter process narrow ($\Gamma < 100$ MeV) states with masses between 500 and 1000 MeV were searched for and a limit, $B(J/\psi \rightarrow \gamma X \rightarrow \gamma\pi^0\pi^0) < 1.3 \cdot 10^{-5}$, 95% C.L., was placed on their occurance. In contrast, the known radiative decays of J/ψ generally have branching ratios of the order 10^{-3}[18]. For further discussion, see Ref. 31.

$\overline{\Gamma}_{\gamma\gamma} \cdot B_{\pi\pi}$

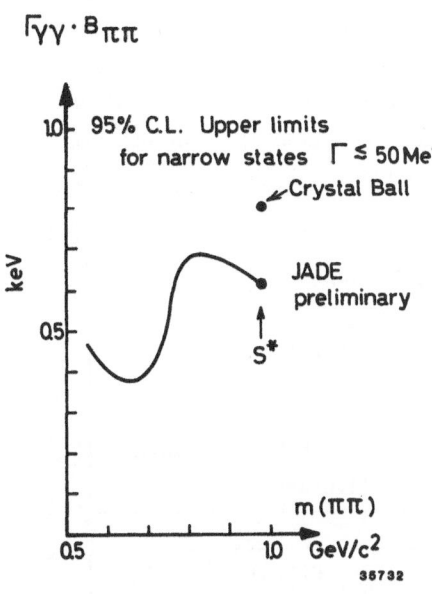

35732

Fig. 14

95% C.L. limit on the product of $\Gamma_{R\gamma\gamma} \cdot B_{\pi\pi}$, as function of the mass of the resonance R. The width of R is $\lesssim 50$ MeV and $B_{\pi\pi}$ is the branching ratio into $\pi\pi$.

Limits on the production of broad (i.e. with widths $\gtrsim$ a few hundred MeV) states below 1 GeV/c^2 in $\gamma\gamma$ reactions are not yet available. They would be of interest for the question of ε(or σ) production, e.g. in the DM1 experiment[32] at DCI, a broad low mass ε is invoked to explain the excess of $\pi^+\pi^-$ events above the expectation from the Born term calculation. For further discussion of $\gamma\gamma$ production of scalar states, see Refs. 16, 24, 26, 33 and 34.

The few events around 2 GeV/c^2 in Fig. 9 have been used by the JADE group to obtain a limit on $\gamma\gamma$ production of the h(2040) meson. This limit is given as

$$\Gamma_{h\gamma\gamma} \cdot B(h \rightarrow \pi\pi) < 0.22 \text{ keV} \quad 95\% \text{ C.L.} \quad \text{(preliminary)}$$

corresponding to $\sigma(e^+e^- \rightarrow e^+e^-h) \cdot B(h \rightarrow \pi\pi) \lesssim 60$ pb. Here isotropic decay of the h meson was assumed. In Ref. 34 $\sigma(e^+e^- \rightarrow e^+e^-f*)$, f* being the f recurrence, is estimated to ~ 0.4 nb.

The production of glueballs in $\gamma\gamma$ production is naively expected to be suppressed, compared with $q\bar{q}$ meson production. This is because the photons do not couple directly to the gluons, but only via an intermediate quark loop,

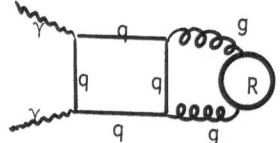

Since an extra quark loop is involved, one could expect an order of magnitude suppression, corresponding to the square root of the Okubo-Zweig-Iizuka-factor[35], $\sqrt{OZI} \sim 10$. This is at the limit of present statistics at PETRA and PEP. Nevertheless, glueballs have been searched for in $\gamma\gamma$ collisions. Of particular interest are the states $\iota(1440)$ and $\Theta(1640)$, which have been seen in radiative J/ψ decays[36] and which have no natural place in the SU(3) nonets. Since J/ψ radiative decays are a likely source of glueball production[37], both ι and Θ are now counted as candidates for such states. A number of models have been put forth that mix these states with the isoscalar members of the pseudoscalar and tensor nonets[38]. In such models, gluonium states may acquire larger $\gamma\gamma$ widths due to the mixing with $q\bar{q}$ states. Definite predictions may be made about decay modes for all the involved states, and also for widths of the decays into $\gamma\gamma$. Thus the results of the experimental searches are important, although presently only upper limits are given.

TABLE 3

Upper limits for $\gamma\gamma$ widths of ι and Θ

Experiment (Ref)	Decay Mode X	$\Gamma_{\iota\gamma\gamma} \cdot B(\iota \to X)$ 95% C.L.
MARK II[9]	$K\bar{K}\pi$	< 8.0 keV
TASSO[39]	$K\bar{K}\pi$	< 7.0 keV (preliminary)
TASSO[40]	$\rho^0\rho^0$	< 1.0 keV

Experiment (Ref.)	Decay Mode X	$\Gamma_{\Theta\gamma\gamma} \cdot B(\Theta \to X)$ 95% C.L.
Crystal Ball[31]	$\eta\eta$	< 0.3 keV
MARK II[41]	$K\bar{K}$	< 0.4 keV
TASSO[39]	$K\bar{K}$	< 0.3 keV (preliminary)
TASSO[40]	$\rho^0\rho^0$	< 1.2 keV

Several decay modes of ι and Θ have been investigated and the present limits are summarized in Table 3. Some of the data underlying these limits are shown in the talks by H. Spitzer and H. Kolanoski, these proceedings.

$K\bar{K}\pi$ is so far the only decay mode of ι which has been seen. If it indeed is the major decay mode, i.e. $B(\iota \to K\bar{K}\pi) \gtrsim 50\%$, then $\Gamma_{\iota\gamma\gamma} \lesssim 15$ keV. This is not yet restrictive although some models do predict a $\gamma\gamma$ width in this range. The limit involving $B(\iota \to \rho^0\rho^0)$ contradicts the large estimate in the model by Milton, Palmer and Pinski[38]. For Θ, the only known decay modes are $\eta\eta$ and $K\bar{K}$. A possible third decay mode is $\Theta \to \rho^0\rho^0$. Recently a resonance like enhancement in the radiative decay $J/\psi \to \gamma\rho^0\rho^0$, with the branching ratio $(1.25 \pm 0.35 \pm 0.40)\cdot 10^{-3}$, was reported[42]. If interpreted as a Breit-Wigner resonance, the mass and width of this state are very similar to those of $\Theta(1640)$. If these decay modes dominate, then the limits in Table 3 places limits on $\Gamma_{\Theta\gamma\gamma}$ that are in the same range as current predictions, $\lesssim 1$ keV.

The $\gamma\gamma$ production of $\eta_c(2980)$ is heavily suppressed due to the large mass, since the cross section is proportional to m_η^{-3}, cfr. eq.(5). However, theoretical estimates[43] give η_c a rather large $\gamma\gamma$ width, ~ 6 keV. This corresponds at PETRA and PEP beam energies to $\sigma(e^+e^- \to e^+e^-\eta_c) \sim 50$ pb and thus to ~ 4000 events presently at each PETRA interaction point. Small detection efficiencies and the smallness of the known decay modes of η_c again puts it at the limit of detection with the present statistics. The theoretical estimates are based on two different approaches : in the non-relativistic potential models[43],

$$\Gamma_{\eta_c\gamma\gamma} = \frac{12\alpha^2\,Q_c^4\,|R_s(0)|^2}{m_{\eta_c}^2} \qquad\qquad \Gamma_{J/\psi\,e^+e^-} = \frac{4\alpha^2\,Q_c^2\,|R_s(0)|^2}{m_{J/\psi}^2}$$

where $R_s(0)$ is the s-wave radial wave function at the origin and Q_c the charm quark charge. Since the mass difference of η_c and J/ψ is small, the assumption that the two wave functions are equal is made and

$$\Gamma_{\eta_c\gamma\gamma} = \frac{4}{3}\Gamma_{J/\psi\,e^+e^-} \quad \text{or} \quad \Gamma_{\eta_c\gamma\gamma} \simeq 6 \text{ keV} .$$

More careful considerations of the difference in $R_s(0)$ modify this estimate to 5.5 keV. Similar estimates are obtained with the QCD based dispersion sum rule calculations[43,44], in the range 4 – 7 keV. In view of the success with which the sum rule calculations have been applied to the charmonium system, a precise measurement of $\Gamma_{\eta_c\gamma\gamma}$, as well as the $\gamma\gamma$ widths of the heavier charmonium states $\chi_0(3415)$, $\chi_2(3555)$ and $\eta_c'(3590)$, will be an important test. The measurement of $\Gamma_{\eta_c\gamma\gamma}$ is therefore of great interest and almost all known decay modes of η_c[45] have been used for the searches. The results are summarized in Table 4. To give an idea of the present experimental sensitivity, Fig. 15 shows the $4\pi^{\pm}$ spectrum from the TASSO group, together with the gaussian fit which gives the present limit in this decay mode. Similarly the JADE data in Fig. 16 shows the $2\pi^{\pm}2\pi^0$ spectrum. Here all events in the η_c region were used to set the upper limit. One may conclude from Table 4 that both more data and better known decay modes of η_c are needed for the eventual measurement of $\Gamma_{\eta_c\gamma\gamma}$. The experimental limits, together with the poorly known branching ratios of η_c, give corresponding limits on $\Gamma_{\eta_c\gamma\gamma}$ of the order of 50-100 keV or more. The lowest limit so far is obtained by the Crystal Ball group[47], from the decay $J/\psi \to 3\gamma$: $\Gamma_{\eta_c\gamma\gamma} \lesssim 20$ keV, which is still well above the theoretical predictions.

It should be mentioned here that the decay $\chi_2(3555) \to \gamma\gamma$ has been observed by the

Crystal Ball group in the reaction $\psi' \to 3\gamma$ [46]. The reported values for the branching ratio, $B(\chi_2 \to \gamma\gamma) = (6 \pm 2) \cdot 10^{-4}$, and absolute width of $\chi_2(3555)$, $\Gamma_\chi = 2.1 {}^{+1.0}_{-0.7}$ MeV [46], correspond to a value $\Gamma_{\chi_2\gamma\gamma} = 1.3 {}^{+0.7}_{-0.6}$ keV. From the dispersion sum rule calculations the estimate 1.7 – 2.3 keV is given (Novikov et al. [43]). Similarly, for $\chi_0(3415)$ a limit can be obtained, $\Gamma_{\chi_0\gamma\gamma} < 9.2$ keV (90% C.L.). Here the values 4.6 – 5.4 keV have been predicted (Novikov et al. [43]).

Table 4
Upper Limits for $\gamma\gamma$ Production of η_c

Experiment (Ref)	Decay Mode X	$\Gamma_{\eta_c\gamma\gamma} \cdot B(\eta_c \to X)$ keV (preliminary)	$B(\eta_c \to X)$ [45,46] %
TASSO [39]	$p\bar{p}$	< 0.4 95% C.L.	$0.29 {}^{+0.30}_{-0.16}$
TASSO [39]	$K\bar{K}\pi$	< 27 95% C.L.	$3 \cdot (5.4 {}^{+3.3}_{-2.4})$
TASSO [39]	$2\pi^+2\pi^-$	< 0.7 95% C.L.	$2.0 {}^{+1.5}_{-0.9}$
JADE [39]	$\pi^+\pi^-2\pi^0$	< 4.2 95% C.L.	$(1-2) \cdot (2.0 {}^{+1.5}_{-0.9})$
JADE [39]	$\eta\pi^+\pi^-$	< 2.3 95% C.L.	$2.6 {}^{+1.8}_{-1.7}$

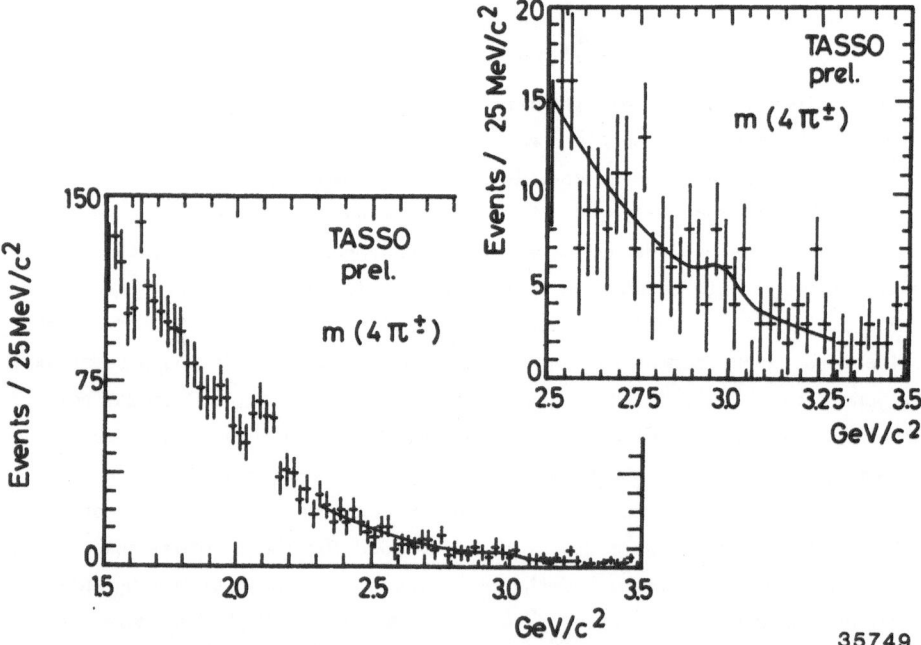

35749

Fig. 15 Mass spectrum of $4\pi^\pm$ in exclusive $4\pi^\pm$ events. The insert shows the same data in magnified scale. The fit is a smooth background with a gaussian resolution function corresponding to the known $4\pi^\pm$ mass resolution at the η_c mass.

Although the above results are still meager, it is clear that future larger data samples will provide important information on these radiative widths. Continued studies of the charmonium states and their decays, at the J/ψ and ψ' resonances, will be as important in this respect.

This review closes with an example of the possibility to use the large integrated luminosities for the study of rare decay modes of well known C = +1 resonances. The JADE group presents at this conference a search for the decay f → π⁺π⁻2π⁰. While the corresponding all charged decay f → 4π± is well known, 2.8 ± 0.4%[18], the partly neutral decay is poorly known. Expected is B(f → π⁺π⁻2π⁰) = (1-2)·B(f→4π±), depending on the isospin structure of the decay. The data are shown in Fig. 16. In these data, no background subtraction (from π⁰ sidebands) has yet been done and the word preliminary should be stressed. A more detailed presentation of these data, in particular concerning the presence of a ρ⁺ρ⁻ component, is given by H. Kolanoski elsewhere in these proceedings. In Fig. 15, events which are candidates for the process γγ → ρ⁺ρ⁻ have been removed and the decay f → ρ⁺ρ⁻ is not considered. All events below 1.6 GeV/c² are taken as candidates for the f decay. The rate is compared with the rate of e⁺e⁻ → e⁺e⁻f → e⁺e⁻π⁰π⁰, which was observed in the same experiment, and this gives the limits

$$B(f \rightarrow \pi^+\pi^-2\pi^0) \qquad\qquad < \ 0.12 \quad 95\% \ \text{C.L.} \quad \text{(preliminary)}$$
$$B(f \rightarrow \pi^+\pi^-2\pi^0)/(B(f \rightarrow \pi\pi) < \ 0.15 \quad 95\% \ \text{C.L.} \quad \text{(preliminary)} \ .$$

The latter value can be compared with the previous measurement [48], 15 ± 7%.

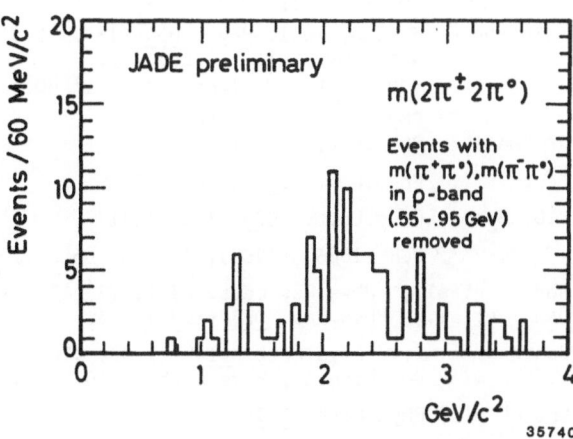

Fig. 16 Mass of 2π±2π⁰ in exclusive π⁺π⁻4γ events. All events with two π⁺π⁰ combinations in ρ bands have been removed. The γγ combinations in a π⁰ band were adjusted to the π⁰ mass. A cut, $p_t(2\pi^+2\pi^0) < 0.2$ GeV/c, has been applied. The data correspond to 77 pb⁻¹.

CONCLUSIONS

Resonance production in $\gamma\gamma$ reactions continues to provide much useful information to test our ideas about the quark structure of mesons. Large amounts of data have already been collected at SPEAR, PETRA and PEP and these data are being analyzed. To resolve questions about the $\gamma\gamma$ couplings of scalars, glueballs and heavy mesons like η_c, much higher statistics is needed. This will take some time to come, at least at PETRA, since present plans foresee running at highest possible energies, with corresponding lower luminosities. Much work however remains to be done with the existing data, not only in investigating new exclusive final states but also in the better understanding of the detectors and of the physics involved.

Acknowledgement

I wish to thank all my colleagues at DESY and SLAC for giving freely of their time and knowledge and for providing the material for this review. Thanks are also due to the organizers of this workshop, for providing such nice surroundings for a successful meeting.

REFERENCES

1. G. Bellettini et al.,   Nuovo Cim. 66A (1970) 243
    V.I. Kryshkin et al.,   JETP 30 (1970) 1037
    A. Browman et al.,   Phys.Rev.Lett. 33 (1974) 1400

2. C. Bemporad et al.,   Phys.Lett. 25B (1967) 380
    A. Browman et al.,   Phys.Rev.Lett. 32 (1974) 1067

3. C.N. Yang,   Phys.Rev. 77 (1950) 242

4. G. Bonneau, M. Gourdin and F. Martin, Nucl. Phys. B54 (1973) 573
    J. Field,   Nucl.Phys. B168 (1980) 477,
                                 and Erratum Nucl.Phys.B176 (1980) 545

5. R.B. Curtis,   Phys.Rev. 104 (1956) 211
    R.H. Dalitz and D.R. Yenni, Phys.Rev. 105 (1957) 1598

6. F.E. Low,   Phys.Rev. 120 (1960) 582

7. I.F. Ginzburg and V.G. Serbo, Phys.Lett. 109B (1982) 231

8. A. Rittenberg,   UCRL-18863 (1969), Thesis, unpublished

9. MARK II Coll., G. Abrams et al., Phys.Rev.Lett. 43 (1979) 477
    MARK II Coll., P. Jenni et al.,   Phys.Rev. D27 (1983) 1031

10. CELLO Coll., H.J. Behrend et al., Phys.Lett. 114B (1982) 378,
                                 and Erratum, Phys.Lett. 125B (1983) 518

11. JADE Coll., W. Bartel et al., Phys.Lett. 113B (1982) 190

12. F. Close,   An Introduction to Quarks and Partons
                                 (Academic Press, London, 1979).

13. G. von Dardel et al.,   Phys.Lett. 4 (1963) 51
    G. Bellettini et al.,   Nuovo Cim. 40A (1965) 1139.

14. CERN/SPSC/82-35 and CERN/SPSC/82-40   (Proposal)

15. PLUTO Coll., presented by M. Zachara in the parallel sessions,
    see H. Spitzer, these proceedings.

16. S. Matsuda and S. Oneda,   Phys.Rev. 187 (1969) 2107
    S. Okubo,   in Symmetries and Quark Models, ed. R. Chand
                                  (Gordon and Breach, New York, 1970)
    H. Suura, T.F. Walsh and B.-L. Young, Lett. Nuovo Cim 4 (1972) 505
    A. Bramon and M.Greco,   Phys.Lett. 48B (1974) 137
    F. Gault et al.,   Nuovo Cim. 24A (1974) 259
    A. Kazi, G. Kramer and D.H. Schiller, Lett. Nuovo Cim. 15 (1976) 120
    N. Isgur,   Phys.Rev. D13 (1976) 129
                                  and Erratum, Phys.Rev. D23 (1981) 817
    Etim-Etim and M. Greco,   Nuovo Cim. 42 (1977) 124
    V.M. Budnev and A.E. Kaloshin, Phys.Lett. 86B (1979) 351
    M. Chanowitz,   Phys.Rev.Lett. 35 (1979) 977
            ibid.,   Phys.Rev.Lett. 44 (1980) 59

17. D. Binnie et al.,   Phys.Lett. 83B (1979) 141

18. Particle Data Group,   Phys.Lett. 111B (1982) 1

19. A.T. Filippov,   Sov. J. Nucl.Phys. 29 (1979) 534

20. W.D. Apel et al.,   Sov. J. Nucl.Phys. 25 (1977) 300

21. M. Chanowitz,   Proc. SLAC Summer Institute on Particle Physics,
                                  1981

22. PLUTO Coll., Ch. Berger et al., Phys.Lett. 94B (1980) 254
    TASSO Coll., R. Brandelik et al., Z. Phys. C10 (1981) 117

23. MARK II Coll., A. Roussarie et al., Phys.Lett. 105B (1981) 304

24. Crystal Ball Coll., C. Edwards et al., Phys.Lett. 110B (1982) 82

25. PLUTO Coll., Ch. Berger et al., Nucl.Phys. B202 (1982) 189

26. G.Mennessier,   Z. Phys. C16 (1982) 241

27. TASSO Coll., M. Althoff et al., Phys.Lett. 121B (1983) 216

28. J.L. Rosner,   Phys.Rev. D24 (1981) 1347
    J.F. Donoghue,   Phys.Rev. D25 (1982) 1875

29. A rather complete list is found in
    PLUTO Coll., Ch. Berger et al., Phys.Lett. 94B (1980) 254
    In addition, and of newer date are, e.g.
    G.M. Radutskii,   JETP Lett. 6 (1967) 336
    Z. Kunszt, R.M. Muradyan and V.M. Ter-Atonyan, Budna Report E2-5424 (1970)
    D. Faiman, H.J. Lipkin and H.R. Rubinstein, Phys.Lett. 59B (1975) 269
    P. Grassberger and R. Kögerler, Nucl.Phys.B106 (1976) 451
    H. Krasemann and J.A.M. Vermaseren, Nucl. Phys. B184 (1981) 269
    P. Singer,   Phys.Lett. 124B (1983) 531
    L. Bergström, G. Hulth and H. Snellman, Z. Phys. C16 (1983) 263

30. H.J. Lipkin,   Nucl.Phys. B7 (1968) 321
        ibid.,   Proc. EPS Int. Conf. on High Energy Physics,
                                  Palermo 1975, p. 609
    D. Faiman et al.,   Ref. 29

31. K. Wacker,   Proc. of the XVIIIth Rencontre de Moriond,
                                  La Plagne, March 13-19, 1983

32. A. Coureau et al.,   Phys.Lett. 96B (1980) 402
    A. Falvard et al.,   Paper No. 48 submitted to the Int. Symposium
                                  on Lepton and Photon Interactions at High Energies,
                                  Bonn 1981

33. V.M. Budnev, A.N. Vall and V.V. Serebryakov, Sov. J. Nucl.Phys. $\underline{21}$ (1975) 531
    A.E. Kaloshin and V.V. Serebryakov, Tph-125, Novosibirsk 1981
    D.H. Lyth,   University of Lancaster preprint May 1982

34. B. Schrempp-Otto, F. Schrempp and T.F. Walsh, Phys.Lett. $\underline{36B}$ (1971) 463
                                       and private communication

35. S. Okubo,   Phys.Lett. $\underline{5}$ (1963) 105
    G. Zweig,   CERN Report 8182/TH.401 (1964)
    J. Iizuka et al.,   Prog. Theor. Phys. $\underline{35}$ (1966) 1061

36. Crystal Ball Coll., C. Edwards et al., Phys.Rev.Lett.$\underline{49}$ (1982) 259
                                       and Erratum, Phys.Rev.Lett. $\underline{50}$ (1983) 219
        ibid.,   Phys.Rev. Lett. $\underline{48}$ (1982) 458

37. J.D. Bjorken, Proc. Int. Conf. on High Energy Physics, Geneva 1979, p.245
    S. Meshkov,   Proc. Orbis Scientiae, Coral Gables, 1980, p.43
    M. Chanowitz,   Proc. SLAC Summer Institute, Stanford, 1981

38. C. Rosenzweig, A. Salomone and J. Schechter, Phys.Rev. $\underline{D24}$ (1981) 2545
    S. Iwao,   Lett. Nuovo Cim. $\underline{35}$ (1982) 481
    J.F. Donoghue,   Proc. XVIth Int. Conf. on High Energy Physics,
                                      Paris, 1982, p. C3-89
    K. Senba and M. Tanimoto,   Phys.Rev. $\underline{D25}$ (1982) 792
        ibid.,   Phys.Rev. $\underline{D26}$ (1982) 3270
    T. Teshima and S. Oneda,   Phys.Lett. $\underline{123B}$ (1983) 455
        ibid.,   Phys.Rev. $\underline{D27}$ (1983) 1551
    S. Ono and O. Pène,   Phys.Lett. $\underline{109B}$ (1982) 101
    K.A. Milton, W.F. Palmer and S.S. Pinsky, Proc. of the XVIIth Rencontre de
                                      Moriond, Les Arcs, France, March 20-26, 1982
    W.F. Palmer and S.S. Pinsky,   Phys.Rev. $\underline{D27}$ (1983) 2219
    J. Schechter,   Phys.Rev. $\underline{D27}$ (1983) 1109
    B. Li et al.,   Beijing Institute of High Energy Physics,
                                      preprint, BIHEP Th-82-10, June 1982
    J.L. Rosner,   Phys.Rev. $\underline{D24}$ (1981) 1347
        ibid.,   Phys.Rev. $\underline{D27}$ (1983) 1101
    J.L. Rosner and S.F. Tuan,   Phys.Rev. $\underline{D27}$ (1983) 1544
    E. Kawai,   Phys.Lett. $\underline{124B}$ (1983) 262
    J.F. Donoghue and H. Gomm,   Phys.Lett. $\underline{121B}$ (1983) 49
    S.-C. Chao,   University of Oregon preprint,OITS 205,
                                      Dec. 1982
    N. Aizawa, Z. Maki and I. Umemura, Kyoto University preprint, July 1982

39. TASSO and JADE Collaborations, data submitted to this conference

40. TASSO Coll., M. Althoff et al., Z. Phys. $\underline{C16}$ (1982) 13

41. D.L. Burke,   Proc. XVIth Int. Conf. on High Energy Physics,
                                      Paris, 1982, p. C3-513

42. MARK II Coll., D.L. Burke et al., Phys.Rev.Lett. $\underline{49}$ (1982) 632

43. V.A. Novikov et al.,   Phys.Reports $\underline{41C}$ (1978) 1
    T. Appelquist, R.M. Barnell and K.D.Lane, Ann.Rev.Nucl.Part.Sci. $\underline{28}$ (1978) 387
    M.A. Shifman,   Z.Phys. $\underline{C4}$ (1980) 345
    L. Bergström, H. Snellman and G. Tengstrand, Phys.Lett. $\underline{82B}$ (1979) 419

44. V.A. Novikov et al.,   Phys.Rev.Lett.$\underline{38}$ (1977) 626
        ibid.,   Phys.Lett. $\underline{67B}$ (1977) 409
    R. Kirschner and A. Schiller,   Z. Phys. $\underline{C16}$ (1982) 141
    L.J. Reinders, H.R. Rubinstein and S. Yazaki, Phys.Lett. $\underline{113B}$ (1982) 411

45. MARK II Coll., T.M. Himel et al., Phys.Rev.Lett. $\underline{45}$ (1980) 1146
    Crystal Ball Coll., R. Partridge et al., Phys.Rev.Lett. $\underline{45}$ (1980) 1150

46. M. Oreglia,   Proc. of the XVth Rencontre de Moriond,
                                      Les Arcs, France, March 15-21, 1980
    J.E. Gaiser,   Proc. of the XVIIth Rencontre de Moriond,
                                      Les Arcs, France, Jan.24-30, 1982

E. Bloom,   Proc. "Physics in Collision", Stockholm,
                                   1982

47. K. Königsmann,   Proc. XVIIth Rencontre de Moriond, Les Arcs,
                                   France, March 14-20, 1982

48. Y. Eisenberg et al.,   Phys.Lett. 52B (1974) 239.

Discussion

Q. P. SINGER (Haifa) : Concerning the new result for $\eta \rightarrow \gamma\gamma$, I wonder what happened
    with the very first measurement of Bemporad et al., which gave a value close
    to 1 keV? It seems that the Crystal Ball value comes inbetween the old DESY
    value and the Cornell value.

A. OLSSON : Browman et al. (Cornell 1974) state in their paper that their fit co-
    efficients are compatible with the DESY data from 1967 (Bemporad et al.), although
    the sets of coefficients are different in the two experiments. They suggest
    that the DESY data may not be sensitive enough to distinguish the two solutions,
    mainly due to limited energy range. The DESY measurement is not included by
    PDG in a mean value for $\Gamma_{\eta\gamma\gamma}$.

Q. J.A.M. VERMASEREN (NIKHEF) : What is known about the f in the tagged data ?

A. OLSSON : The TASSO group showed data on tagged f production in Paris 1981, but
    no signal could be claimed. No data were submitted to this review talk, although
    the PLUTO group will show some results in the parallel sessions. Again, no signal
    is seen.

    VERMASEREN : This shows also that the form factors are very strong.

Q. .S. BRODSKY (SLAC) : I don't believe it is appropiate to use a unitarized Born-
    term model based on point-like pions and kaons to estimate the background to
    resonance production in $\gamma\gamma \rightarrow \pi^+\pi^-$ and $\gamma\gamma \rightarrow K^+K^-$. The fall-off of the meson form
    factors should give an index of the expected suppression of the continuum con-
    tribution due to meson compositeness.

A. OLSSON : Experimentalists are of course happy to have at least one model that
    seems to explain the data, including the mass shift of the f meson and the large
    low mass continuum. But for the details of your question I must refer you to
    Mennessier.

    G. MENNESSIER (Montpellier) : To describe the nearby left hand cut, one needs
    Born terms. Notice that because of gauge invariance, one cannot introduce a
    true form factor for the pion (or kaon) exchange. Of course farther singularities
    are expected to become more important with higher energies and should be included.
    Several exchanges have been tried, with or without form factors. It turns out
    that a correct description of the data up to the f mass can be achieved with
    the pure point-like Born term.

    I agree that a model for the t and u channel singularities which would inter-
    polate between the Born behaviour at low energies and the scaling behaviour
    expected at high energies would be more satisfactory from a theoretical point
    of view. It does not seem necessary within the present status of experimental
    data.

Resonance Production in $\gamma\gamma$ Collisions

F. M. RENARD[*]

Stanford Linear Accelerator Center
Stanford University, Stanford, California 94305

1. Introduction

The first motivation[1] for $\gamma\gamma$ collisions was indeed the study of $C = +1$ resonances. After the pioneer works listed in Ref. 1, one can trace back from the four preceeding $\gamma\gamma$ workshops[2-5] and other high energy physics conferences[6] how the subject developed since this time. The theoretical concern progressively evolved from kinematical considerations (properties of the $e^+e^- \rightarrow e^+e^-X$ processes) to dynamical ones, namely resonances, soft hadronic processes, current algebra constraints, sum rules, duality relations and more recently hard (point-like) processes and QCD tests. Several phenomenological studies on the possible production of exotic particles (like technihadrons, Higgs bosons, supersymmetric particles, excited fermions, ...) have also been made but this is outside the scope of our review.

The processes $\gamma\gamma \rightarrow$ hadrons can be depicted as follows. One photon creates a $q\bar{q}$ pair which starts to evolve; the other photon can either (A) make its own $q\bar{q}$ pair and the $(q\bar{q}q\bar{q})$ system continue to evolve or (B) interact with the quarks of the first pair and lead to a modified $(q\bar{q})$ system in interaction with $C = +1$ quantum numbers. The main lines of evolution are, in case (A):

— each $q\bar{q}$ pair forms a vector meson V and both V's interact like in hadronic collisions with formation of resonances at low energy and with diffractive and peripheral scattering at high energy.

— the $q\bar{q}$ pairs make a quark rearrangement and hadronize (this can be described by perturbative QCD when the arrangement is due to hard gluons, i.e., when one has a large momentum transfer).

in case (B):

— the $q\bar{q}$ pair has some probability of making a bound quarkonium,

— it can quasi-freely evolve and make independent quark jets if the momentum transfer between the two photons (or the two quarks) is large enough.

*Work supported by the Department of Energy, contract DE-AC03-76SF00515 and by Centre National de la Recherche Scientifique, France.

— it can also hadronize into a small number of hadrons (this can again be described by QCD if the momentum transfer is large).

— it can annihilate into two or more gluons and make glueballs or gluon jets.

The theoretical challenge is the computation of the probabilities of each of these possibilities. I am supposed to review the recent theoretical activity concerning resonance production and related problems. I think that since the last (1981) $\gamma\gamma$ workshop the new aspect of the theoretical works precisely concerns the resonance spectroscopy and the description of exclusive processes. In particular a special attention has been paid to the unusual states like four quark, mixed quark and gluon bound states and glueballs. As a consequence we organize our review as follows:

Sec. 2: Hadronic $C = +1$ spectroscopy ($q\bar{q}$, $qq\bar{q}\bar{q}$, $q\bar{q}g$, gg, ggg bound states and mixing effects).

Sec. 3: Exclusive $\gamma\gamma$ processes (generalities, unitarized Born method, VDM and QCD).

Sec. 4: Total cross section (soft and hard contributions).

Sec. 5: q^2 dependence of soft processes (soft/hard separation, $1^{\pm+}$ resonances).

Sec. 6: Polarization effects.

Sec. 7: Conclusion.

2. Hadronic $C = +1$ Spectroscopy

2.1 STANDARD $q\bar{q}$ BOUND STATES

The low lying $C = +1$ states are $0^{-+}(^1S_0)$, $0^{++}(^3P_0)$, $1^{++}(^3P_1)$ and $2^{++}(^3P_2)$. Predictions for the $\gamma\gamma$ decay rates of the light quark bound states have been given a long time ago[1-8]. It is striking how many of the recent experimental results tend to favor the simplest picture based on the non-relativistic quark model. In this short section we just want to quote the recent calculations concerning light and heavy quark bound states and the new assignments for the 0^{++} nonet. The case of the 1^{++} states which cannot couple to two real photons will be discussed in Sec. 5.

0^{-+} states: $\pi^0, \eta, \eta', \eta_c, \eta_b \ldots$. Theoretical and experimental values of $\Gamma_{\gamma\gamma}$ are given in Table 1. $\Gamma_{\pi \to \gamma\gamma}$ is computed from the triangle anomaly in terms of f_π.[7] $\Gamma_{\eta \to \gamma\gamma}$ and $\Gamma_{\eta' \to \gamma\gamma}$ are then obtained by nonet symmetry ($f_\pi = f_{\eta_1} = f_{\eta_8}$).[8] Heavy quarkonia decay widths have been computed using the relativistic corrections calculated in Ref. 9 who give a factor 0.5 for η_c and 0.6 for η_b with respect to the non-relativistic case.[12]

0^{++} states. In Table 2 we tentatively assign $\epsilon(1425)$, $X^0(1770)$, $\vec{X}(1300)$ to the light scalar nonet[10,13] instead of the old choice $\sigma(700)$, $S^*(980)$, $\vec{\delta}(980)$. θ is the unknown mixing angle in this nonet. The $\Gamma_{\gamma\gamma}$ normalization has been fixed applying the factor $\frac{15}{4}$ to the quark model prediction for the 2^{++} nonet (see below). The heavy quarkonia decay widths have also been computed according to the results of Ref. 9 who give a correction factor 0.4 for $\chi_0(c\bar{c})$ and 0.6 for $\chi_0(b\bar{b})$ with respect to the non-relativistic case.

For completeness we want to quote old computations giving[8] 6.0 to 22.0 keV for $\sigma(700)$ and[11] $12.8 \left(sin\theta - \frac{cos\theta}{2\sqrt{2}}\right)^2$ keV for $S^*(980)$ and 4.8 keV for $\delta^0(980)$. However see Sec.2.2 for the four quark interpretation of these states.

2^{++} states: $f, f', A_2^0, \chi_2(c\bar{c}), \chi_2(b\bar{b}), \ldots$. Many different predictions have been given for the f meson (using tensor dominance of the energy-momentum tensor, finite energy or superconvergence sum rules, quark models $\ldots$)[8] which range from 1.0 to 12.0 keV. In Table 3 we took for reference the expressions of Ref. 11. With a mixing angle $\theta = 24^o$ or 35.3^o one would get 2.4 or 2.3 keV and 0.014 or 0.18 keV for $\Gamma_{f \to \gamma\gamma}$ and $\Gamma_{f' \to \gamma\gamma}$ respectively. There is also a recent calculation[14] on the basis of a Veneziano-type dual meson-meson amplitude which gives $\Gamma_{f \to \gamma\gamma} = 2.66 \pm 0.45$, $\Gamma_{f' \to \gamma\gamma} B_{f' \to K\bar{K}} = 0.141 \pm 0.039$, $\Gamma_{A_2 \to \gamma\gamma} = 0.90 \pm 0.36$ keV in good agreement with experiments. However a precise comparison with experimental results in the case of the f has to face the problems of mass and width shifts (see Sec. 3.2).

Heavy quarkonia predictions also come from Ref. 9, the relativistic correction factors being now 0.45 for $\chi_2(c\bar{c})$ and 0.75 for $\chi_2(b\bar{b})$.

Table 1
$\gamma\gamma$ Decay Widths of 0^{-+} Mesons

	π^0	η	η'	η_c	η_b
TH	7.6 eV	0.39 keV	6 keV	3 keV	0.25 keV
EXP	7.95 ± 0.55	$\begin{cases} 0.324\pm0.046 \\ 0.56\pm0.12 \end{cases}$	5.3 ± 0.6		

Table 2
$\gamma\gamma$ Decay Widths of 0^{++} Mesons

	$\epsilon(1425)$	$X^0(1770)$	$X^0(1300)$	$\chi_0(c\bar{c})$	$\chi_0(b\bar{b})$
TH	$(sin\theta + 2\sqrt{2}\,cos\theta)^2$ keV	$(cos\theta - 2\sqrt{2}\,sin\theta)^2$ keV	3 keV	1.4 kev	25 eV
EXP	$\Gamma_{\gamma\gamma}B_{\pi\pi} < 1.5$ keV				

Table 3
$\gamma\gamma$ Decay Widths of 2^{++} Mesons

	$f(1270)$	$f'(1515)$	$A_2^0(1320)$	$\chi_2(c\bar{c})$	$\chi_2(b\bar{b})$
TH	$0.28(sin\theta + 2\sqrt{2}\,cos\theta)^2$ keV	$0.28(cos\theta - 2\sqrt{2}\,cos\theta)^2$ keV	0.83 keV	0.5 keV	8.0 eV
EXP	2.8 ± 0.2	$\Gamma_{\gamma\gamma}B_{KK} = 0.11\,^{\pm0.02}_{\pm0.04}$ keV	0.82 ± 0.30		

<u>Excited states:</u>   In the light quark spectroscopy there are several candidates for orbital excitations:

$$2^{-+} \; : \; A_3(1680) \; , \; X(1820)$$

$$4^{++} \; : \; A_2(2000) \; , \; h(2040)$$

and for radial excitations:

$$0^{-+} \; : \; \pi'(1270) \; , \; \xi(1275)$$

$$2^{++} \; : \; (\pi f)_{1700}$$

$$2^{-+} \; : \; A_3^1(2100) \; .$$

No estimation of the $\gamma\gamma$ widths of these kinds of states has been given. One can expect that they will decrease with the degree of radial excitation like $\Gamma_{V\to e^+e^-}$ (one could for example use the ratios $\frac{\Gamma_{\rho'\to e^+e^-}}{\Gamma_{\rho\to e^+e^-}}$ and $\frac{\Gamma_{\psi'\to e^+e^-}}{\Gamma_{\phi\to e^+e^-}}$).

There are well-known $(\chi', \chi'', \ldots)$ excitations in the $c\bar{c}$ and $b\bar{b}$ spectroscopy. Predictions for their $\gamma\gamma$ decay widths have been given in Ref. 9.

2.2 $(qq\bar{q}\bar{q})$ States

The classification of these four quark states has been given by Jaffe.[13] Low lying (S-wave) states are obtained by coupling color and spin taking into account the exclusion principle and the mass shifts due to gluon exchanges. A distinction is made between exotic states $E(J^{PC}, \underline{n})$ and cryptoexotic states $C(J^{PC}, \underline{n})$. The first ones have Y, I_3 values which cannot be obtained in a $(q\bar{q})$ nonet; they necessarily pertain to flavor $SU(3)$ representations $\underline{n} > \underline{9}$ (i.e., $\underline{18}, \underline{18}^*, \underline{36}, \ldots$). The second ones pertain to any $\underline{n}$ but have Y, I_3 values that one can find in a nonet. The lowest states are the cryptoexotic nonets because they are the states which maximize the negative mass shifts due to gluon exchanges.

The lowest one is $(0^{++}, \underline{9})$ whose contents are: $u\bar{u}d\bar{d}; \frac{1}{\sqrt{2}}s\bar{s}(u\bar{u}+d\bar{d}); s\bar{s}d\bar{u}, \frac{1}{\sqrt{2}}s\bar{s}(u\bar{u}-d\bar{d}), s\bar{s}u\bar{d}; d\bar{s}u\bar{u}, u\bar{s}d\bar{d}; s\bar{d}u\bar{u}, s\bar{u}d\bar{d}$. Candidates are $\sigma(650), S^*(980), \delta(980)$ and $\kappa(900)$. These states should decay by fall-apart (quark rearrangement) in Pseudoscalar-Pseudoscalar channels and very little in Vector-Vector channels (see Table 4). This is why one gets large widths for $\sigma \to \pi\pi$ and $\kappa \to K\pi$; S^* and δ widths are limited $K\bar{K}$ phase space.

The second $(0^{++}, \underline{9}^*)$ nonet should be higher in mass (1450-1800 MeV) because of different recouplings of the four quark states which also favor VV decays with respect to PP (see Table 4).

There is also a $(2^{++}, \underline{9})$ nonet in the range (1650-1950 MeV) which should only decay into VV channels (Table 4).

Table 4
$qq\bar{q}\bar{q}$ States and $\gamma\gamma$ Widths[17,18]

	M(GeV)	Γ (GeV)	Recoupling PP	Recoupling VV	$\Gamma_{\gamma\gamma}$ (keV)
$(0^{++}, \underline{9})$	$\sigma(0.65)$ $S^*(0.98)$ $\delta(0.98)$	0.6 0.04 0.05	0.743	-0.041	8.? 0.27 0.27
$(0^{++}, \underline{9}^*)$	1.45 1.8 1.8	0.07 0.23 0.18	-0.177	0.644	1.7 0.8 0.1
$(2^{++}, \underline{9})$	1.65 1.8 1.95	0.04 0.57 0.58	0	$\sqrt{2/3}$	1.7 0.04 0.35
$(2^{++}, \underline{36})$	1.65 1.65 1.65 1.95 1.95 2.25	0.2 0.2 0.19 0.29 0.29 0.36	0	$\sqrt{1/3}$	1.26 1.23 0.3 0.02 0.17 0.02
$(0^{++}, \underline{36}^*)$	1.8 1.8 1.8 2.1 2.1 2.35	> 2 GeV	0.041	0.743	> 2 keV

Next higher representations with exotic and cryptoexotic states are $(0^{++}, \underline{36})$, $(0^{++}, \underline{36}^*)$ and $(2^{++}, \underline{36})$. There are also several 1^+ states: $(1^{+-}, \underline{9})$ and $(1^{+-}, \underline{36})$ which cannot couple to 2 photons because of $C = -1$ and $(1^+, \underline{18})$, $(1^+, \overline{18})$, $(1^+, \underline{18}^*)$, $(1^+, \overline{18}^*)$ which can mix into $C = \pm 1$ states; the $C = +1$ states could appear in the case of virtual photons.[15] States with high values of orbital momentum have been discussed in the framework of dual models and in connection with possible baryonium states.[16]

The $\gamma\gamma$ decay widths of these various four quark states have been computed[17,18] using VDM and their strong VV fall-apart decays. Results are summarized in Table 4. The first $(0^{++}, \underline{9})$ nonet is only weakly coupled to $\gamma\gamma$ because of its small VV fall-apart. This may explain why S^* and δ^0 are only weakly or not at all observed in present experiments. The case of the $\sigma(650)$ is not yet clear (see Sec. 3.2). The other cryptoexotic states get larger $\gamma\gamma$ widths because of their strong VV fall-apart. However simultaneously their large total width may render their identification difficult. The $\rho\rho$ enhancement may be partly due to such states.[17,18] Predictions for other VV channels have also been given. The $\theta(1640)$ could be tentatively assigned to a 4-quark state although its copious production by the 2-gluon channel in $\psi \to \gamma X$ would not be likely.

It is important to notice that with a recent non-relativistic potential model Weinstein and Isgur concluded[19] that only the lowest 4-quark states could exist and in fact as meson-meson bound states. Because of strong color mixing forces no other resonance state should be observed. The interpretation of $S^*(980)$ and $\delta(980)$ as $K\bar{K}$ bound states also came out from analyses of $\pi\pi$ and $K\bar{K}$ scattering amplitudes.[20]

2.3 $(q\bar{q}g)$ States

The existence of valence gluons is controversial. So far the description of these kinds of states has mainly been attempted with the bag model.[21] Such states get various names such as hermaphrodite,[22] hybrid[23] or meikton.[24] One expects them to be rather stable because of large color magnetic forces between the $(q\bar{q})$ octet and the gluon octet. The lowest gluon states in the bag correspond to transverse electric $TE(J^{PC} = 1^{+-})$ and transverse magnetic $TM(1^{--})$ radiations. Masses are computed by minimizing the total energy contribution to the bag. A recent overall fit of $(q\bar{q})$, $(q\bar{q}g)$ and (gg) states has been done by Chanowitz and Sharpe[24] including the important effects of gluon self-energy and interpreting the $\iota(1440)$ as a 0^{-+} glueball.

The low lying spectrum is obtained with $(q\bar{q})$ in S-wave ($^1S_0$ and $^3S_1$) and either a $TE(1^{+-})$ gluon giving 1^{--}, 0^{-+}, 1^{-+}, 2^{-+} states or a $TM(1^{--})$ gluon giving 1^{+-}, 0^{++}, 1^{++}, 2^{++} states (see Table 5). Decays should proceed in a first step by $g \to q\bar{q}$ and in a second step by $(q\bar{q}q\bar{q})$ hadronization.[23] A distinction is expected[23] between TE and TM states. The first ones should have smaller widths because in the non-relativistic limit the $q\bar{q}$ state issued from the gluon is in a P-wave and leads to small overlap integrals. The second ones should have normal hadronic widths because they are in S-wave.

$\gamma\gamma$ decays can be estimated with *VDM* and the strong *VV* channels in the case of 0^{++} and 2^{++} *TM* states (Table 5).

These $(q\bar{q}g)$ states could also be reasonably produced in $\psi \rightarrow \gamma X$ decays. The rates should be half way between those of $(q\bar{q})$ states and those of (gg) glueballs. $\iota(1440)$ and $\theta(1640)$ do not fit with $(q\bar{q}g)$ states because of their too large experimental widths.

The appearance of exotic states (1^{-+}) in the $(q\bar{q}g)$ spectrum has been especially advertised[22] but they are not easily producible in $\gamma\gamma$ collisions (see Sec. 5 and Ref. 15).

Table 5
$(q\bar{q}g)$ States and $\gamma\gamma$ Widths[23,24]

	M(GeV)	Γ (GeV)	Modes	$\Gamma_{\gamma\gamma}$
1^{--}	1.83	?	PP	–
0^{-+}	1.41	0.015	PS	small
1^{-+}	1.61	0.003	PA, PB	(virtual γ's) small
2^{-+}	1.97	0.002	PT	small
1^{+-}	2.0	?	PP, PS, PA, PT	–
0^{++}	1.6	0.2	PP, VV, PB	$\simeq 0.4$ keV
1^{++}	1.8	0.06	PV, PS, PA, PT	(virtual γ's) small
2^{++}	2.2	0.15	VV	$\simeq 1.2$ keV

Each state appears in a nonet of flavor-like ρ, ω, ϕ, K^* with mass differences due to strange quarks. In this table we give the mass value for the ρ, ω-like states. Chanowitz and Sharpe[24] described the *TE* states. We placed the *TM* states about 0.2 GeV higher in mass because of the higher magnetic gluon radiation energy.

2.4 (gg) AND (ggg) GLUEBALLS

Pure gluonic bound states are predicted in QCD by lattice calculations and have been studied phenomenologically in bag models and in non-relativistic potential models. Results have also been obtained by ITEP sum rules.

Progress in lattice calculations has for example been reviewed by Berg.[25] There are still many quantitative uncertainties however the mass of the lowest state (0^{++}) generally falls around or below 1 GeV and the next states $(0^{-+}, 1^{+-}, 2^{++})$ come out between 1 and 2 GeV.

ITEP sum rules[26] give quite different results with the 2^{++} at 1.5 GeV but the 0^{-+} around $2 \div 2.5$ GeV and the 0^{++} around 4 GeV.

A systematic classification of glueballs has been attempted with the bag model.[27-29] With the TE and TM gluon radiation states (defined in Sec. 2.3) and Bose statistics one gets the lowest states:

$$TE \times TE \; : \; 0^{++}, \; 2^{++} \qquad\qquad TE \times TM \; : \; 0^{-+}, \; 2^{-+} \; .$$

Identifying the 0^{-+} with $i(1440)$ Chanowitz and Sharpe[24] obtained the spectrum: $0^{++}(1.2 \pm 0.5)$, $2^{++}(2.15 \pm 0.4)$ and $2^{-+}(2.30)$. Three gluon states were predicted with 0^{++}, 1^{+-}, 3^{+-} around 1.45 GeV and 0^{-+}, 1^{-+}, 2^{-+}, 2^{--}, 1^{--}, 3^{--}, 3^{-+} around 1.8 GeV[29] when gluon-self energy effects are neglected.

Confining potential models predict additional states formed with longitudinal components for massive (effective) valence gluons; for example there are now 1^{-+}, 2^{-+} and 3^{-+} states for (gg) glueballs and many others for (ggg) glueballs.[30] It was noticed that the level ordering is much dependent upon the type of potential which is used (and equivalently upon the boundary conditions used in the bag model).[31]

With an effective gluon mass $m_g = 500$ MeV Cornwall and Soni[32] obtained the following (gg) spectrum: $m(0^{++}) \simeq 1200$ MeV, $m(0^{-+}) \simeq 1400$ MeV, $m(2^{++}) \simeq 1900$ MeV, $m(1^{-+}) \simeq 1500$ MeV, $m(2^{-+}) \simeq 1800$ MeV and a (ggg) 0^{-+} bound state at 2400 MeV.

The decay process of glueballs is still controversial. A somewhat standard claim is that because $gg \rightarrow q\bar{q}$ involve an α_s factor the corresponding width should be halfway between an OZI forbidden decay and a normal hadronic decay, i.e., a few tens of MeV for a 1.5 GeV glueball. This is not obvious because it is not clear what value of Q^2 and then what value of $\alpha_s(Q^2)$ will control these decays. Cornwall and Soni[32] claim that the process $(gg) \rightarrow g + g$ is perfectly allowed. If gluons then hadronize non-perturbatively (for example, $gg \rightarrow q\bar{q}q\bar{q}$) one can get a normal hadronic decay width (50-200 MeV). There are also possibilities of strong mixing with other hadronic states which could drive these decays. Decay modes should a priori be flavor singlet but large quark mass effects in the basic amplitude and in hadronization as well as phase space effects will spoil this property. Lipkin[33] argues for a similarity with $\psi \rightarrow 3g \rightarrow$ hadrons; however with a lower mass the effects quoted above could be more important than for the ψ. Also mixing with nearby non-singlet states will completely modify the basic pattern.[34,40]

Consequently $\gamma\gamma$ decays of glueballs are very model dependent and variable from one state to the other; widths can lie between the two extreme values $\sqrt{\dfrac{\Gamma_{OZI}}{\Gamma_{allowed}}} \; \Gamma_{(q\bar{q}) \rightarrow \gamma\gamma}$ and $\Gamma_{(q\bar{q}) \rightarrow \gamma\gamma}$. If the decay of a glueball into two vector mesons is known obviously VDM can be applied in order to get the $\gamma\gamma$ decay.

Glueball candidates are $i(1440)$ and $\theta(1640)$, the 0^{-+} and 2^{++} states observed in $\psi \rightarrow \gamma X$.[35]

They are not (yet) observed in $\gamma\gamma$ experiments which only give upper limits:[6,36]

$$\Gamma_{\gamma\gamma} B_{i \to KK\pi} < 8\,\text{keV}$$

$$\Gamma_{\gamma\gamma} B_{i \to \rho\rho} < 1.0\,\text{keV}$$

$$\Gamma_{\gamma\gamma} B_{\theta \to \eta\eta} < 5\,\text{keV}$$

$$\Gamma_{\gamma\gamma} B_{\theta \to \rho\rho} < 1.2\,\text{keV}$$

$$\Gamma_{\gamma\gamma} B_{\theta \to K\bar{K}} < 0.3\,\text{keV}$$

The very different pattern of $\psi \to \gamma X$ and of $\gamma\gamma \to X$ certainly is an important hint for the glueball interpretation. Broad enhancements have been observed in $\pi^- p \to (\phi\phi) + n$ just above the $\phi\phi$ threshold. It has been proposed to interpret[37] them as due to the presence of 2^{++} states $g_t(2160)$, $g_t'(2310)$ which could be glueballs.

In addition the problems associated with the $f(1270)$ production in $\gamma\gamma$ collisions (mass and width shifts, small $\gamma\gamma$ width) and the strong $\rho\rho$ production is suggestive for the mixing of the f with a nearby state which could be a glueball.[27,28,40]

2.5 MIXING EFFECTS

The $\gamma\gamma$ landscape corresponding to the $q\bar{q}$, $qq\bar{q}\bar{q}$, $q\bar{q}g$, gg and ggg spectroscopy is depicted in Fig. 1. It appears to be rather rich but also rather intricate especially because of the large widths of the $qq\bar{q}\bar{q}$ states. In addition the basic states described in the previous sections do not necessarily correspond to the physical states because of mixing effects. Basic mixing terms are:

— quarkonia mixing ($q\bar{q} \leftrightarrow q'\bar{q}'$); for example the ones responsible for the deviation to ideal structure (strange/non-strange mixing).

— $q\bar{q} \leftrightarrow gg$ mixing. Such a term is important for hadronic decay of glueballs. It has also been invoked for explaining anomalous properties of isosinglet pseudoscalar mesons (η, η', η_c) by giving them some pure glue component.

— $qq\bar{q}\bar{q} \leftrightarrow gg$ mixing. This is another possibility for hadronic decay of glueballs.

— $q\bar{q}g \leftrightarrow q\bar{q}q\bar{q}$. It is the basic term for hadronic decays of ($q\bar{q}g$) mesons.

The mixing amplitudes may be due to simple point-like couplings between quarks and gluons (the so-called annihilation diagrams) or to intermediate single or multi-body hadronic states (the so-called unitary corrections with complex amplitudes). The second case can always "accidently" happen when two nearby states have the same quantum numbers and some common decay channels which induce these "unitary" terms. The rich spectroscopy expected between 1 and 2 GeV certainly offers many such possibilities.

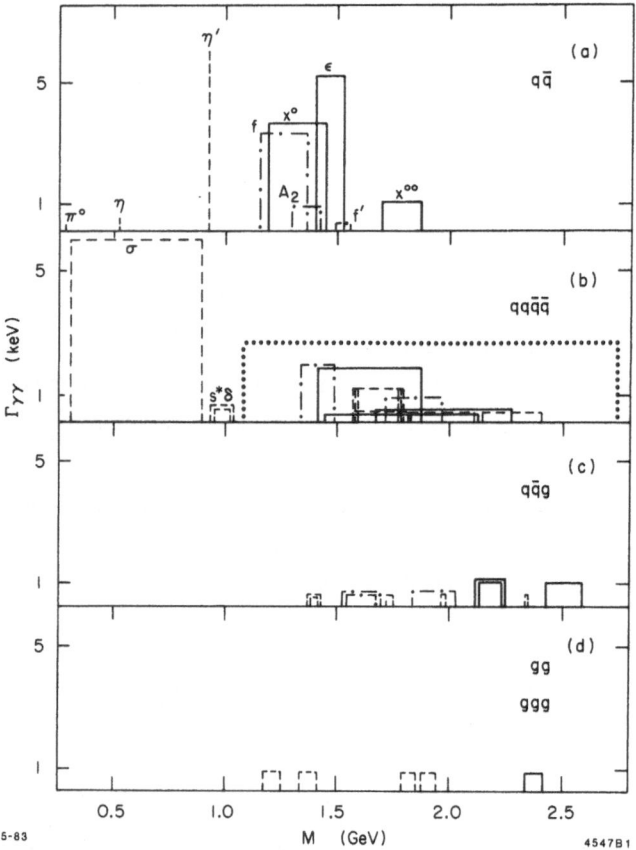

Fig. 1. Predictions for $\gamma\gamma$ widths of: a) $q\bar{q}$ states, b) $qq\bar{q}\bar{q}$ states, c) $q\bar{q}g$ states, d) gg and ggg states.

Fundamental aspects of mixing are on the one hand to allow couplings a priori forbidden for one of the states (i.e., $s\bar{s} \to$ non-strange hadrons; $(gg) \to$ hadrons, ...) and on the other hand to forbid some couplings by destructive interference between components. This last property can always be accidental for a particular channel but it is systematic for the dominant channels of mixed degenerate states (decoupling theorem[28,38] well known in quantum mechanics and particle physics).

Applications have been done for glueballs by several groups. An $\eta-\eta'-G$ mixing was tried for explaining the η, η' anomalies in ψ decays and recently G was identified with $i(1440)$. Fishbane et al.,[39] recently showed that this does not work with a 3×3 orthogonal mass mixing, the minimal value for the 0^- glueballs mass being 2 GeV. An $f - f' - G$ mixing was also used because of the problems in f, f' production in ψ decay; here the 2^+ G state was taken as the $\theta(1640)$. This also fails to work as recently explained by Rosner and Tuan.[40-42] The modes $f' \to \pi\pi$, $K\,K$, $\gamma\gamma$ are inconsistent with the smallness of the mode $\theta \to \pi\pi$ as being due to a destructive interference between $q\bar{q}$ and gg components. It was observed that an ad-hoc $f - \theta$ mixing could work while the f' is left unmixed; it is however difficult to understand why a non-strange $q\bar{q} \leftrightarrow gg$ term would exist whereas a strange $s\bar{s} \leftrightarrow gg$ would not.

It may well be that these applications are oversimplified again because of the rich spectroscopy in this energy range. The mixing formalism could be improved by considering s-dependence and imaginary parts[38,43] but at least a qualitative understanding should appear before going into very technical and perhaps artificial solutions. It is possible that the θ is not a glueball (it could be a $q\bar{q}g$ or a $qq\bar{q}\bar{q}$ state) or is not the glueball which mixes with f and f'. Other narrower states could both or separately mix with them. It is also possible that the spectroscopy in the θ region is more complex; new results coming from ψ decays and from $\gamma\gamma$ processes could clarify this situation.

3. Exclusive $\gamma\gamma$ Processes

3.1 GENERALITIES

Although in both cases one deals with a non-hadronic initial state there is a great difference between $\gamma\gamma \to$ hadrons and $e^+e^- \to \gamma^* \to$ hadrons. Not only are the J^{PC} different but the general analytic structures of the amplitudes are completely different. For example, a two-body amplitude $\gamma\gamma \to A + B$ has both left-hand cuts (due to t- or u-channel exchanges like Born terms) and right-hand cuts (final state interactions including possible resonances). Any reliable analysis of a particular $\gamma\gamma \to$ hadrons process should consider these various contributions. Resonances may come from rescattering in the final state or may also have a direct coupling to $\gamma\gamma$. A multi-channel analysis (K-matrix type) can disentangle these possibilities. A very important related question is the role of threshold effects. They may be more important than in $e^+e^- \to$ hadrons because of enhancements due to light particle exchanges (for example pions) in the crossed channels. Again analyticity should tell us how to separate a threshold enhancement from a true resonance. An example of such analysis has been recently given by Mennessier[44] for $\gamma\gamma \to \pi\pi$ at low energy. Earlier works in this domain can be found in Brodsky's review at the Paris colloquium in 1973.[2] It would be interesting to try similar studies for other simple channels like $\gamma\gamma \to VP$ and $\gamma\gamma \to VV$ where there are many new states to look for. At least we would like to ask for a minimal caution when fitting experimental cross-sections with broad resonances especially near below or above a threshold (for example $\gamma\gamma \to \rho\rho$). Modified Breit-Wigner forms including finite width effects consistent with analyticity and unitarity should be used.[38,45] The output parameters should be very sensitive to the correctness of the treatment. In these cases a comparison of the resonance effects in different channels (with different threshold locations, like $\rho\rho$ and $\rho\gamma$) should be very instructive.[46]

In the soft domain there are several properties to check. There are low energy theorems coming from gauge invariance, current algebra and soft pions. For example, there exist precise predictions for the $\gamma\gamma \to n$ pion amplitudes just above the thresholds.[47] Cross sections are rather high in the case of even n but low for odd n. Contact terms like $\gamma\gamma\rho\pi$, $\gamma\gamma\rho\pi\pi$, $\gamma\gamma A_2\pi$ should give some threshold enhancements.[48] $\gamma\gamma$ amplitudes should agree with VDM and the hadron-like behavior of the photons. Tests can be done either in the resonance region (when $VV - R$ couplings are known) or outside (when VV scattering amplitudes are known).

When the momentum transfer Q between the two photons increases these "soft" contributions should decrease like powers of $\frac{m^2}{Q^2}$ where m is an hadronic mass. One enters in the "hard domain" defined by large W and Q where point-like contributions should dominate. QCD results are reviewed elsewhere at this conference.[49] Here we only want to quote a few contributions for exclusive channels $\gamma\gamma \to M\bar{M}$[50] and $\gamma\gamma \to B\bar{B}$[51] in order to compare the behaviors of the cross sections in soft and hard domains. We want to underline the fact that the transition between these two domains is not well understood.

3.2 $\gamma\gamma \to M\bar{M}$ AT LOW ENERGY; UNITARIZED BORN METHOD

Mennessier[44] proposed a method for treating the strong interaction effects in the $M\bar{M}$ channel according to analyticity and unitarity constraints. The model for $\gamma\gamma \to M\bar{M}$ consists in:

— "Born terms": gauge invariant (P, V) meson exchanges and contact terms.

— final $M\bar{M}$ scattering in agreement with $\pi\pi$, πK and $K\bar{K}$ phase shifts.

— possible direct couplings of resonant states to $\gamma\gamma$.

This method allows a very interesting discussion of the respective role of background (Born terms), of resonances in the final state and of states directly coupled to $\gamma\gamma$. A first application [44] to $\gamma\gamma \to \pi\pi$ in the DCI-SPEAR-PEP-PETRA energy range $W \leq 1.4$ GeV led to the following conclusions:

— A state with $M \simeq \Gamma \simeq 600$ MeV (the $\sigma(600)$?) is required with a direct coupling corresponding roughly to $\Gamma_{\gamma\gamma} \simeq 8$ keV.

— Only a weak $S^*\gamma\gamma$ coupling is necessary (this would agree with the interpretations for the S^* to be either a $(qq\bar{q}\bar{q})$ or a $(K\bar{K})$ bound state).

— The $f^0\gamma\gamma$ direct coupling is approximately the standard one. An apparent mass shift appears from the interference with the non-resonant amplitudes. However this mass shift is always larger (by -30 MeV) in the $\pi^+\pi^-$ channel than in $\pi^0\pi^0$. If the disagreement with experiment would persist it would be a signal for the need of another nearby state.

3.3 VDM DESCRIPTIONS

For real photons the VDM hypothesis consists in replacing the photon state $|\gamma >$ by the series $\Sigma_V \frac{eg_{V\gamma}}{m_V^2}|V >$ where the summation extends to ρ, ω, ϕ in the "restricted" version and to series ρ, ρ', ρ'', ... in the "extended" version. For any $\gamma\gamma$ process we would write:

$$R(\gamma\gamma \to F) = \sum_{V,V'} \frac{e^2 g_{V\gamma} g_{V'\gamma}}{m_V^2 m_{V'}^2} R(VV' \to F) \ .$$

Such a formula supposes an extrapolation procedure from $q^2 = m_V^2$, $q'^2 = m_{V'}^2$ to $q^2 = q'^2 = 0$. The "VDM assumption" generally consists in taking a gauge invariant amplitude which coincides for onshell vector mesons with the strong $VV' \to F$ amplitude and which has the weakest q^2, q'^2 dependence. There may nevertheless be some ambiguities. There are also constraints from Bose statistics; for recent applications see for example Refs. 15 and 18. Clear VDM predictions and tests correspond to cases where the $VV' \to F$ amplitudes are well known either from experiment or from a reliable model. The usually quoted results concern the asymptotic behavior of $\gamma\gamma \to VV$. Using either additive quark model relations for cross sections or factorization properties for diffractive scattering one gets the relation:

$$\sigma(\gamma\gamma \to VV) = \frac{[\sigma(\gamma p \to Vp)]^2}{\sigma(pp \to pp)} \ .$$

It is not satisfied by experiment in the case of $\gamma\gamma \to \rho^0\rho^0$ for $W < 2$ GeV. This is not so surprising. Its use at low energy is not reasonable for two kinds of reasons, kinematical and dynamical ones. Firstly the t-channel exchange processes are strongly dependent on the mass extrapolations from m_V, $m_{V'}$ to $m_\gamma = 0$ for example through t_{min} effects. Alexander et al.,[52] proposed to use the above relation at fixed $p_{c.m.}$ (instead of fixed $c.m.$ energy) in order to reduce the kinematical effects. Because of flux corrections this has the result of enhancing largely the predictions for $\sigma(\gamma\gamma \to VV)$ at low energy. Secondly, there may exist typical $VV' \to F$ resonances (for example the states discussed in Sec. 2) which locally enhance $\gamma\gamma \to F$ and which are not reproduced by the above relation. In other words as long as we do not know the low energy behavior of $VV' \to F$ we cannot make a reliable prediction for $\gamma\gamma \to F$ by VDM.

There is a remarkable exception to this assertion in the case of a broad resonance which decays dominantly into VV' channels such that it saturates unitarity for this partial wave. In this case one gets

$$\sigma(\gamma\gamma \to VV') \simeq \frac{8\pi(2J+1)}{W^2} B_{\gamma\gamma} \qquad \text{for } W^2 \simeq M^2 \ ,$$

with $B_{\gamma\gamma}$ just given by the VDM couplings:

$$B_{\gamma\gamma} \simeq \frac{e^4 g_{V\gamma}^2 g_{V'\gamma}^2}{m_V^4 m_{V'}^4} \ .$$

This limiting case gives a VDM test independently of the hadronic VV'-Resonance coupling. Such broad resonances can be found in $q\bar{q}$ or $qq\bar{q}\bar{q}$ spectroscopy (see Sec. 2) and this prediction can apply to $\gamma\gamma \to \rho\rho$. The experimental amplitude analysis[53] suggests a peak around 1.2-1.3 GeV with possible additional contributions at 1.45 and 1.65 GeV. These contributions should interfere constructively below 1.6 GeV but destructively above in the case of $\rho^0\rho^0$. $I = 0$ and $I = 2$ contributions are required in order to explain the $\frac{\rho^+\rho^-}{\rho^0\rho^0}$ ratio. A multi-state structure ($q\bar{q}$, $qq\bar{q}\bar{q}$, ...) could give such features.[18] Tests of this picture can be obtained by looking to related processes (VV' and $V\gamma$: $\rho\gamma$, $\omega\gamma$, $\phi\gamma$, $\rho\omega$, $\omega\omega$, $\phi\phi$, $\phi\rho$, ...).[18,46] By the way the comparison of $\gamma\gamma \to W + M$ with $\gamma\gamma \to \gamma + M$ gives a good VDM test. One can also use the few experimental results existing in $p\bar{p} \to \rho\rho$, $\rho\omega$ at low energy to predict $\gamma\gamma \to p\bar{p}$ and compare with TASSO results. The order of magnitude turns out to be good.[54]

3.4 EXCLUSIVE $\gamma\gamma$ PROCESSES IN QCD

At large W and large momentum transfer (i.e., fixed angle above the $C = +1$ resonance region) it is expected[50] that exclusive processes will be dominated by the QCD diagrams of Fig. 2. The amplitude is expressed in terms of an hadronic distribution amplitude $\phi(x_i, Q)$ for each final hadron and a hard scattering amplitude T_H for $\gamma\gamma \to$ valence quarks. Such a factorized form reproduces the results of dimensional counting[55] $R_{fi} \simeq (Q^2)^{\frac{4-n}{2}} f(\theta)$, up to $log \frac{Q}{\Lambda}$ factors, where n is the total number of valence constituents in the initial and final states. Q is the typical

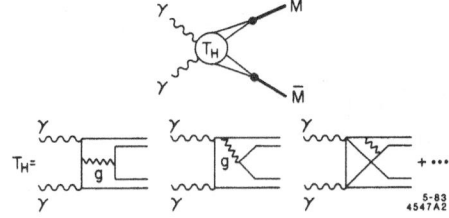

Fig. 2. QCD description of the
exclusive process $\gamma\gamma \to M\bar{M}$.

momentum transfer in the process. The detailed properties of $f(\theta)$ depend upon the unknown
form of $\phi(x_i, Q)$ which come from non-perturbative effects. However the normalization is fixed by
the meson leptonic decay constant $\int_0^1 dx \phi_M(x, Q) = \frac{f_M}{2\sqrt{3}}$. Using a similar analysis of the meson
form factors Brodsky and Lepage[50] were able to reabsorb all $\alpha_s(Q^2)$ factors in the expression:

$$\frac{d\sigma}{dt}(\gamma\gamma \to M\bar{M}) = 16\pi^2\alpha^2 \left|\frac{F_M(W^2)}{W^2}\right|^2 G_M(\theta) .$$

$G_M(\theta)$ depends upon $\phi(x_i, Q)$ and has a different expression for helicity zero and for helicity ± 1
mesons. Several extreme choices for the amplitudes $\phi(x_i, Q)$ have been tried.[50] The normalization
condition and the known values of f_M (i.e., $f_\pi = 93$ MeV, $f_K = 112$ MeV, $f_\rho = 154$ MeV, $f_\omega =$
158 MeV, $f_\phi = 161$ MeV) already give interesting predictions for the asymptotic magnitudes of the
cross section. See Table 1 of Ref. 50 for a comparison of $\gamma\gamma \to \pi\pi$, $K\bar{K}$, $\pi\eta$, $\eta\eta$, $\rho\rho$, $\rho\omega$, $\omega\omega$, $\phi\phi$
and $\mu^+\mu^-$. For example:

$$\frac{d\sigma}{dt}(\gamma\gamma \to \rho_\perp^+\rho_\perp^-) \simeq 8\,\frac{d\sigma}{dt}(\gamma\gamma \to \pi^+\pi^-) \simeq \frac{5\,\text{GeV}^4}{W^4} \cdot \frac{d\sigma}{dt}(\gamma\gamma \to \mu^+\mu^-)$$

for $\theta \simeq \frac{\pi}{2}$. The same method has been applied to $\gamma\gamma \to B\bar{B}$ by Damgaard.[51] In this case the
normalization has been fixed by the known decay rate $\psi \to 3g \to p\bar{p}$ which involve the same
$\phi_B(x_i, Q)$ amplitude. The cross section $\frac{d\sigma}{dt}(\gamma\gamma \to p\bar{p})$ which behaves for large W and θ like
$\frac{1}{W^{12}} f(\theta)$ (up to $\alpha_s(Q^2)$ factors) has tentatively been compared to the preliminary TASSO results
between $2 \le W \le 3$ GeV. Although a detailed comparison should not be valuable so close to the
threshold the right order of magnitude seems to be obtained.

On another hand attention has been driven to possible enhancements of exclusive processes
involving both light and heavy quarks. Ecclestone and Scott[56] pursuing an idea laready used for
Z^0 decays considered the processes $\gamma\gamma \to M(Q\bar{q}) + M(\bar{Q}q)$ where M is a charmed or bottomed
or topped meson. From diagrams of Fig. 3 they expected a strong enhancement (governed by
the $\frac{m_Q}{m_q}$ ratio) due to the light quark propagator. However it is merely possible that this effect is
completely washed out by strong final state interactions.

Fig. 3. QCD diagrams for heavy
mesons production.

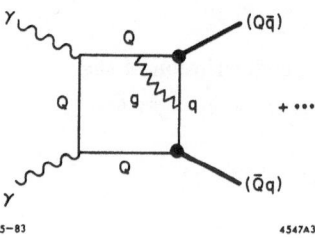

5–83   4547A3

4. Total Hadronic Production at Low W

The asymptotic behavior of $\sigma(\gamma\gamma \to \text{hadron})$ is generally thought to be dominated by the hadronic component of the photon.[1-8] VDM <u>and</u> Pomeron exchange with factorization or quark model relations[1-8] give:

$$\sigma \to \sigma_0 \simeq 0.24 \mu b \ .$$

The same assumptions for $\alpha = \frac{1}{2}$ Regge trajectory exchanges give the W dependence:

$$\sigma \to \sigma_0 + \sigma_1 = 0.24 \mu b + \frac{0.27 \mu b \, \text{GeV}}{W} \ .$$

On the experimental side estimations of the total cross section have been given by the PLUTO and TASSO groups.[57,58] Both agree for $W \geq 2.5$ GeV and give larger values than the above relations. For $1.7 \leq W \leq 2.5$ GeV the results disagree. PLUTO results are especially high and can be represented by the expression:

$$\sigma = 0.97 \left(0.24 \mu b + \frac{0.27 \mu b \, \text{GeV}}{W} \right) + \frac{2.25 \mu b \, \text{GeV}^2}{W^2} \ .$$

However the TASSO group[59] has shown that experimental estimations are very dependent upon the model used for event topology. Nevertheless several theoretical explanations for high values of σ can be found. Low lying Regge trajectories like pion exchange may not satisfy the factorization assumption. For example π exchange seems to be more important in electromagnetic interactions (photo-production) than in purely hadronic interactions. A better description would then be obtained with:[52]

$$\sigma \to \sigma_0 + \sigma_1 + \sigma_2 \qquad \text{with } \sigma_2 = \frac{A^\pi}{W^2} \ .$$

EVDM descriptions also predict higher values of σ_0 and σ_1. For example with a series (Veneziano-type) of vector mesons with masses $m_n^2 = m_0^2(1 + 2n)$, photon couplings $g_{\gamma n}^2 = g_{\gamma 0}^2$ and cross sections satisfying the relations $\sigma_{\gamma n}(W)/\sigma_{\gamma 0}(W) = (m_0^2/m_n^2)^\alpha$ with $\alpha = 1$ for Pomeron (i.e., σ_0) and $\alpha = \frac{1}{2}$ for other Regge exchanges (i.e., σ_1) one obtains:[60]

$$\sigma \to 0.29 \mu b + \frac{0.46 \mu b \, \text{GeV}}{W} \ .$$

Point-like contributions may also give additional terms. Greco and Srivastava suggested many years ago[61] on the basis of duality relations (Regge-Resonances) that an additional term is needed and proposed the contribution of the Box diagram $\gamma\gamma \to q\bar{q}$:

$$\sigma_{\gamma\gamma}^{box} \simeq \frac{4\pi\alpha^2}{W^2} \sum_q e_q^4 \, log \, \frac{W^2}{m_q^2} \ .$$

Such a term is similar to the above other $\frac{1}{W^2}$ terms in the low W region.

In the low W region, i.e., for $W \le 2$ GeV where there are large resonance effects, the above formulae can only be considered as giving averaged descriptions. Also one should probably not add all the above contributions. For example EVDM is an alternative to QCD (point-like contributions). There probably is fewer overlap between other "hadronic" and "point-like" components which should respectively correspond to two different domains when one integrates over transverse momenta inside photon structure: $\int_0^{\mu^2} dp_T^2$ for the hadronic part and $\int_{\mu^2}^{\infty} dp_T^2$ for the point-like part where μ is of the order of hundreds of MeV.[62]

A way to disentangle the various components of the total cross section is to look to the q^2-dependences with slightly virtual photons. It should be steeper for hadron-like parts than for point-like parts (or equivalently for low mass vector mesons than for high mass vector mesons in EVDM series). At low W the contributions to the cross section can also be separated into exclusive processes where threshold effects (Born terms, contact terms), standard $q\bar{q}$ resonances and unusual resonances ($qq\bar{q}\bar{q}$, $q\bar{q}g$, gg, ggg) contribute. It would be interesting to reconstruct their respective averaged W and q^2 behaviors and to compare them to the above formulae. In Fig. 4 we draw a very tentative sum of resonance contributions from the results of Sec. 2. This at least shows that the so-called unusual part may be as important as the usual one.

Fig. 4. Total $\gamma\gamma \to$ hadrons cross-section at low energies. Usual contribution refers to standard $q\bar{q}$ resonances and unusual to ($qq\bar{q}\bar{q}$), ($q\bar{q}g$) and glueball states discussed in Sec. 2.

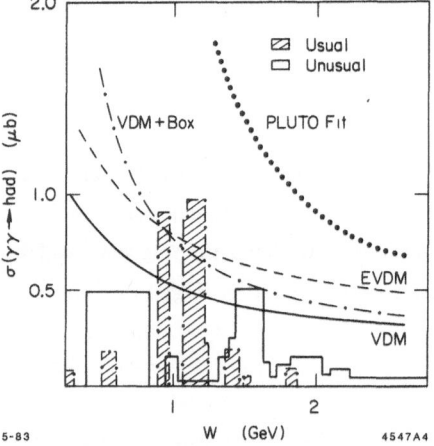

5. q^2 Behavior of Soft Processes

When one or both photons becomes slightly virtual soft processes should evolve according to VDM:

$$R(\gamma\gamma^* \to F) = \sum_V \frac{eg_{V\gamma}}{m_V^2 - q^2} \, R(\gamma V \to F) \ .$$

This should be applicable both to non-resonant and to resonant processes (as long as longitudinal γ^* amplitudes can be neglected). For example, when one vector meson (like ρ) is dominant one expects for the resonance decay width:

$$\Gamma_{R \to \gamma\gamma^*} \simeq \left(\frac{m_V^2}{m_V^2 - q^2}\right)^2 \Gamma_{R \to \gamma\gamma} \ .$$

This behavior should reflect in the total cross section (because ρ and ω with $m_V^2 \simeq 0.6 \text{ GeV}^2$ give the same effect) and this seems to be the case up to $-q^2 \simeq 1 \text{ GeV}^2$. For higher q^2 either point-like processes (QCD) or higher vector meson states (EVDM) begin to contribute and lead to a less decreasing cross section. In fact this transition from soft to hard processes is a difficult theoretical problem. Its experimental and phenomenological approach through the q^2 dependence is certainly most interesting. First it has been shown[50] that both VDM and QCD predict that the resonance contribution to $\gamma\gamma^*$ should fall off like $(1 + q^2/0.68\,\text{GeV}^2)^{-1}$. A change in the dominant helicity amplitudes when passing from low $q^2(R \to \gamma\gamma)$ to high $q^2(R \to \gamma\gamma^*$ or $R \to \gamma^*\gamma^*)$ is also expected on the basis of the point-like behavior.[63] For example in f^0 case, $\lambda = 0$ should dominate instead of $\lambda = 2$ for real photons.

On the other hand in the hard domain (high momentum transfer) $\gamma\gamma^* \to M\bar{M}$ or $B\bar{B}$ should be rather insensitive to q^2 provided that $q^2 \ll W^2$.[50] These properties could be well illustrated by the behavior of the real photon structure functions in the low q^2 domain and especially in the exclusive limit.

Another interesting feature of $q^2 \neq 0$ $\gamma\gamma^*$ collisions is the possibility of exciting $1^{\pm+}$ states forbidden for two real photons by Bose statistics. For one real photon and one virtual photon one in general has two independent couplings corresponding to $(\pm, \pm)$ and $(\pm, 0)$ helicity amplitudes.[15,63] The first ones vanish like q^2 for $q^2 \to 0$ because of (Bose) symmetrization effects. The second ones vanish like $\sqrt{-q^2}$ as expected for longitudinal amplitudes. 1^{++} states exist in $(q\bar{q})$, $(qq\bar{q}\bar{q})$, $(q\bar{q}g)$, (gg) and (ggg) spectroscopy. 1^{-+} states are exotic and exist in $(q\bar{q}g)$, (gg) and (ggg) spectroscopy. Predictions have been given recently for these various kinds of states.[15] On the basis of VDM one can expect that for low q^2 the decay widths will be of the order of:

$$\Gamma_{\gamma\gamma^*} \simeq \frac{|q^2|}{m^2} \, \Gamma_{\gamma\gamma^*}^R$$

where m is an hadronic mass and $\Gamma_{\gamma\gamma^*}^R$ stands for the decay width of a state (like $0^{\pm+}$, $2^{\pm+}$) normally allowed to decay into two real photons and whose q^2 dependence is only the one due to

VDM poles. For example $D(1285)$ and several $(qq\,\bar{q}\,\bar{q})$ predicted 1^{++} states are expected to have especially large decay widths. In $e^+e^- \rightarrow e^+e^- + X$ this additional $\frac{|q^2|}{m^2}$ factor will cancel the $\frac{1}{q^2}$ factor coming from the photon propagator so that one will lose one $log\frac{s}{m_e^2}$ enhancement in the non-tagged cross section. However tagged experiments with $|q^2| \geq m_V^2$ should allow to observe these resonances. Looking to specific channels like $P + V$ or $P + A$ could also help. For 1^{-+} exotic states the channels $\pi\eta$, $\pi\eta'$, $\eta\eta'$ were especially noticed in Ref. 24. Polarization (see the next section) could also help to disentangle the new helicity components.

In the whole the $1^{\pm+}$ contributions to the γ structure functions for $|q^2| \geq m_V^2$ are expected to be comparable to those of other partial waves.

6. Polarization Effects

The cross section for $e^+e^- \to e^+e^- + X$ with polarized $e^\pm$ beams has been given in the most general case in Ref. 64. When T and P invariance hold, this form reduces to the expressions more frequently written.[65,69] Let us consider the final $e^\pm$ distributions which like in the unpolarized case are expressed in terms of $\gamma\gamma$ luminosity coefficients times $\gamma\gamma$ cross sections for various helicity combinations. In the quasi-real $\gamma\gamma$ limit one gets:[64]

$$\frac{\ell_1^0 \ell_2^0 d\sigma}{d_3\ell_1 d_3\ell_2} = \frac{\alpha^2}{2\pi^4} \cdot \frac{|\vec{q}|W}{sq_1^2 q_2^2}$$

$$\left\{ K_{TT}\sigma_{TT} + K'_{TT}Re\tau_T - Q'_{TT}Im\tau_T + P_L P'_L K''_{TT} \,\tilde\sigma_{TT} + P_L \tilde V_{TT} \,\sigma_T^1 + P'_L V_{TT}\sigma_T^2 \right\} \ .$$

The kinematical coefficients are explicitly given in Ref. 64. We just notice that K'_{TT} and Q'_{TT} are proportional to $cos2\phi$ (the azimuthal angle between e^+e^- and $\gamma\gamma$ planes). When T-invariance holds $Im\tau_T = 0$ and when P-invariance holds $\sigma_T^1 \equiv \sigma_T^2 \equiv 0$ (there is no P_L or P'_L dependence in the cross section). Hence we are left with:

$\sigma_{TT} = \frac{1}{2}(\sigma_\parallel + \sigma_\perp) = \frac{1}{2}(\sigma_0 + \sigma_2)$, the unpolarized $\gamma\gamma$ cross section,

$\tilde\sigma_{TT} = \frac{1}{2}(\sigma_0 - \sigma_2)$ which will appear with longitudinally polarized $e^\pm$ beams,

$Re\tau_T = \sigma_\parallel - \sigma_\perp$, the linear correlation which will appear in the azimuthal $(cos2\phi)$ distribution without $e^\pm$ polarization.

Measurements of these three quantities give interesting independent informations about the dynamics. In resonance physics they allow to separate the contributions of the different couplings. For a $2^\pm$ resonance one can separate $\lambda = 0$ from $\lambda = 2$ contributions[65,66]; (for $0^\pm$ resonances $\sigma_2 \equiv 0$ and $\sigma_{TT} \equiv \tilde\sigma_{TT} \equiv \pm\frac{1}{2}Re\tau_T$). For high W soft (hadron-like) processes one can separate various Regge terms.[66] $Re\tau_T$ and $\tilde\sigma_{TT}$ are given by unnatural parity exchanges which should decrease faster with W than σ_{TT}. This fact has been used in Ref. 67 for studying the sum rules

$$\int Re\tau_T(W^2)dW^2 = \int \tilde\sigma_{TT} (W^2)dW^2 = 0 \ ,$$

their resonance saturation and the possibility of fixed j-plane singularities associated to point-like processes. For example the $\gamma\gamma \to f\bar{f}$ contributions are:

$$\sigma_{TT} = \frac{4\pi\alpha^2}{W^2} R_{\gamma\gamma}\left[\left(1 + \frac{4m^2}{W^2} - \frac{8m^4}{W^4}\right)L - \left(1 + \frac{4m^2}{W^2}\right)\sqrt{1 - \frac{4m^2}{W^2}} \right]$$

$$\tilde\sigma_{TT} = -\frac{4\pi\alpha^2}{W^2} R_{\gamma\gamma}\left(L - 3\sqrt{1 - \frac{4m^2}{W^2}} \right)$$

$$Re\tau_T = -\frac{16\pi\alpha^2}{W^2} R_{\gamma\gamma} \frac{m^2}{W^4}\left(2m^2 L + W^2 \sqrt{1 - \frac{4m^2}{W^2}} \right)$$

with

$$L \equiv 2\, log\left(\frac{W}{2m} + \sqrt{\frac{W^2}{4m^2} - 1}\,\right)$$

and

$$R_{\gamma\gamma} = Q_f^4 \ .$$

Notice that $\frac{\sigma_2}{\sigma_0} \simeq log\, \frac{W}{m}$.

It has also been pointed out[68] that the very interesting process $\gamma\gamma \to gg$ which gives $\sigma_2 \simeq \sigma_0$ could be separated from the $f\bar{f}$ background by using longitudinally polarized $e^\pm$ beams.

The advent of unusual contributions ($qq\,\bar{q}\,\bar{q}$, $q\,\bar{q}\,g$, gg, ggg) could give further motivations for polarization. If these contributions turned out to be important they should play a role in these various duality sum rules.

7. Conclusion

Photon-photon collisions offer another example of the power of electromagnetic interactions for studying hadronic structure. Owing to important experimental progress during these last years it is now a major field in particle physics. It can be compared to e^+e^- annihilation, lepton-hadron deep inelastic scattering and quarkonia physics. $\gamma\gamma$ collisions have genuine features which make them complementary to these other fields. The presence of two photons leads to a greater sensitivity to the point-like structure. One also has the possibility of tuning the q^2 value and in this way we can look at the soft and hard components with variable weights. The soft/hard transition actually is an unsolved theoretical problem. The present descriptions still need a lot of phenomenological inputs like vector meson properties, mesonic wave functions, which interfer with quark and gluon properties at short distances (i.e., VDM versus QCD). Not only the high energy scattering processes but also the formation of standard and unusual $C = +1$ states appear to be very sensitive to this double aspect of the strong interactions. Already it would be an important contribution from $\gamma\gamma$ collisions if this unusual spectroscopy could be confirmed and at least if the old problems of the 0^{++} mesons were clarified. The 1983 $\gamma\gamma$ Workshop is certainly a step on this way.

Acknowledgements

We thank Prof. S. D. Drell for his hospitality at SLAC Theory Group and S. Brodsky, D. Burke, F. Gilman for discussions.

9. References

1. F. Low, Phys. Rev. <u>120</u> (1960) 582;

   F. Calogero and C. Zemach, Phys. Rev. <u>120</u> (1960) 1860;

   N. Arteaga-Romero, A. Jaccarini and P. Kessler, C.R.A.S. <u>269B</u> (1969) 153, 1129;

   N. Arteaga-Romero, A. Jaccarini, P. Kessler and J. Parisi, Lett. Nuov. Cim. <u>4</u> (1970) 933; Phys. Rev. D <u>3</u> (1971) 1569; Phys. Rev. D. <u>4</u> (1971) 2927;

   V. E. Balakin, V. M. Budnev and I. F. Ginzburg, JETP Lett. <u>11</u> (1970) 388;

   V. M. Budnev and I. F. Ginzburg, Phys. Lett. <u>37B</u> (1971) 320;

   S. Brodsky, T. Kinoshita and H. Terazawa, Phys. Rev. Lett. <u>25</u> (1970) 972; <u>27</u> (1971) 280; Phys. Rev. D <u>4</u> (1971) 1532.

2. Proc. Int. Colloq. $\gamma\gamma$ Collisions, Paris; Jour. Phys. Suppl. <u>35</u> (1974).

3. Proc. Int. Conf. on Two-Photon Interactions, Lake Tahoe, ed., J. F. Gunion (1979).

4. Proc. Int. Workshop on $\gamma\gamma$ Collisions, Amiens, eds., G. Cochard and P. Kessler, Springer Verlag <u>134</u> (1980).

5. Proc. 4th Int. Colloq. on $\gamma\gamma$ Interactions, Paris, ed., G. W. London (1981).

6. S. Cooper, Talk at 2nd Int. Conf. on Physics in Collision, Stockholm 1982; DESY 82-050.

   D. L. Burke, Talk at XXI Int. Conf. on High Energy Physics, Paris 1982, SLAC-PUB-2988.

7. S. L. Adler, Phys. Rev. <u>177</u> (1969) 2426.

   J. S. Bell and R. Jackiw, Nuov. Cim. <u>60A</u> (1969) 47.

8. M. Greco, Ref. 4.

9. L. Bergström et al., Phys. Lett. <u>82B</u> (1979) 419; Z. Phys. <u>C8</u> (1981) 363.

10. M. S. Chanowitz, Proc. SLAC Summer Inst. 1981; SLAC Report No. 245 (1982).

11. V. M. Budnev and A. E. Kaloshin, Phys. Lett. <u>86B</u> (1979) 351.

12. F. Gilman, Ref. 3.

13. R. L. Jaffe, Phys. Rev. D <u>15</u> (1977) 267.

14. P. Singer, Phys. Rev. D. <u>26</u> (1982) to appear; preprint Technion-Phys-83-2 (1983).

15. F. M. Renard, SLAC-PUB-3126.

16. W. W. Buck, C. B. Dover and J. M. Richard, Ann. Phys. <u>121</u> (1979) 47.

17. B. A. Li and K. F. Liu, SLAC-PUB-2783.

18. N. N. Achasov, S. A. Devyanin and G. N. Shestakov, Phys. Lett. <u>108B</u> (1982) 134.

19. J. Weinstein and N. Isgur, Phys. Rev. D. <u>27</u> (1983) 588.

20. A. C. Irving, A. D. Martin and P. J. Done, preprint DAMTP.

21. T. Barnes, Nucl. Phys. <u>B158</u> (1979) 171;

    P. Hasenfratz et al., Phys. Lett. <u>95B</u> (1980) 299.

    F. deViron and J. Weyers, Nucl. Phys. <u>B185</u> (1981) 391.

22. T. Barnes and F. E. Close, Phys. Lett. <u>116B</u> (1982) 365.

23. M. Tanimoto, Phys. Lett. <u>116B</u> (1982) 198.

24. M. Chanowitz and S. Sharpe, LBL-14865 (1982).

25. B. Berg, CERN TH-3327 (1982).

26. V. Novikov et al., Nucl. Phys. B191 (1981) 301.

27. R. Jaffe and K. Johnson, Phys. Lett. 60B (1976) 201.

28. J. Donoghue, K. Johnson and B. A. Li, Phys. Lett. 99B (1981) 416.

29. J. F. Donoghue, Talk at APS meeting, Santa Cruz 1981; UMHEP-157.

30. J. Coyne, P. Fishbane and S. Meshkov, Phys. Lett. 91B (1980) 259.

31. P. M. Fishbane, Talk at 1981 Orbis Scientiae.

32. J. M. Cornwall and A. Soni, Phys. Lett. 120B (1983) 431.

33. H. J. Lipkin, ANL-HEP-PR-81-35 (1981).

34. J. F. Donoghue and H. Gomm, Phys. Lett. 121B (1983) 49.

35. E. D. Bloom, Talk at Int. Conf. on High Energy Physics, Paris 1982.

36. P. Jenni et al., SLAC-PUB-2758.

    TASSO Coll., H. Kolanoski, Proc. Renc. Moriond 1982.

37. S. J. Lindenbaum, Talk at Int. Conf. on High Energy Physics, Paris 1982.

38. K. E. Lassila and P. V. Ruuskanen, Phys. Rev. Lett. 19 (1967) 762.

    F. M. Renard, Nucl. Phys. B82 (1974) 1; Lett. Nuov. Cim. 21 (1978) 15.

39. P. Fishbane et al., UCLA/82/TEP/18 (1982).

40. J. L. Rosner, Phys. Rev. D 24 (1981) 1347.

41. J. L. Rosner and S. F. Tuan, Phys. Rev. D (to appear).

42. H. J. Schnitzer, Nucl. Phys. B207 (1982) 131.

43. J. F. Donoghue, Phys. Rev. D 25 (1982) 1875.

44. G. Mennessier, Z. Phys. C16 (1983) 241.

45. N. A. Törnqvist, Phys. Rev. Lett. 49 (1982) 624.

46. J. Layssac and F. M. Renard, Nuov. Cim. 70A (1982) 1.

47. H. Terazawa, Rev. Mod. Phys. 45 (1973) 615.

48. A. Bramon and F. Cornet, Phys. Lett. 83B (1979) 235.

    Y. Goldschmidt and H. R. Rubinstein, Nucl. Phys. B91 (1975) 445.

    C. Ayala, A. Bramon and F. Cornet, Phys. Lett. 107B (1981) 235.

49. J. Stirling, Talk at this Workshop.

50. S. J. Brodsky and G. P. Lepage, Phys. Rev. D. 24 (1981) 1808.

51. P. Damgaard, Nucl. Phys. B211 (1983) 435.

52. G. Alexander and U. Maor, Phys. Rev. D 26 (1982) 1198.

53. TASSO Coll., M. Althoff et al., Z. Phys. C16 (1983) 13.

54. J. Layssac, private communication.

55. S. J. Brodsky and G. R. Farras, Phys. Rev. Lett. 31 (1973) 1153; Phys. Rev. D 11 (1975) 1309.

    V. A. Matveev et al., Lett. Nuov. Cim. 7 (1973) 719.

56. R. E. Ecclestone and D. M. Scott, DAMTP/82/17.

57. PLUTO Coll., Ch. Berger et al., Phys. Lett. 99B (1981) 287.

58. TASSO Coll., E. Hilger et al., DESY 80/75 (1980).

59. H. Kolanoski, report at Montpellier meeting, December 1982.

60. U. Maor and E. Gotsman, TAUP-1080-82 (1982).

61. M. Greco and Y. Srivastava, Nuov. Cim. 43A (1978) 88.

62. S. J. Brodsky, Ref. 5.

63. G. Köpp, T. F. Walsh and P. Zerwas, Nucl. Phys. B70 (1974) 461.

64. N. S. Craigie, K. Hidaka, M. Jacob and F. M. Renard, Phys. Rep. (to appear).

65. V. M. Budnev et al., Phys. Rep. 15C (1975) 183.

66. V. M. Budnev et al., Phys. Lett. 96B (1980) 387.

67. I. F. Ginzburg and V. G. Serbo, Phys. Lett. 103B (1981) 68.

68. B. L. Combridge, Nucl. Phys. B174 (1980) 243.

69. F. Martin, Thesis, Paris 1974.

SPEAKER: F. Renard

NAME OF QUESTIONER: H. Kolanoski

INSTITUTION: University of Bonn

QUESTION: Can you comment on the differences in the predictions for the coupling of 4-quark states to $\rho^+\rho^-$ (compared to $\rho^0\rho^0$)?

ANSWER: It is essentially due to the uncertainty in the interference effects between $I = 0$ and $I = 2$ resonances. As you can see from the picture of the $qq\bar{q}\bar{q}$ spectrum there are several broad resonances which overlap in this energy region. So results are very sensitive to the choices of couplings, masses and widths. It would be very interesting to know the energy dependence of the $\frac{\rho^+\rho^-}{\rho^0\rho^0}$ ratio and to check if it is only a local destructive interference effect or a more general property.

SPEAKER: F. Renard

NAME OF QUESTIONER: U. Maor

INSTITUTION: Tel-Aviv University

QUESTION: Concerning $\gamma\gamma \to VV$ I wish to draw attention to a recent factorizable VDM calculation done by Alexander Williams and myself. We do reproduce the $\gamma\gamma \to \rho^0\rho^0$ data as well as the new results on $\gamma\gamma \to \rho^+\rho^-$.

ANSWER: It is easy to understand how the use of VDM at fixed $p_{c.m.}$ give larger predictions than in the case of fixed $E_{c.m.}$. However it is still not clear to me what is the local dynamical structure of the $\gamma\gamma \to VV$ amplitudes in the $\rho^0\rho^0$ enhancement region ($W \simeq 1.2 - 1.7$ GeV).

STATUS OF QCD

R. Petronzio
Theory Division
CERN, Geneva, Switzerland

ABSTRACT

Some aspects of the present status of the perturbative and of the lattice approach to QCD are briefly reviewed and discussed.

The entire perturbative QCD relies on its asymptotic freedom,
which implies that, in a laboratory with a diameter of about 0.1 Fermi,
strong interactions can be analyzed, very much like the electroweak ones,
by a perturbative expansion in the running coupling constant.

However, there is a long way to go from such a small box to the
actual size of a laboratory and the natural question is: how can we
believe that the results of the prediction will not be affected by such
an extrapolation ? After all, it is known that something very relevant
must happen in between, i.e., the confinement of coloured sources.

Theorists have answered, or tried to answer, this question by introducing the concept of "infra-red" sensitivity, which is just a statement about the smoothness of the infinite volume limit. Unfortunately,
very few predictions are in fact infra-red insensitive, a well-known
example of them being the total cross-section of e^+e^- into hadrons.

In many other cases, and in particular when hadrons are present
in the initial state, the infra-red sensitivity manifests itself through
the appearance of "mass singularities". Quarks and gluons, when squeezed
into a small box, go very far off-shell: the smoothness of the infinite
volume limit translates into a smooth dependence upon this off-shellness.
When this is not the case, such as when the dependence is logarithmic,
mass singularities appear.

Factorization techniques[1] allow us to convert the dependence upon
the box size (or the off-shellness) into the one upon a scale where new
basic ingredients of perturbative QCD, the parton densities, are normalized. One can see that, even if the infinite volume limit is outside
the perturbative range, the effects of local change of the small box
size are still perturbative. This leads to Altarelli-Parisi types of
equations[2] for the parton densities, where the perturbative expansions

apply to the kernels (the probabilities), regulating the <u>variation</u> of the parton densities by a local change of the scale.

The procedure of removing the infra-red cut-off, as well as that of removing the ultra-violet one, is not free of ambiguities when one performs a calculation beyond the leading order. Renormalization and factorization scheme dependences may render the interpretation of next-to-leading corrections to the basic QCD processes rather cumbersome. The principle of minimal sensitivity (PMS) seems to be a good guideline for determining the "suitable" scheme[3]; one should have a more systematic use of these techniques when comparing higher order calculations with experimental data. The universality of the "Λ" parameter among different reactions is not lost when using an optimization procedure; only the "natural" scale of the process is adapted to the particular case, always measured in units of a fixed Λ parameter. In other words, different reactions will lead to a series expansion in $\alpha(\mu_1^2/\Lambda^2)$, $\alpha(\mu_2^2/\Lambda^2)$ where μ_1^2, μ_2^2 have been determined by the PMS: the fit of the theoretical formulae to the experiment will provide a value for $\alpha(\mu_1^2/\Lambda^2)$ which, knowing μ_1^2, will lead in turn to a determination of a "universal" Λ.

The principle of minimal sensitivity provides a way, based on the renormalization group, to resum some of the terms of the perturbative expansion. Some other resummations are performed, not by renormalization group methods, but rather by explicit diagrammatic techniques. This is the case for those terms in the perturbative expansion which are produced by soft gluon emissions[4]. Improvements of this type are normally relevant only at the edge of the phase space, when the whole gluon bremsstrahlung is forced to be soft, and therefore rather difficult to test experimentally. In the same framework, one can mention the resummation techniques leading to a perturbative description of low $p_\perp$, high mass muon pairs produced in Drell-Yan or back-to-back jet distributions in e^+e^- [5].

In the latter case, these methods allow us to promote to the perturbative domain some effects which were ascribed to the non-perturbative behaviour of the theory, and therefore parametrized with phenomenological quantities like the intrinsic $k_\perp$.

The role of this parameter is very crucial in all reactions aiming to measure a usually steep transverse momentum distribution: even a small shift in the value of $p_\perp$ may produce large effects on the absolute normalization and considerably alter any test of predictions based on a perturbative expansion.

The influence of non-perturbative phenomena can also considerably affect the perturbative behaviour for reactions which are officially infra-red insensitive, like the three-jet cross-section in e^+e^-, where different hadronization models lead to different values for α[6]. This, in fact, happens because the definition of jets is clear at the level of the elementary reaction in terms of gluons and quarks, but it becomes obscured by the fragmentation process which must take place in the conversion from partons to hadrons.

For this problem, a complete QCD calculation is unavailable: the only way out at present is to generate the parton cascade down to off-shellness of the order of 1 GeV^2 and then hope that the remaining dressing will not spoil the pattern established perturbatively: one comes to the edges of the small box where perturbation theory applies and hopes for the best. Already at perturbative level however, there are various problems which have to be solved. In particular, the simple rule of the jet calculus, where the fragmentation would follow a strong ordering in the invariant masses of fragmenting partons, does not hold in the case of soft gluon emissions[7]. There, also, a strong ordering in the angle of emissions is required: terms which are formally of higher order with respect to the basic ladder-type structure provide a destructive interference for the emissions which do not obey this rule.

Recently, a new Monte Carlo generation of parton cascade has been made, respecting the angular ordering, and therefore including both the leading collinear singularities and the leading infra-red singularities[8]. In Fig. 1 a typical event appears for an incoming jet energy of 300 GeV. The numbers on the lines are the energies; those at the right of the vertices are the value of a parameter ξ regulating the opening angle of the emission and defined as follows·

$$\xi = q_1 \cdot q_2 / (\omega_1 \omega_2) \sim 1 - \cos\theta_{12} \quad \text{if} \quad q_1^2, q_2^2 \ll q_0^2 \tag{1}$$

where q_1^2, q_2^2 are the invariant masses of the decay partons. The ordering condition reads:

$$\xi_1, \xi_2 < \xi < \xi_{MAX} \leq 1 \tag{2}$$

where ξ_1 and ξ_2 are the angular variables of the eventual branchings of gluons 1 and 2, while ξ_{max} is the angular variable of the vertex at which the incoming gluon was produced.

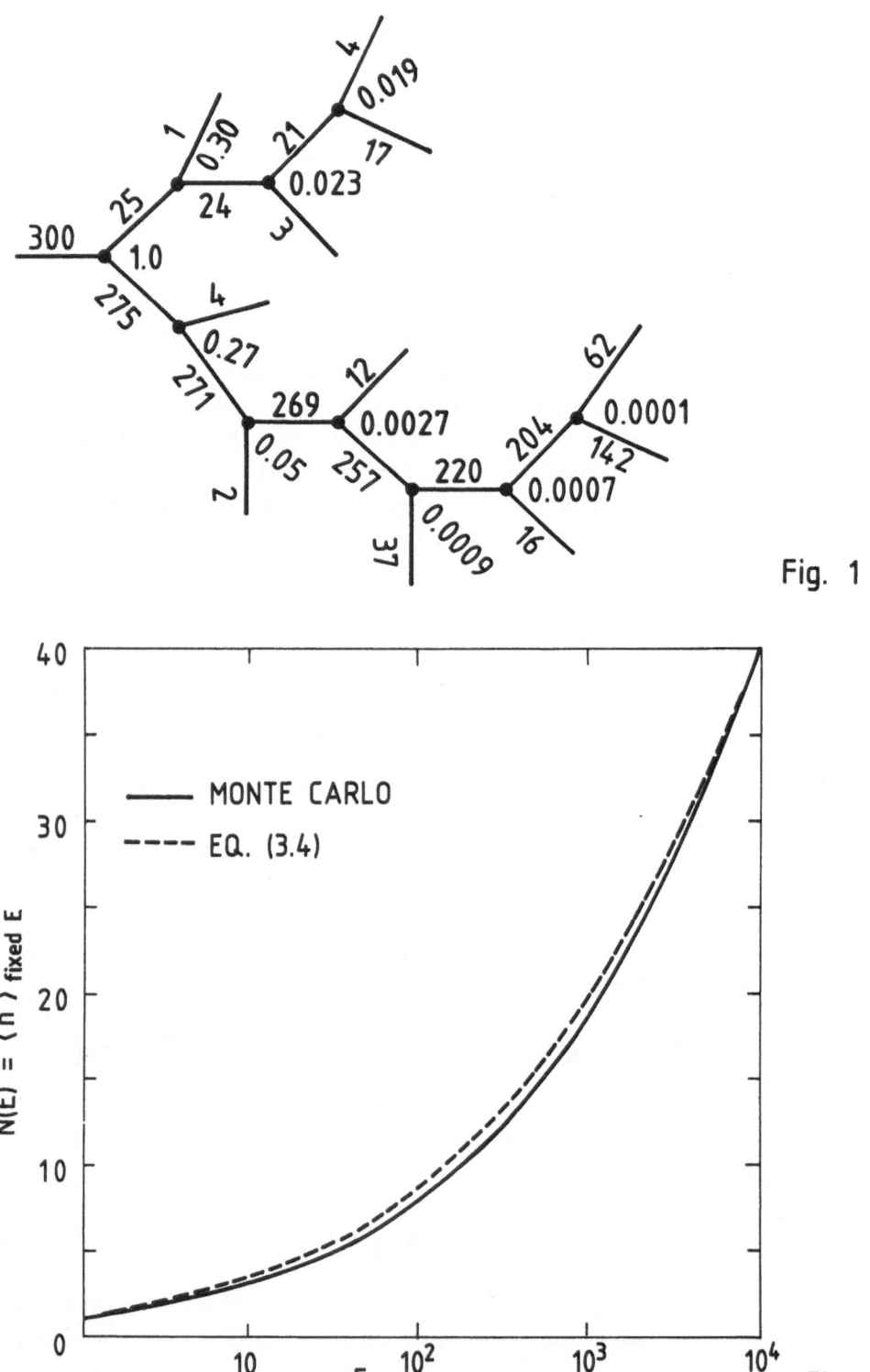

Fig. 1

Fig. 2

Any Monte Carlo calculation introduces several additional next-to
leading contributions with respect to the leading analytic result: it
may then serve "theoretically" to estimate the validity of asymptotic
formulae. In Fig. 2 the asymptotic behaviour of the average multiplicity is confronted with the Monte Carlo results: the differences are
quite small. Of course, the real test of whether the QCD-inspired
Monte Carlo approach to the jet fragmentation is a way out from the
problem of hadronization, is the sensitivity of the results to the
type of the explicit hadronization model which must be attached at the
end of the cascade.

The complexity of QCD as a quantum field theory limits the possibility of really testing the simple perturbative regime in many other
places. The most popular effect is that generally ascribed to "higher
twist" terms, i.e., the contributions which die as inverse powers of
large kinematical invariants in the perturbative expansion. One can
distinguish two types of higher twists: the inclusive ones and the
exclusive ones. The first type occurs in inclusive reactions, like
deep inelastic scattering, and is related, generally speaking, to the
inclusion of effects like the off-shellness or the $k_\perp$ of interacting
partons, which are normally neglected in the impulse approximation. It
also includes the case when more than a single parton per hadron ini-
tiates the hard process. In Fig. 3, one can see an example of a diagram giving rise to higher twists. The possibility of finding a simple
physical picture for higher twists, like the generalized parton model
for the leading ones, is complicated by the fact that the description
in terms of higher twists is not unique. In fact, because of the equations of motion, operators containing derivatives of the fields can be
converted into operators where a higher number of fields appear or,

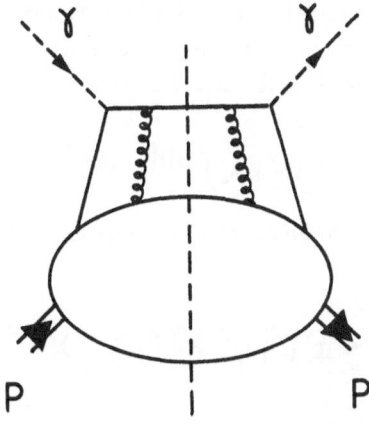

Fig. 3

in simpler terms, where the off-shellness can be traded for more incoming partons. Various operator bases have been presented in the
literature: one suggested by Politzer[9] and fully analyzed by Jaffe
and Soldate[10] consists of a systematic reduction of off-shellness and
transverse momentum in favour of collinear partons. However, the lowest
order coefficient function is rather complicated, and depends upon integrals of many independent operators. A simpler answer is obtained by
working in what is called the transverse basis, where the transverse
degrees of freedom are retained[11]. A simple argument which supports
the use of this basis for higher twist analysis is the following. Let
us consider the expansion of the matrix element of the Wilson operators
$\bar{\psi}\, \gamma_{\mu_1}\, D_{\mu_2}, \ldots, D_{\mu_N}\psi$ up to terms of $O(\Lambda^2)$. One has:

$$\langle p | \bar{\psi}\gamma_{\mu_1} D_{\mu_2} \cdots D_{\mu_N} \psi | p \rangle = A_0(N)\, p_{\mu_1} p_{\mu_2} \cdots p_{\mu_N} + \frac{1}{2} \sum_{i \neq j} B_{ij}(N)\, g^{\mu_i \mu_j}. \qquad (3)$$

$$\cdot (p_{\mu_1} \cdots p_{\mu_N})/(p_i p_j) + O\left(\frac{\Lambda^4}{Q^4}\right)$$

Dimensional arguments say that $A_0(N)$ controls the leading twist while
$B_{ij}(N)$ controls the next-to-leading twist. The leading twist part can
be isolated by contraction with a light-like vector n_μ; the identification of n_μ with the gauge fixing vector leads to the expression:

$$A_0(N) = \langle p | \bar{\psi}\not{n}\, x^{N-1} \psi | p \rangle \qquad (4)$$

where $x = in^\mu \partial_\mu$ in configuration space.

The above relation is the starting point for the parton model
picture when x is identified with the fraction of momentum that the
parton carries and, consequently, the anti-Mellin transform of A(N) is
called the parton (momentum) distribution. Therefore, the natural
question one may ask to establish a parton language for the twist four
sector is: what is the projector which isolates a given term $B_{ij}(N)$
in Eq. (6.1) ? $B_{ij}(N)$ are projected by the tensor:

$$d^{\mu_i \mu_j}\, n_{\mu_1} \cdots n_{\mu_N}/(n_{\mu_i} n_{\mu_j}) \quad \text{with} \quad d^{\mu\nu} = -g_{\mu\nu} + \frac{p_\mu n_\nu + p_\nu n_\mu}{(p \cdot n)} \qquad (5)$$

where only transverse components of the covariant derivatives appear:

$$B_{ij}(N) = \frac{1}{2} \langle p | \bar{\psi}\not{n}\, x^{i-2} D_\perp^\mu\, x^{3-i-1}\, D_{\perp\mu}\, x^{N-j} | p \rangle \qquad (6)$$

The answer for the transverse structure function of deep inelastic scattering is then:

$$F_\perp(x_B, \frac{\Lambda^2}{Q^2}) = A_0(\xi) + \frac{\Lambda^2}{Q^2}\left[4T_1(\xi) - \xi\int dx_2\, dx_1\, \frac{\delta(\xi-x_2)-\delta(\xi-x_1)}{x_2-x_1}T(x_2,x_1)\right] \tag{7}$$
$$+ O(\Lambda^4/Q^4)$$

where

$$A_0(x) = \frac{1}{4}\int \frac{d\lambda}{2\pi}\, e^{i\lambda x}\langle p|T[\bar\psi(0)\not\!n\,\psi(\lambda)]|p\rangle$$

$$\Lambda^2 T_1(x) = \frac{1}{4}\int \frac{d\lambda}{2\pi}\, e^{i\lambda x}\langle p|T[\bar\psi(0)\gamma_\mu \not\!n\,\gamma_\nu\, D_\perp^\mu(0)\, D_\perp^\nu(\lambda)\,\psi(\lambda)]|p\rangle$$

$$\Lambda^2 T_2(x_1,x_2) = \frac{1}{4}\int d\lambda\, d\eta\, e^{i\eta(x_2-x_1)+i\lambda x_1}\cdot\langle p|T\{\bar\psi(0)\gamma_\mu \not\!n\,\gamma_\nu\, D_\perp^\mu(\eta)\, D_\perp^\nu(\eta)\,\psi(\lambda)\}|p\rangle$$

$A_0(x)$, $T_1(x)$, $T_2(x_2,x_2)$ are the Fourier transform with respect to the light-cone position λ of bilocal or trilocal operators on the light-cone. In the above expression, the contribution of operators involving four fermions has not been included[*]. Nevertheless, the introduction of the two additional structure functions, besides the one appearing at leading twist level, makes it already very difficult to imagine that, as in the leading twist case, experimental data by themselves may fix the magnitude and the shape of the new structure functions. Models are required to reduce the number of unknowns and to make the comparison with experiment really useful.

Exclusive higher twists originate when hadrons, either in the initial or in the final state, are "directly" coupled to the hard scattering[12]: the physics of these reactions is very close to that occurring for the e.m. form factor at large momentum transfer. Similar methods are then applied to the description of these effects. In general, for the form factors, one can write[13]

$$F_\pi(Q^2) = \int_0^1 dx \int_0^1 dy\, \phi_\pi^*(x,Q^2)\, T(x,y\,;Q^2)\, \phi_\pi(y,Q^2) \tag{8}$$

----

[*] For the N = 1 moment the contribution of the four fermion operators is zero.

where

$$\phi_\pi(x,Q^2) = \int^{Q^2} \frac{d^2k_\perp}{16\pi^3} \, \psi_{q\bar{q}/\pi}(x, \vec{k}_\perp)$$

is the amplitude for finding valence q and $\bar{q}$ in a pion collinear up to a scale Q with light-cone fractions x and (1-x). The same distribution amplitude controls the exclusive processes like $\gamma q \to Mq$ (cf. Fig. 4) whose matrix element behaves like:

$$\mathcal{M} \propto \phi_\pi \otimes T \tag{9}$$

A systematic analysis of these effects has still to come: they look particularly interesting for the experimentally rather clear signature that they offer.

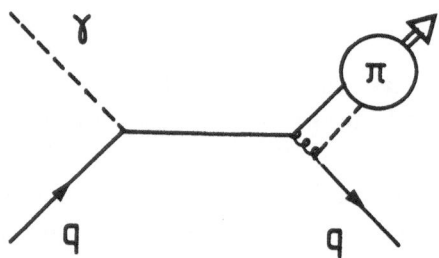

Fig. 4

Higher order calculations in the perturbative sector and diagrammatic resummation techniques will never allow us to extend the predictive domain of QCD much beyond the small box we were talking about at the beginning. Genuine non-perturbative calculations are needed in order to make contact with the hadron world. The simplest approach to gain insight into the non-perturbative aspects of the theory is one of the so-called "ITEP sum rules"[14]. The idea consists of relating the property of resonances like the width or the masses, which are due to non-perturbative effects, to the properties of the "non-perturbative" QCD vacuum, where chiral breaking occurs and non-vanishing expectation values of operators like $G_{\mu\nu}G^{\mu\nu}$ can be found. The method can be exemplified by a simple case: take the vacuum expectation values of two vector currents at different points in space-time:

$$\langle 0| \, T\left[\, J^\mu(0) \, J^\nu(x)\,\right] |0\rangle \qquad\qquad (10)$$

Its Fourier transform is related to the vacuum polarization of the photon propagator, which can be related to the experimental total cross-section of e^+e^- into hadrons via dispersion relations. On the other hand, at short distances, the operator product in Eq. (10) can be Wilson-expanded in terms of vacuum expectation values of local operators. The information about the non-perturbative characteristics of the total cross-section in the resonance region is then related to those vacuum expectation values (VEV). Typically, one finds

$$\langle 0| \sqrt{\alpha_s} \; G^\mu \, G_{\rho\nu} \, |0\rangle \;\sim\; (400 \; \text{MeV})^4$$

and

$$\left(\langle 0| \, \bar{q}q \, |0\rangle\right)^{1/3} \;\sim\; .3 \; \text{GeV}$$

Once some basic VEVs are determined, thanks to the universality of the operators appearing in the Wilson expansion, the same procedure can be applied to other current-current correlation functions in order to determine the mass of the states dominating the correlation at large distances[15]. Hadron masses have been determined in this way in satisfactory agreement with the particle data table. These methods are powerful because they exploit the internal consistencies of a local quantum field theory and relate different non-perturbative quantities among themselves.

However, in principle, the quark masses and the value of Λ are the only inputs needed to make any predictions in QCD: the only hope we have today of making calculations which require the minimal number of inputs is through the lattice approach[16].

On the lattice, we are still in a box, but of much larger size, say of about 1 Fermi. However, a finite lattice spacing "a" is provided to make the total number of degrees of freedom finite. The physics which can be learned is believed to be independent of the lattice spacing when this goes to zero. This limit is achieved by letting the (bare) coupling constant go to the ultra-violet attractive point $g \to g_c$: the system then approaches a second order phase transition, where the correlation length measured in units of the lattice spacing goes to infinity and becomes then independent of the particular way the lattice discretization is performed. Sufficiently close to the "continuum" limit, one must observe a definite scaling law expressing

how a change of the lattice spacing "a" can be compensated by a readjustment of the bare coupling constant. This is just the usual renormalization group behaviour, followed by the running coupling constant as
a function of the renormalization scale:

$$a\Lambda = [\frac{8\pi^2}{33}\beta]^{51/121} exp[-4\pi^2\beta/33] \quad where \quad \beta = \frac{6}{g_0^2} \tag{11}$$

Physical dimensionful quantities, measured in lattice spacing units
must then show an exponential behaviour in β. When this behaviour
is observed, one is confident that predictions will be credible. In
other words, QCD on the lattice is a theory with an ultra-violet cut-
off: renormalizability implies a cut-off independency for the relation
between physical quantities. This can only be believed if the expected
renormalization group behaviour of Eq. (11) is manifest.

Up to now, the only tests which have been made of the expected
scaling laws concern the pure gauge sector but the results cannot be
considered conclusive[17]. An approximate scaling behaviour has recently
been found by Kogut for the deconfinement temperature T_c, the physical
quantity characterizing the phase transition from the hadron phase to
that of the quark-gluon plasma[18].

Perhaps the most interesting results expected in the pure gauge
sector concern the estimates of the glueball mass. Various groups
have been working actively in the last years[19]: by now there is a
certain agreement about the lowest lying state, while the detailed
analysis of higher excitation is still in progress. The actual value
of the glueball mass needs the knowledge of the lattice spacing value
at a given β in physical units. This question has not been completely
settled in my opinion: the existing analysis deduces this quantity
from the scaling law behaviour of the string tension. (The coefficient
of the linearly rising $q\bar{q}$ potential.) This, in turn, is obtained by
observing the expectation value of the "Wilson" loop: the still rather
limited size of lattice available does not allow us to explore Wilson
loops of sizes much greater than 3-4 at $\beta \cong 6$, where one is not gua-
ranteed that the linearly rising part of the potential is really being
tested.

An independent way of determining the lattice size in physical units
is by trying to fit the hadron spectrum obtained by introducing fermions
on the lattice. As in the case of the pure gauge sector, there are
various ways of defining a fermionic action on the lattice, all of them
being equivalent in the continuum limit.

Unfortunately, only results in the "quenched" approximation are available today, where the contribution of virtual fermion loops is neglected[20]. Various good reasons have been presented to support such an approximation mainly based on the observed Zweig rule of hadron physics. The error due to the quenching is estimated to be of the order of a few percent, while the present statistical and systematical errors of simulations done within the quenching approximation are still of the order of 10%.

The method used is very simple and consists of the calculation of a (Euclidean) correlation function between operators carrying definite quantum numbers. The calculation is made by i) producing the quark propagators in the presence of a given field configuration; ii) using it to form the correlation function of operators bilinear or trilinear in the fermion fields; iii) averaging the result over many field configurations. When the distance between the operators becomes large, the correlation is exponentially damped with a scale corresponding to the lowest mass bound state, carrying the same quantum numbers of the operator. In formulae:

$$\int d^3x \; \langle o| \; O(o) \; O(\vec{x},t) |o\rangle \underset{t \; large}{\sim} c \, e^{-m_o t} \qquad (12)$$

A fit to the correlation function at large t provides a value for m_o. The main source of fluctuations is due to the inadequacy of the volume of the lattice. In Fig. 5, one can see the fluctuations among different Monte Carlo configurations of the "ρ" mass, with "current" quark masses of about 300 MeV calculated on a $5^3 \times 10$ or a $10^3 \times 20$ lattice: fluctuations are drastically reduced by increasing the volume[21]. A systematic error also comes from the limited distance in the time direction which, on lattices of order 5-6 at $\beta = 6$, introduces a large contamination due to radially excited states: the estimate of the masses then turns out to be systematically higher on smaller lattices. Another source of errors is the impossibility of calculating the quark propagator with values for the current masses too close to the actual values of light quarks (few MeV). The present limit is a bit less than 100MeV and is obtained on a $10^3 * 20$ lattice. If the light quark spectroscopy remains out of reach, the strange quark hadron spectrum does not require any extrapolation of the values obtained. A comparison of the results with the real spectrum is in this case very successful. One can estimate errors of the order 5-7% for the standard mesons and of $\sim$10-15% for the baryons. If an extrapolation is used to extend the predictions also to the light-quark world, one finds the same quality of results for the meson but a rather unsatisfactory scenario for the baryons: the masses are still too heavy and the spin-spin splittings definitely too small. A rather cumbersome result is that the value of the lattice spacing in physical units obtained by the analysis of hadron masses is smaller by about 30% than the one found by the scaling behaviour of the string tension. The question is not settled.

The experimental value of the total cross-section of $e^+e^- \to$ $\to$ hadrons can be used to provide an "experimental" value of the vector current-vector current correlation function in the co-ordinate space. Indeed, given $\rho(s)$ which is the imaginary part of the photon vacuum polarization, one can relate this to the real part by dispersion relations: the Fourier transform of the result is the desired correlation function. In formulae [I neglect overall constants of terms which behave like $\delta(0)$]

$$\delta^{ij}\, \Pi_{ij}(Q^2) \equiv \int dt\, e^{iq_0 t} \int d^3x\, <0|\,T[\, J_i(0)\, J^i(\vec{x},t)]|0> =$$

$$= Q^2 \int \frac{ds\, \rho(s)}{(s+Q^2)} \tag{13}$$

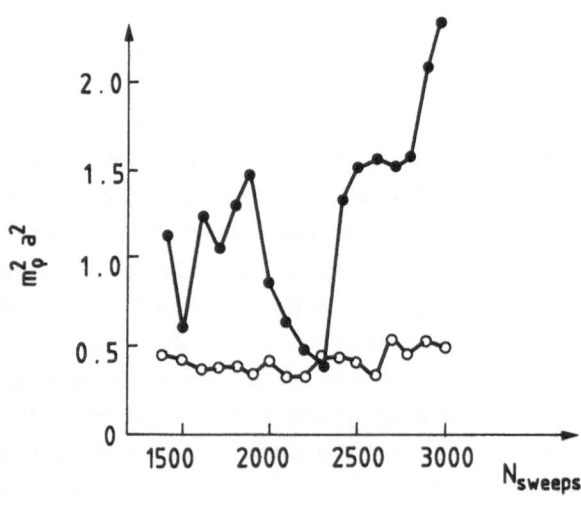

Fig. 5

from which follows:

$$C_E(t) \equiv \int d^3x \; \langle o | J_i(0) \, J^i(\vec{x}, t) | o \rangle_E =$$

$$= \int dq_0 \; e^{-q_0 t} \left\{ Q^2 \int ds \, \rho(s) / (s + q^2) \right\}$$

where the Euclidean rotation of the time has been performed. Injecting into Eq. (13) a phenomenological form for $\rho(s)$ of the type (we work out the case of strange quarks):

$$\rho(s) = 1.8 \, m_\phi^2 \, \delta[s - m_\phi^2] + \Theta[s - 1.8 \, m_\phi^2]$$

one can perform the integral. In Fig. 6 the "experimental" value of $C_E(t)$ (full line) is compared (up to an overall normalization) to the Monte Carlo results (dashed line): the agreement is rather good.

The short term development of lattice QCD will probably involve i) the analysis of larger lattice sizes made possible by the performances of the present vector machines, and ii) a systematic use of improved actions. In fact, as I already said, the definition of the lattice action is rather arbitrary: the universality character of the phase transition occurring in the continuum limit allows arbitrary definitions, differing by powers of the lattice spacing and, possibly, by inverse powers of β. As usual, the real continuum limit is a mathematical concept: in practice one has to deal with a finite lattice spacing and the precocity of the approach to the continuum may then crucially depend upon the choice of the action. (That reminds one of the ambiguity in the choice of the "correct" scale for the perturbative expansion: "asymptotically" it does not matter, but it may turn out to be very important at moderate energies.)

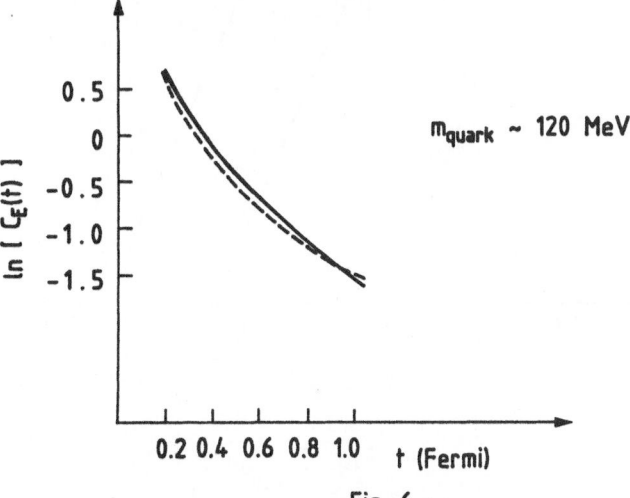

Fig. 6

The improvement consists of modifying the action by adding those terms which will make the approach faster. The solution to this problem can be found order-by-order in perturbation theory[22], by requiring the absence of corrections of the type $(a/\xi)^2 \ln \xi/a$, $(a/\xi)^4 \ln \xi/a$, ..., to the renormalization group equations of the n-point Green functions where ξ is the correlation length of the theory. An application of these ideas, due originally to Symanzik, to the case of the O(3) non-linear σ model, is given in Fig. 7[23]. One compares the behaviour with β of the inverse of the correlation length in lattice spacing units (the mass gap) for the standard or the improved actions: the straight lines are the scaling curves and the faster approach to the continuum of the modified action is evident.

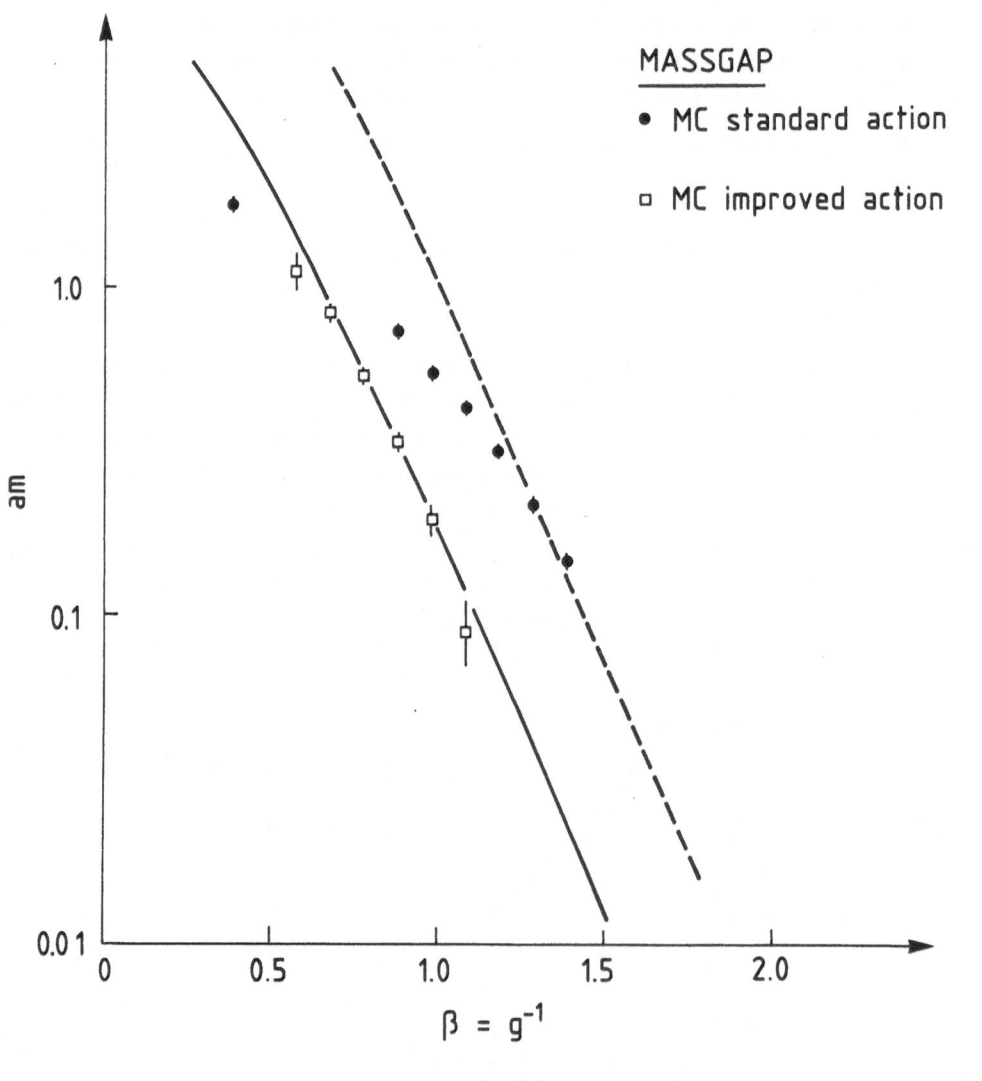

Fig. 7

Various open questions remain both in the perturbative as in the lattice approach to QCD: for example, in the former, i) a complete analysis of second order corrections of the available hard processes is still lacking, as, for instance, in the case of large $p_\perp$ production, so relevant in the collider energy region; ii) the role of power law corrections, and in particular the role of the intrinsic transverse momentum, remain to be elucidated; iii) exclusive higher twist should be looked for given the relevant insight that they provide into the shape of hadronic wave functions; iv) a QCD inspired hadronization model remains to be built which shows itself to be insensitive to possible phenomenological readjustment at least as far as the jet analysis is concerned.

In the case of the lattice, the renormalization group behaviour of calculations involving fermions must be proved, the possibility of defining improved actions must be exploited, the quenching approximation must be removed, and larger volumes must be reached.

The two approaches are complementary and they are both needed to give in the near future more rigorous support to the widespread idea that QCD is the theory of strong interactions.

REFERENCES

1) A.H. Mueller, Phys. Rev. D19 (1974) 963 and D18 (1978) 3705;
   R.K. Ellis, H. Georgi, M. Machacek, H.D. Politzer and G.G. Ross,
      Phys. Lett. 78B (1978) 281; Nucl. Phys. B152 (1979) 285;
   D. Amati, R. Petronzio and G. Veneziano, Nucl. Phys. B140 (1978) 54;
      B146 (1978) 29.

2) G. Altarelli and G. Parisi, Nucl. Phys. B126 (1977) 298.

3) P.M. Stevenson, Phys. Lett. 100B (1981) 61; Nucl. Phys. B203
      (1982) 472.

4) S. Ellis, M. Fleishon and W. Sterling, Phys. Rev. D24 (1981) 1386;
   J. Kodaira and L. Trentadue, SLAC preprint SLAC-PUB-2862 (1981).

5) Yu.L. Dokshitzer, D.I. D'yakonov and S.Y. Troyan, Phys. Rep. 58
      (1980) 269;
   G. Parisi and R. Petronzio, Nucl. Phys. B54 (1979) 427;
   J. Collins and D. Soper, Nucl. Phys. B193 (1981) 381 and B194 (1982)
      445;
   P. Chiappetta and M. Greco, Phys. Lett. B106 (1981) 219.

6) H.J. Berends et al., "The influence of fragmentation models on the
      determination of α_s", CELLO collaboration, DESY preprint (1982).

7) A.H. Mueller, Phys. Lett. 104B (1981) 161;
   A. Bassetto, M. Ciafaloni, G. Marchesini and A.H. Mueller, Nucl.
      Phys. B207 (1982) 189.

8) G. Marchesini and B.R. Webber, CERN preprint TH.3525 (1983).

9) H.D. Politzer, Nucl. Phys. B172 (1980) 349.

10) R. Jaffe and M. Soldate, Phys. Rev. D26 (1982) 49.

11) R.K. Ellis, W. Furmanski and R. Petronzio, Nucl. Phys. B207 (1982)
      1 and B212 (1983) 29.

12) For a recent review see,
    S.J. Brodsky, T. Huang and G.P. Lepage, SLAC preprint SLAC-PUB-
      2868 (1982).

13) S.J. Brodsky and G.R. Farrar, Phys. Rev. Letters 31 (1973) 1153
      and see also Ref. 12).

14) M. Shifman, A. Vainshtein and V. Zakharov, Nucl. Phys. B147 (1979)
      385.

15) B.L. Ioffe, Nucl. Phys. B188 (1981) 317;
    N.V. Krasnikov, CERN preprint TH.3422 (1982).

16) K.G. Wilson, New Phenomena in Subnuclear Physics, ed. A. Zichichi,
      Erice (1975), (Plenum Press, New York, 1977).

17) M. Creutz, Phys. Rev. Lett. 43 (1979) 553; 45 (1980) 313;
    G. Bhanot and C. Rebbi, Nucl. Phys. B188 (1981) 469;
    E. Pietarinen, Nucl. Phys. B190 (FS3) (1981) 349;
    M. Creutz and K.J.M. Moriarty, Phys. Rev. D26 (1982) 2166.

18) J. Kogut et al., University of Illinois preprint ILL-TH-82-5 and
    ILL-TH-82-39.

19) See for example the talk by
    B. Berg at the 21st International Conference on High Energy Physics,
    Paris, (1982) and references therein;
    For a recent analysis, see:
    B. Berg and A. Billoire, Saclay preprint Sph.T/42 (1983).

20) E. Marinari, G. Parisi and C. Rebbi, Phys. Rev. Lett. 47 (1981)
    1978;
    M. Hamber and G. Parisi, Phys. Rev. D28 (1983) 247;
    D. Weingarten, Nucl. Phys. B215 (FS7) (1983) 1;
    F. Fucito, G. Martinelli, C. Omero, G. Parisi, R. Petronzio and
    F. Rapuano, Nucl. Phys. B210 (FS6) (1982) 407;
    G. Martinelli, C. Omero, G. Parisi and R. Petronzio, Phys. Lett.
    117B (1982) 434;
    A. Hasenfratz, P. Hasenfratz, C.B. Lang and Z. Kunszt, Phys. Lett.
    117B (1982) 81;
    R. Gupta and A. Patel, Caltech preprint CALT-68-966 (1982);
    C. Bernard, T. Draper and K. Olynyk, UCLA preprint UCLA/82/TEP/10
    (1982);
    K.C. Bowler, E. Marinari, G.S. Pawley, F. Rapuano and D.J. Wallan,
    Edinburgh preprint 82/236 (1982).

21) H. Lipps, G. Martinelli, R. Petronzio and F. Rapuano, CERN preprint
    TH.3548 (1983).

22) K. Symanzik, Mathematical Problems in Theoretical Physics, eds.
    R. Schrader et al., Conf. Berlin (1981), (Springer Verlag, 1982),
    Lectures Notes 153;
    G. Martinelli, G. Parisi and R. Petronzio, Phys. Lett. 114B (1982)
    251;
    P. Weisz, DESY preprint 82-044 (1982).
    B. Berg, S. Meyer, I. Montvay and K. Symanzik, DESY preprint
    (1983);
    M. Falcioni, G. Martinelli, M.L. Paciello, B. Taglienti and
    G. Parisi, University of Rome preprint (1983).

23) B. Berg et al., Ref. 22).

DISCUSSION

W. Frazer, Univ. of Calif.: How useful is the resummation technique in improving the QCD perturbation series for $x \to 1$? Isn't it true that only the leading singularity as $x \to 1$ is eliminated, and the resulting series still converges poorly for x near 1?

R. Petronzio, CERN: The next-to-leading terms are normally down by a power of α: one could find a region where $\alpha \ln^2(1-x) \sim 1$ but $\alpha \ln(1-x) \ll 1$. Admittedly, this region is rather small. Alternatively, one can try to construct diagrammatic techniques to resum also some next-to-leading corrections [4].

S. Brodsky, SLAC: 1) Many 'direct' or 'exclusive higher twist' high p_T processes can be calculated in terms of meson form factors without ambiguity, and I agree these are very important to study. In addition the leading higher twist contribution to the meson structure function can be related to the meson form factor.

2) Initial state interactions in hadron-hadron reactions add to the complexity of QCD phenomenology leading to breakdown of factorization for long targets, a new source of k_T fluctuations, central region radiation, etc. Factorization of the Drell-Yan process in leading twist has so far only been established to two-loop order.

3) The scale fixing procedure for perturbative expansions apparently is still controversal. Lepage, Mackenzie and I have noted that there is no ambiguity in setting the scale in the corresponding QED problem, and (for any renormalization scheme) thus automatically leads to an unambigous scale-setting method for QCD to the first non-trivial order. Thus the method gives sensible results for all processes including the change of scale for high n moments. The PMS method is inconsistent when applied to QED reactions,e.g. positronium decay.

JET PRODUCTION AND HIGH p_T PHENOMENA IN PHOTON-PHOTON REACTIONS[*]

NORBERT WERMES[†]

Stanford Linear Accelerator Center

Stanford University, Stanford, California 94305

ABSTRACT

The status of experimental investigations of high p_T phenomena and jet production in photon-photon collisions is reviewed. Taking the challenging questions on hard scattering processes in $\gamma\gamma$ reactions as a guide, the experimental approach to these questions is summarized. Results from the PETRA experiments CELLO, JADE, PLUTO, and TASSO are presented including preliminary results on the Q^2-dependence of jet cross sections. Experimental limitations and background problems are discussed.

Zusammenfassung — Es wird ein Überblick über den Stand experimenteller Untersuchungen von high-p_T Phänomenen und der Erzeugung von Jets in Photon-Photon Wechselwirkungen gegeben. Anhand der herausfordernden Fragen bezüglich harter Streuprozesse in $\gamma\gamma$ Reaktionen wird der experimentelle Ansatz zur Beantwortung dieser Fragen zusammengefasst. Ergebnisse der PETRA Experimente CELLO, JADE, PLUTO und TASSO werden presentiert. Vorläufige Untersuchungen der Q^2-Abhängigkeit von Jet-Wirkungsquerschnitten sowie experimentelle Einschränkungen und Untergrundprobleme werden diskutiert.

[*] Work supported by the Department of Energy, contract DE-AC03-76SF00515.

[†] Alexander von Humboldt Fellow.

1. INTRODUCTION

This talk reviews experimental results on hard scattering reactions in $\gamma\gamma$ collisions via high p_T phenomena.

The talk is divided into three parts. First I shall try to list the physics challenges for photon-photon experiments, in the context of hard scattering processes at high p_T. In the main part the experimental approach to these challenging questions is discussed. It will be explained why we believe that hard scattering processes do exist. Then the explicit jet-searches performed by the different experiments are reviewed. Finally I shall conclude with an attempt to assess to what extent the challenging questions can be answered.

Many theoreticians[1] believe that photon-photon collisions and particularly the high p_T hard scattering phenomena are perhaps the cleanest laboratory for testing QCD. The arguments are that there are only fundamental particles involved and the processes are computable in QCD. Also, when approaching $\gamma\gamma$ scattering through $e^+e^- \rightarrow e^+e^- +$ hadrons in e^+e^- storage rings the corresponding structure functions in the "hard scattering expansion" are relatively simple compared to hadron-hadron interactions. There are no spectator jets accompanying the leading order $\gamma\gamma \rightarrow q\bar{q}$ reaction as in pp scattering which makes the experimental situation much cleaner. Last, but not least, the photon itself is a very direct probe of matter and the fact that the photon's Q^2 can be varied in e^+e^- collisions is very useful to disentangle hadronic and pointlike reactions in $\gamma\gamma$ collisions.

In the following I want to list a not neccessarily complete set of explicit tasks being formulated by theory and challenging the experiments.

- $\gamma\gamma \rightarrow q\bar{q}$ scattering at high transverse momenta allows one to test the quark propagator at large p^2. This should be much cleaner in $\gamma\gamma$ collisions than in e^+e^- reactions via $e^+e^- \rightarrow q\bar{q}g$ because no uncertainties introduced through the strong coupling constant α_s confuse the issue. In the same context there is a question whether current or constituent quark masses appear in the quark propagator.

- In their epochal paper[2] in 1971 Berman, Bjorken and Kogut pointed out that for pointlike hard scattering processes like $\gamma\gamma \rightarrow q+X$ the jet (quark)-trigger cross sections should scale as

$$E \frac{d\sigma}{d^3p} (\gamma\gamma \rightarrow \text{jet} + X) \propto \frac{1}{p_T^4} f(x_T, \theta_{c.m.}) \quad , \tag{1}$$

where $x_T = 2p_T/\sqrt{s}$, $\theta_{c.m.} = $ center-of-mass angle of jet, $p_T = $ transverse jet-momentum with respect to the $\gamma\gamma$ axis. This typical scaling behaviour should be tested in $\gamma\gamma \rightarrow$ jet $+ X$ cross sections and in inclusive particle cross sections at high p_T.

- $\gamma\gamma \rightarrow q\bar{q}$ scattering provides a useful tool to test the quark charges and resolve the question of fractionally or integrally charged quarks. Let me spend a few words on why $\gamma\gamma$ scattering is unique for this. Quark charges in models which satisfy the modified Gell-Mann-Nishijima relation

$$Q = \left(T_3 + \frac{Y}{2}\right) + \frac{1}{3} t \tag{2}$$

where T_3 and $Y/2$ are flavour generators and t is an arbitrary number which may depend on colour, can be assigned in a very general way[3]

$$
\begin{array}{c|cccc}
 & u & d & s & c \\
\hline
R & z_R & z_R - 1 & z_R - 1 & z_R \\
B & z_B & z_B - 1 & z_B - 1 & z_B \\
Y & z_Y & z_Y - 1 & z_Y - 1 & z_Y \\
\end{array}
\tag{3}
$$

where R,B,Y denote different colours and the constraint that $z_R + z_B + z_Y = 2$, which follows from the fact that the $\Delta^{++}(u_R, u_B, u_Y)$ has charge +2, has to be obeyed. In the fractionally charged quark model (FCQ), also called Gell-Mann-Zweig model,[4] one has $z_R = z_B = z_Y = 2/3$; in integrally charged quark models (ICQ) originally invented by Han and Nambu[5] the assignment is $z_R = 0$, $z_B = z_Y = 1$. In this nomenclature the photon can be considered as

$$"\gamma" \sim (e_f q\,\bar{q},\ R\bar{R} + B\bar{B} + Y\bar{Y}) + \left(\left(z_R - \frac{2}{3}\right)q\,\bar{q},\ R\bar{R} - Y\bar{Y}\right) + \left(\left(z_B - \frac{2}{3}\right)q\,\bar{q},\ B\bar{B} - Y\bar{Y}\right) \tag{4}$$

where e_f = flavour-charge of quarks (2/3,-1/3). The notation $(q\,\bar{q}, R\bar{R}+B\bar{B}+Y\bar{Y})$ means $(q_R\,\bar{q}_R + q_B\,\bar{q}_B + q_Y\,\bar{q}_Y)$ and it is easy to see that in (4) the photon is decomposed into a flavour and a colour part. Expressing (4) in terms of flavour and colour multiplets we find in the particular case of the ICQ-model

$$"\gamma" \sim (\{NS\}_F,\ \{1\}_C) - (\{1\}_F,\ \{8\}_C) \tag{5}$$

where F,C denote flavour and colour and NS represents 'non-singlet' ($\{NS\}_F = \{8\}_F$ in case of $SU(3)_F$). Note that for FCQ models $z_\alpha - (2/3) = 0$ for $\alpha = R, B$ and therefore there is no colour octet piece of the photon in these models. The colour octet part of the photon is responsible for a different value of $R_{\gamma\gamma}$ in the two quark model alternatives. While in one photon annihilation (Fig. 1) the final state colour singlet can only be generated by a colour singlet photon, in two photon reactions (Fig. 2) two colour octet photons can produce a colour singlet final state ($\{8\} \otimes \{8\} = \{1\} \oplus \{8\} \oplus \{8\} \oplus \{10\} \oplus \{\overline{10}\} \oplus \{27\}$). Thus[6]

$$R_{\gamma\gamma} = \frac{d\sigma/dt(\gamma\gamma \to q\,\bar{q})}{d\sigma/dt(\gamma\gamma \to \mu\mu)} = \left.\begin{cases} 3 \cdot \Sigma_i e_i^4 \\ 1/3 \cdot \Sigma_i (\Sigma_\alpha e_{i\alpha}^2)^2 \end{cases}\right\} = \begin{cases} 34/27 & \text{for FCQ} \\ 10/3 & \text{for ICQ} \end{cases} \tag{6}$$

where $e_{i\alpha}$ is the charge of the quark with flavour i and colour α, is different for the different quark models while $R_{1\gamma}$ is not. Therefore it is a real challenge to measure $R_{\gamma\gamma}$ in a clean way.

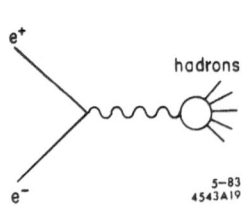

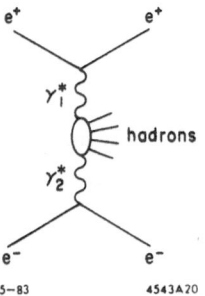

Fig. 1. e^+e^- annihilation into hadrons.   Fig. 2. Two photon production of hadrons.

- $\gamma\gamma$ scattering is a place to study the interplay between hadron-like (VDM) and pointlike coupling of the photon.

- By measuring the different underlying subprocesses like e.g. $\gamma\rho \to q\bar{q}$, $\gamma\gamma \to Mq\bar{q}$, $\rho\rho \to q\bar{q}$, ... one can test the validity of the CIM-model[7] in $\gamma\gamma$ reactions.

- Finally, a strong challenge for the experiments is to test direct $\gamma q \to gq$ and $q\bar{q} \to q\bar{q}$ scattering[8,9,10] via processes as in Fig. 3 where qq-scattering is accompanied by two beam-pipe jets.

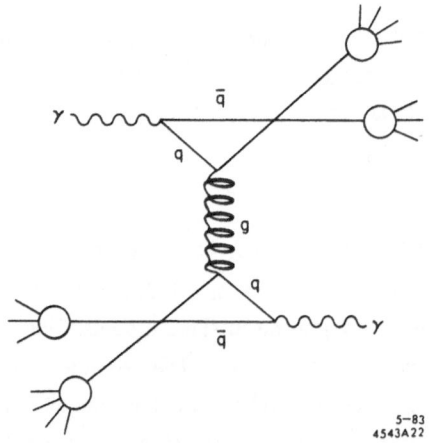

Fig. 3. Direct qq-scattering in a two photon reaction.

2. THE EXPERIMENTAL APPROACH

Let me try to describe the experimental approach to the above questions in three steps:

1. Do we have evidence for hard scattering processes in $\gamma\gamma$ reactions?

2. Can one separate the $\gamma\gamma \to q\bar{q}$ Born process from background processes?

3. To what extent can one answer the challenge questions?

Since two photon physics became accessible to high energy physics there have been six contributions to the subject of hard scattering phenomena via high p_T jets (Table I).

TABLE I

PLUTO (1980)	unpublished 1980	notag + single tag	$2.6\ pb^{-1}$
JADE	Phys. Lett. 107B, 1981	single tag	$9.7\ pb^{-1}$
TASSO	Phys. Lett. 107B, 1981	single tag	$9.0\ pb^{-1}$
TASSO	preliminary 1983	notag	$54\ pb^{-1}$
PLUTO (1982)	preliminary 1983	single tag	$39\ pb^{-1}$
CELLO	preliminary 1983	notag	$6\ pb^{-1}$

The published results from JADE[12] and TASSO[13] using single tag events and the unpublished PLUTO[11] result have been updated by the new PLUTO detector[14], dedicated to two photon physics, with much larger statistics using single tag events in two different Q^2 regions. The problem has also been attacked using notag events by TASSO and CELLO.[14]

Primary evidence for hard pointlike scattering in $\gamma\gamma$-reactions should be seen in:

(a) Jet structure of events eventually seen in single event displays,

(b) $1/W^2$ — contribution in $\sigma^{tot}_{\gamma\gamma \to had}(W)$,

(c) inclusive particle p_T-distributions.

The jet-like character of events seen in $\gamma\gamma$-collisions may well be due to the Lorentz-boost of the $\gamma\gamma$-system which makes the events appear jet-like. The presence of a $1/W^2$-term in the parametrization of the total hadronic $\gamma\gamma$-cross section is still under discussion[15] and its existence as well as its possible origin is not yet established. Therefore we are left with the investigations referring to the last point.

As already mentioned in the Introduction, the authors of Ref.2 predicted scaling for a pointlike scattering process $a + b \to c + d$

$$E_c\ \frac{d\sigma}{d^3 p_c} = \frac{4\pi\alpha^2}{p_T^4}\ f(x_T, \theta_{c.m.}) \quad ; \quad x_T = \frac{2p_T}{\sqrt{s}} \quad . \tag{7}$$

The p_T^{-4}-term comes from a $1/s^2$ in $d\sigma/dt$ and describes the energy dependence. $f(x_T, \theta_{c.m.})$ corresponds to the angular dependence of the process. In order to "see" p_T^{-4}-behaviour of $Ed\sigma/d^3p$

or $d\sigma/dp_T^2$ one must either keep x_T and $\theta_{c.m.}$ fixed or one has to make sure that the angular dependence $f(x_T, \theta_{c.m.})$ does not spoil the p_T-slope too much. An example is given in Fig. 4 for the pointlike process $e^+ e^- \to \mu^+ \mu^-$. The cross section is

$$\frac{d\sigma}{dp_T^2} (ee \to \mu\mu) = \frac{\alpha^2 \pi}{4s^2} \cdot \frac{1 + \cos^2\theta}{\cos\theta} = \frac{1}{p_T^4} f(x_T, \theta_{c.m.}), \tag{8}$$

but the p_T^{-4} term is completely spoiled by the angular dependence. The cross section even exhibits a singularity at $p_T/p_T^{max} = 1$ due to the Jacobian $(\cos\theta)^{-1}$.

Usually the pointlike scattering process revealing the p_T^{-4} is only a subprocess of a more complex reaction as sketched in Fig. 5. The "hard scattering expansion"[16]

$$E_C \frac{d\sigma}{d^3 p_C} (A + B \to C + X) = \int\int dx_a \, dx_b \, G_{a/A}(x_a, p_T^2) \, G_{b/B}(x_b, p_T^2) \, \frac{D_{C/c}}{z_c} (z_c)$$

$$\times \frac{d\sigma}{d\hat{t}} (a + b \to c + d) \cdot \frac{\hat{s}}{\pi} \cdot \delta(\hat{s} + \hat{t} + \hat{u}), \tag{9}$$

$(\hat{s}, \hat{t}, \hat{u} = $ Mandelstam variables of the subprocess, $x_{a,b} = 2p_{a,b}/\sqrt{s}$, $s = $ total center-of-mass energy) parameterizes this reaction in terms of structure functions $G_{x/X}$ and fragmentation functions $D_{Y/y}$ in order to convolute the subprocess and the initial and final state distributions. In the case of $\gamma\gamma \to$ hadron $+ X$ the "outer" process is $e^+ e^- \to e^+ e^- + h + X$. Note that for $\gamma\gamma$ reactions also the processes with more than two quarks in the final state obey the p_T^4 scaling law since the α_s in the subprocess reaction cancels against the α_s in the photon structure function as first noticed by C.H. Llewellyn Smith.[9] Brodsky et al.[8] and Kajantie et al.[10] have performed the convolution of photon-photon flux $G_{\gamma/e}(x)$ and fragmentation function $D_{h/q}(z_h)$ with the subprocess cross section of $\gamma\gamma \to q\bar{q}$. They have shown that the p_T^{-4}-dependence is preserved for moderate x_T

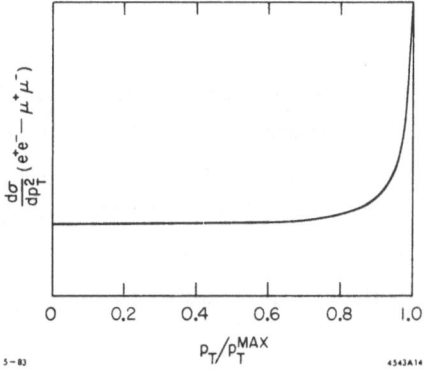

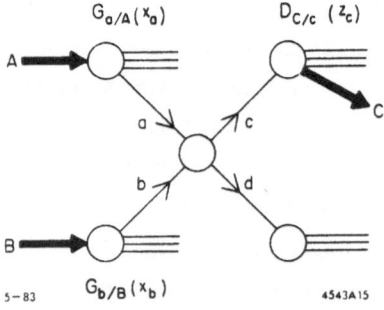

Fig. 4. $d\sigma/dp_T^2$ as function of p_T/p_T^{max} for the pointlike scattering process $e^+ e^- \to \mu^+ \mu^-$.

Fig. 5. The pointlike scattering process $a + b \to c + d$ embedded in the reaction $A + B \to C + D$.

$$\frac{d\sigma}{dp_T^2}(e^+e^- \rightarrow e^+e^- + h^\pm + X) \simeq \frac{1}{p_T^4}\left(log\,\frac{2}{x_T} - \frac{7}{3} + \dots\right) \quad . \tag{10}$$

The angular dependence $f(x_T, \theta_{c.m.})$ only slowly varies with x_T.

This means that the experiments may integrate over $p_\parallel$ not keeping $x_T, \theta_{c.m.}$ fixed without affecting the p_T^{-4}-behaviour too much. Thus we would expect a flattening tail in p_T^2-distributions of $e^+e^- \rightarrow e^+e^- + h + X$ which goes like p_T^{-n}, $n \simeq 4$. Let me remind you that for the QED-process $e^+e^- \rightarrow e^+e^-\mu^+\mu^-$, which only has the photon structure functions in the initial state but no fragmentation in the final state, MARK J[17] and PLUTO[18] have measured a p_T-slope of p_T^{-n}, with $n \sim 4.2 - 4.5$, in agreement with the corresponding QED Monte Carlo calculation.

In Fig. 6 inclusive particle p_T-distributions are shown for the TASSO and PLUTO data. The TASSO data have been selected requiring ≥ 3 charged tracks in the final state. They are corrected for acceptance and compared to different slopes in p_T, normalized to each other at $p_T = 1$ GeV/c. The data at high p_T agree with p_T-slopes of p_T^{-6} and p_T^{-4}. A fit of the type $c_1 exp(-ap_T) + c_2 p_T^{-b}$ yields $a = 7.4 \pm 0.3$ and $b = 3.9 \pm 0.6$. Although this fit result crucially depends on how the fit function is set up and how the transition between low p_T and high p_T data is parametrized, one can, I think, conclude that the data clearly break off from the exponential $exp(-6p_T)$ behaviour, expected from a purely hadronic reaction, at $p_T \sim 1$ GeV/c and the tail at high p_T is consistent with a power law p_T^{-n}, $n \lesssim 6$. In Fig. 6(b) preliminary data from PLUTO plotted as a function of p_T^2 are directly compared to model predictions from a $\gamma\gamma \rightarrow q\bar{q}$ model (see below) and a VDM prediction which includes a limited transverse momentum phase space

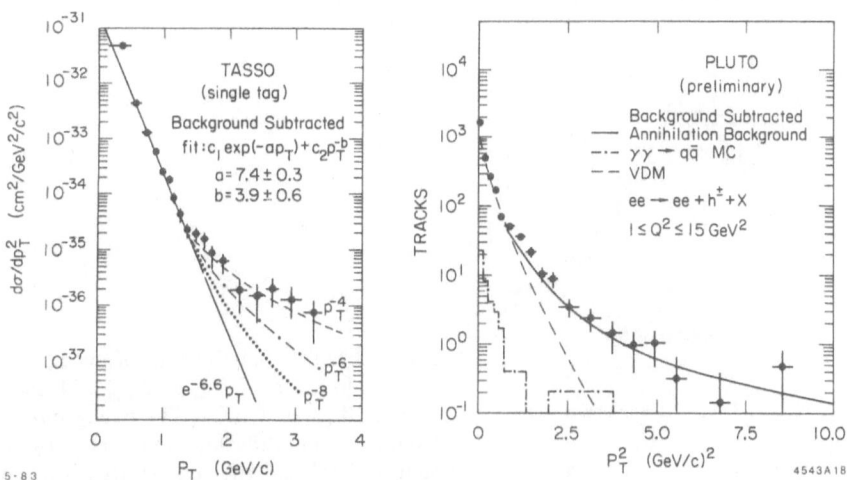

Fig. 6. (a) Inclusive particle $d\sigma/dp_T^2$ cross section observed by TASSO compared to different p_T-slopes (see text). (b) p_T^2-distribution observed by PLUTO (LAT). The curves represent expectations from a $\gamma\gamma \rightarrow q\bar{q}$ model (solid line) and a VDM model (dashed line). The dashed-dotted line represents the 1γ background.

distribution of particles about the $\gamma\gamma$ axis. The matrix element for the latter is proportional to $exp(-6p_T)$. These data require a tag in the PLUTO large angle tagger (LAT) resulting in a Q^2-range from 1 to 15 GeV^2 and are selected demanding at least 4 charged tracks and a visible $\gamma\gamma$-energy, W_{vis}, of more than 4 GeV. The above conclusion is confirmed by the PLUTO data. The high p_T tail cannot be accounted for by the VDM prediction. Since the data are not yet corrected for acceptance no attempt was made to fit the p_T-slope. The data are however consistent with p_T^{-n}, $n \lesssim 5.$[19]

A major concern is possible background from 1γ annihilation which may become very large at high p_T even for single tag events. In Fig. 6 this background has been computed by Monte Carlo and has been subtracted from the data. The subtraction amounts to about 15% for $p_T > 2$ GeV/c in the TASSO data. For the PLUTO data the 1γ background is shown by the histogram to be very small because of the excellent tag identification capability of the new PLUTO detector in the LAT.

For notag data the one photon background is a serious problem as shown in Fig. 7. Plotted are $\sum_i p_i/E_{beam}$ (TASSO) and W_{vis} (CELLO) respectively, with comparisons to Monte Carlo generated 1γ events. For the TASSO data some cuts for selecting high p_T events have been applied. At the high end of the distributions the data are completely explained by the annihilation process $e^+e^- \rightarrow$ hadrons when energy is lost by initial state radiation or if the final state is only partially detected. The total amount of 1γ background is $\sim$13% for the CELLO data but increases to over 50% when high p_T jet events are selected. The TASSO data include $\sim$30% 1γ background when a cut $\sum_i p_i/E_{beam} \leq 0.4$ is applied.

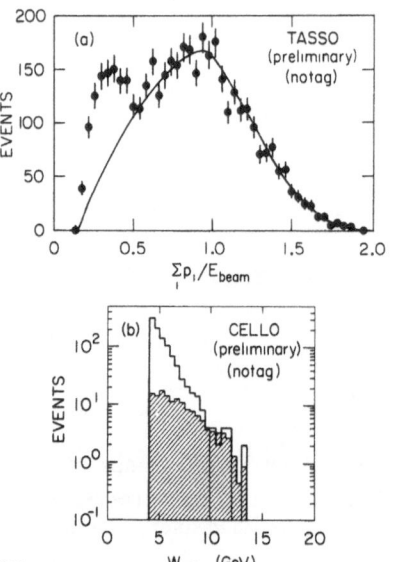

Fig. 7. Energy distributions of notag events plotted as: (a) $\sum p_i/E_{beam}$ (TASSO) and (b) W_{vis} (CELLO). The background from e^+e^- annihilation is indicated by the solid line (a) and the shaded area (b). Note that some cuts for selecting high p_T events have been made in (a).

There is no other choice for notag events than to subtract the computed amount of annihilation background in all distributions. The 1γ subtracted p_T^2-distribution of single particles is shown in Fig. 8 for the CELLO data. Within statistics and despite the above mentioned difficulties the notag data agree with the conclusions drawn from the single tag data p_T^2-distributions.

It is worthwhile to notice that one would expect from a hadronic behaviour of the photons $(\rho\rho \to \rho\rho,\ \gamma\rho \to \gamma\rho)$

$$E\frac{d\sigma}{d^3p}\ (\rho\rho \to \rho\rho) \sim p_T^{-12} \cdot f(x_T, \theta_{c.m.})$$

$$E\frac{d\sigma}{d^3p}\ (\gamma\rho \to \gamma\rho) \sim p_T^{-8} \cdot f(x_T, \theta_{c.m.})$$

as predicted by CIM model counting rules.[20] These p_T-slopes are much steeper than those seen in the data of Figs. 6 and 8.

Let me add two experimental points here. We have required the covered x_T range to be moderate in order to have the fragmentation functions and photon fluxes not affect the p_T dependence. The x_T-range covered by all experiments lies between ~ 0.1 and ~ 0.25 for high p_T tracks, i.e. moderate values. The second point is the question whether detector inefficiencies could change the conclusion. The TASSO data in Fig. 6(a) are corrected for acceptance. But apart from that the detection efficiency for high p_T tracks is almost constant above $p_T \sim 0.5$ GeV/c due to the fact that the detector acceptance is good at large angles where most of the high p_T tracks come from.

Nevertheless, before we definitely conclude that the features discussed so far give evidence for underlying hard scattering processes in $e^+e^- \to e^+e^- + h + X$ we want to directly compare to hadron-hadron data. This has been done in Fig. 9(a). The same data as in Fig. 6(a) taken at

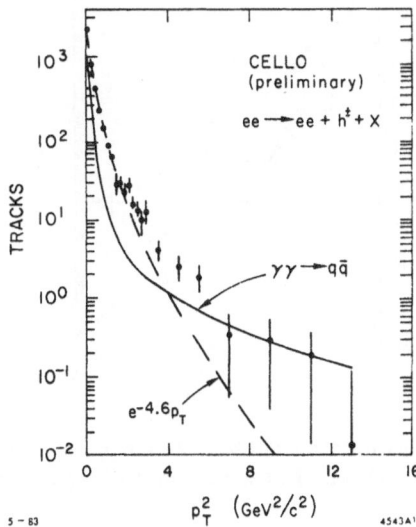

Fig. 8. p_T^2-distribution for the CELLO notag data after subtraction of 1γ events. The curves represent the $\gamma\gamma \to q\bar{q}$ expectation (solid line) and a fit with $exp(-4.6p_T)$ to the data.

5 – 83

4543A13

an average e^+e^- c.m. energy of $\sim$30 GeV are compared to ISR data from $pp \to \pi^\pm + X$ [21] at a c.m. energy of 23 GeV. The two sets of data are normalized at $p_T \sim$ 200 MeV/c. The agreement at low p_T is striking. At $p_T \sim$ 1.5 GeV/c however the e^+e^- data (i.e. $\gamma\gamma \to h + X$) clearly break off from the hadron-hadron slope. In Fig. 9(b) a similar comparison is made for $p\bar{p} \to h + X$ taken by the UA1 detector [22] at a centre-of-mass energy of 540 GeV at the SPS collider. The data are again compared to data from ISR measurements.[21] A similar deviation from the lower energy ISR data is observed revealing a flat high p_T tail which extends to $p_T \sim$ 9 GeV/c. This effect has been attributed to underlying hard scattering subprocesses ($q\bar{q} \to q\bar{q}$) buried in the $p\bar{p}$-reaction which become relevant at high energies and high p_T. In fact the observed p_T-slope for the UA1 data is p_T^{-n}, n $\approx$ 5.[23]

Therefore we conclude that for the $\gamma\gamma$ data the high p_T tail observed in $d\sigma/dp_T^2$ distributions cannot be attributed to hadron-like scattering of the photons. The alternative explanation would be to assume pointlike $\gamma\gamma \to q + X$ subprocesses to be responsible. This would give rise to events showing a jet-topology.

Before I report on jet-searches I would like to add that for a complete understanding of the inclusive particle p_T-spectra one would like to understand the following questions:

1. What is the π, K, p particle decomposition at high p_T? How many particles are ρ's?

2. The highest p_T bins correspond of course to low multiplicity jets due to simple kinematics. How much is just $\gamma\gamma \to \rho\rho$ scattering where the ρ's are interpreted as two particle jets? How much is due to $\gamma\gamma \to \pi + X$ where no jets are formed?

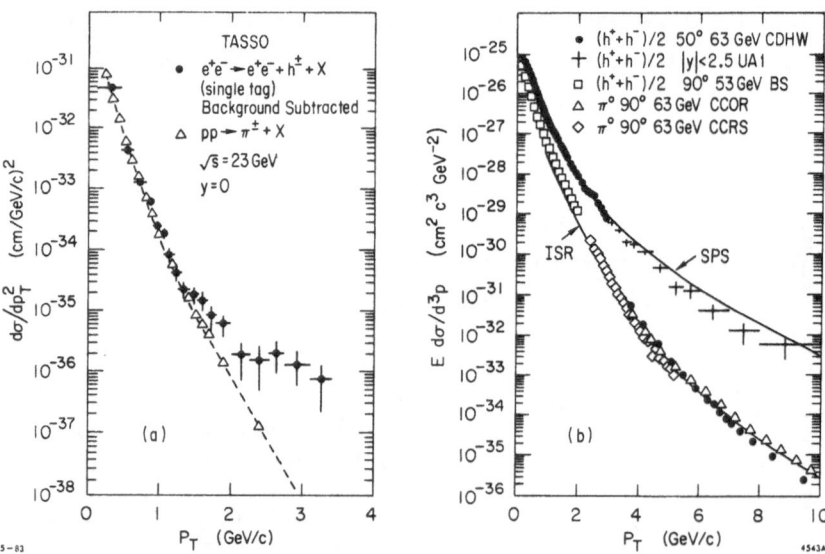

Fig. 9. (a) Comparison of the TASSO single tag data with data from Ref. 21 for $pp \to \pi^\pm + X$. (b) $Ed\sigma/d^3p$ cross sections for UA1-data at $\sqrt{s} = 540$ GeV and ISR-data from Ref. 21 at $\sqrt{s} = 63$ GeV for $pp \to \pi + X$ and $p\bar{p} \to \pi + X$, respectively.

3. Although we concluded that integrating over rapidity or $\cos\theta$ was allowed for $e^+e^- \rightarrow e^+e^- + h + X$, one would nonetheless like to investigate the dependence on rapidity, i.e. measure the full differential cross section $d\sigma/dy_1 d^2 p_T dy_2$.[10]

3. SEARCH FOR JETS

From an experimental and also theoretical point of view it is most interesting to first look for evidence for the Born term process $\gamma\gamma \rightarrow q\bar{q}$ (Fig. 10). One expects about 20-40 events per 10 pb^{-1} for tag and about 200-400 events per 10 pb^{-1} for notag events at $\sqrt{s} = 30$ GeV for $p_T^{quark} \geq 2$ GeV/c after correcting for acceptance.

The four experiments employ two different jet-finding algorithms. PLUTO and TASSO use a 'Thrust' or 'Twoplicity' method[24,13] which always finds two jets per event. JADE and CELLO employ a cluster search algorithm[25,12] which finds 0,1,2,3,... jet events corresponding to the number of distinct clusters found. Both methods provide jets with jet axes strongly correlated to the original quark axes as demonstrated by MC studies in Fig. 11.

There are two major concerns when searching for jets. The first concern is possible background from 1γ annihilation. This background can be reduced by requiring that the detected invariant mass of the final state must not exceed a certain maximum (usually about 40% of the e^+e^- – c.m. energy). A very efficient tool is also to require a tag in the forward spectrometer which is even more powerful the better the tagging particle ($e^\pm$) can be identified. The new PLUTO detector has the capability to associate a tagging $e^\pm$ with its track measured by forward drift and proportional chambers. For a LAT-tag it is also possible to measure the sign of the charge of this track from left or right bending in the forward septum magnet. This drastically reduces possible confusion by hadronic tags or γ conversion in 1γ events.

The second concern is whether one is able to extract the Born process $\gamma\gamma \rightarrow q\bar{q}$ from competing processes such as higher twist reactions[8] and 3- and 4-jet processes.[8,9,10] Bagger and Gunion[26] recently have shown that the normalization for higher twist processes, as quoted in Ref. 8, has to be corrected by a factor $\sim 1/130$. They quote for the ratio of (higher twist/minimum twist) $\sim 10\%$ at $\sqrt{s_{e^+e^-}} = 30$ GeV and $p_T^{jet} \geq 2$ GeV/c. This new understanding makes life easier for the experimentalists.

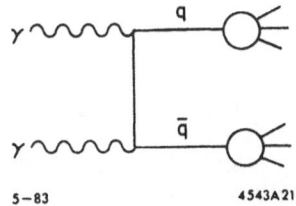

5–83   4543A21

Fig. 10. The Born diagram for
$\gamma\gamma \rightarrow$ jet + X reactions, $\gamma\gamma \rightarrow q\bar{q}$.

Possible identification and separation of 3- and 4-jet reactions which also scale like p_T^{-4} has been discussed by J. Stirling at this conference.[27] From an experimentalist's point of view, a major improvement has been made by adding forward devices to detect two photon events more completely. About 70-80% of the total $\gamma\gamma$ c.m. energy is now seen by most of the detectors for high p_T events. This is demonstrated in Fig. 12 for the PLUTO detector comparing the ratio W_{vis}/W_{true} for the old PLUTO without forward spectrometer and the new PLUTO including the forward spectrometer.

A very nice handle to suppress competing non-Born processes is to make use of large angle tags which corresponds to the photon having high Q^2. This suppresses all kinds of $\gamma\gamma \to$ jet $+$ X background whenever there are form factor-like effects (hadronization) involved at the high Q^2 photon vertex. Naively one would expect to reduce all higher twist and 3,4 jet backgrounds by at least about a factor of 2.

The following basic cuts were applied by the experiments when looking for jets.

- $W_{min} \leq W_{vis} \leq W_{max}$ – This cut limits the detected W_{vis} range in order to suppress 1γ background contributions (W_{max}-cut) and to provide enough energy for the quarks to develop as jets (W_{min}-cut). W_{min} is usually ~ 4 GeV, $W_{max} \sim 40\% \cdot \sqrt{s_{e^+e^-}}$.

- $n_{ch} \geq 4$, $n_{jet} \geq 2$ – This cut limits the jets to contain at least two particles.

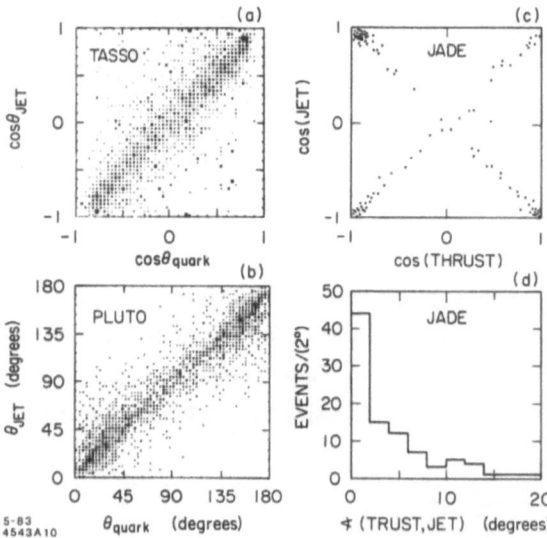

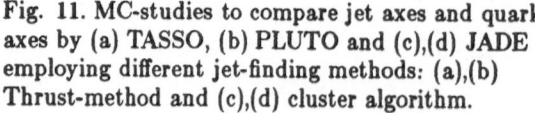

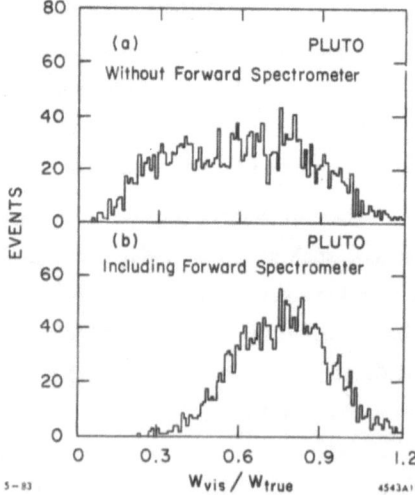

Fig. 11. MC-studies to compare jet axes and quark axes by (a) TASSO, (b) PLUTO and (c),(d) JADE employing different jet-finding methods: (a),(b) Thrust-method and (c),(d) cluster algorithm.

Fig. 12. W_{vis}/W_{true} for the PLUTO data without (a) and including (b) the forward spectrometer.

- Charge balance, p_T balance, p_z balance – Incomplete events especially from the 1γ process tend to be unbalanced in charge, transverse and longitudinal momentum. These cuts are therefore very useful to reduce this background.

- $p_T^{jet} > p_T^{min}$ – At least one detected jet has to have a transverse momentum greater than ~ 2 GeV/c.

- Different tagging requirements – JADE and TASSO analysed single tag events using their small angle taggers. PLUTO distinguishes between small angle tags (SAT) and large angle tags (LAT) and additionally requires that the other electron or positron is scattered under 0°, i.e. is not detected in the opposite forward spectrometer (anti-tag). Data without tag requirement were analysed by CELLO and TASSO.

Note that the first and third cut are crucial for notag data in order to reduce jet-like 1γ background.

Background from beam-gas scattering, 1γ annihilation and $\gamma\gamma \rightarrow \tau^+\tau^-$ has been computed and subtracted in all distributions. The number of events found after cuts is given in Table II.

Table II

Tagging	Experiment		Events	$MC_{1\gamma}$	$\int L \, dt$
SINGLE	JADE		42		$9.7 \; pb^{-1}$
TAG	TASSO		43	4.5	$9.0 \; pb^{-1}$
	PLUTO	SAT	84	5	$28 \; pb^{-1}$
		LAT	71	2	$39 \; pb^{-1}$
NOTAG	TASSO		624	211	$54 \; pb^{-1}$
	CELLO		128	57	$6 \; pb^{-1}$

In Fig. 13 the average transverse momentum of particles with respect to the jet-axis, q_T, is plotted. It is known from e^+e^- annihilation jets at PETRA and PEP that this variable should peak at around 0.3 GeV/c. This is observed for the 2γ jet-events, too.

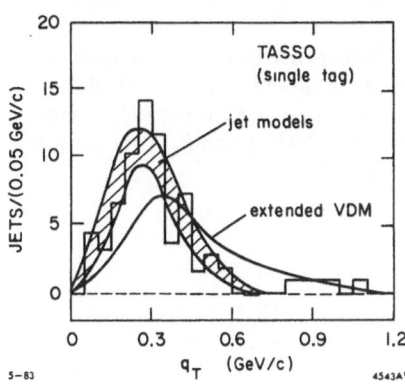

Fig. 13. Average transverse momentum q_T of particles with respect to the jet axis. Expectations from a $\gamma\gamma \rightarrow q\bar{q}$ model with different types of fragmentations (shaded area) and from a VDM model, extended such that it fits the single particle p_T-distributions, are also shown.

In order to simulate the reaction $e^+e^- \rightarrow e^+e^- +$ jet $+$ jet all experiments employ Vermaseren's QED program[28] to generate $e^+e^- \rightarrow e^+e^- + q\bar{q}$ according to QED and then fragment the quarks about the $q\bar{q}$ direction in the $\gamma\gamma$ c.m.-system using the standard Field-Feynman fragmentation algorithm[29] or a phase space model with limited p_T about the quark axis. The latter has been successfully employed when e^+e^- jets were first discovered at SPEAR.[30] The experiments claim that the results are fairly insensitive to the details of the fragmentation. Apart from fragmentation parameters the model has only two free parameters, the explicit quark masses and $R_{\gamma\gamma}$, the effective coupling strength. Using constituent masses for udsc-quarks ($m_q = 300$, 300, 500, $1500 \, MeV/c^2$) one can in turn determine $R_{\gamma\gamma}$ by direct comparison between data and model predictions assuming that for high p_T events the choice of constituent masses over current masses plays a minor role.

In Fig. 14 the thrust distribution as measured by PLUTO is plotted for two different Q^2 ranges corresponding to tags in the SAT and LAT, respectively. The data are compared to the Born term prediction. The agreement is much better for the high Q^2 events indicating a better background suppression for these events as mentioned above.

The most relevant plots are shown in Figs. 15 to 17 which show the direct comparison of the data with the Born term expectation on an absolute scale (i.e. not normalized to each other). The published results from JADE and TASSO (single tag) are shown in Figs. 15(a) and 15(b). In both figures data and model approach each other at high p_T^{jet}. Note however that the statistics is poor at high p_T^{jet}. Fig. 16 shows the results for the notag approach by CELLO and TASSO. Although the same approach of data and model prediction is seen the absolute ratio between data and model is about 2 for the CELLO data whereas it is about 5 for the TASSO data. This discrepancy may be attributed to substantial difficulties in normalizing and subtracting the 1γ

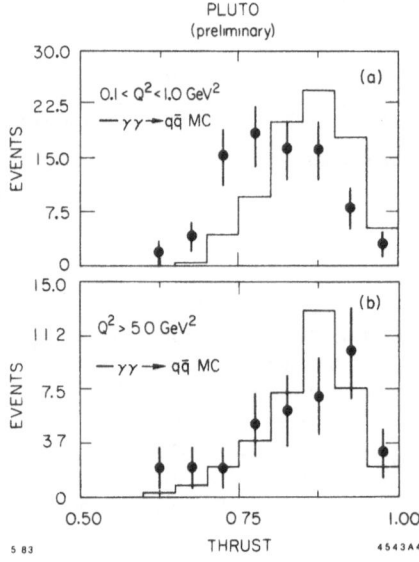

Fig. 14. Thrust distributions as measured by PLUTO compared to a MC calculation of the process $\gamma\gamma \rightarrow q\bar{q}$ for (a) SAT events ($0.1 \leq Q^2 \leq 1 \, GeV^2$) and (b) LAT events ($Q^2 > 5 \, GeV^2$).

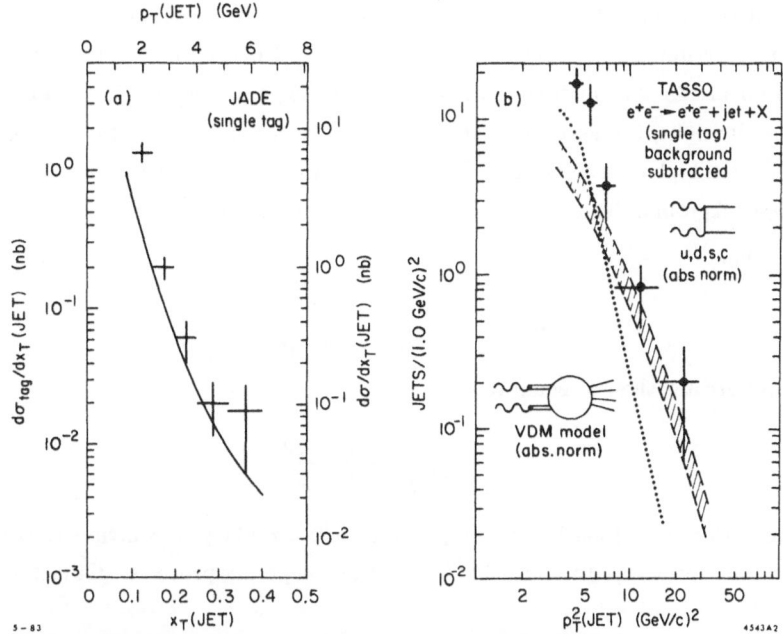

Fig. 15. Jet-p_T cross sections for the JADE (a) and TASSO (b) single tag data, $x_T = 2p_T/\sqrt{s}$. (a) The solid line represents the $\gamma\gamma \to q\bar{q}$ MC calculation. (b) The shaded area represents a $\gamma\gamma \to q\bar{q}$ calculation using for the fragmentation the standard Field-Feynman model and a limited-p_T phase space model with different parameter sets. The dotted line represents the VDM model prediction.

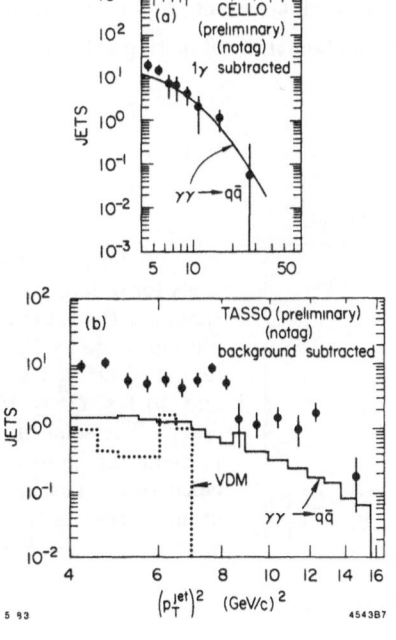

Fig. 16. Jet-p_T^2 distributions for notag events from (a) CELLO and (b) TASSO compared to models.

background. It seems unlikely that the different cuts and event selection methods can account for this effect. Very important progress has been made by PLUTO in Fig. 17 showing the same distributions for events with high (LAT) and low Q^2 (SAT) photons. Again, as already seen for the thrust distributions, the agreement at high p_T between data and the Born term reaction is much better for the high Q^2 events. The data are not yet corrected for acceptance. The indicated amount of 1γ background demonstrates how clean these measurements are.

We now define

$$\tilde{R}_{\gamma\gamma} = \frac{d\sigma/dt(\gamma\gamma \to jet + X)}{d\sigma/dt(\gamma\gamma \to \mu\mu)}$$

which is always greater than or equal to

$$R_{\gamma\gamma} = \frac{d\sigma/dt(\gamma\gamma \to q\bar{q})}{d\sigma/dt(\gamma\gamma \to \mu\mu)}$$

since other (non-Born term) hard scattering $\gamma\gamma$ reactions are likely to contribute. For increasing p_T^{jet} ($x_T \to 1$) we expect $\tilde{R}_{\gamma\gamma}$ to approach $R_{\gamma\gamma}$ as the competing processes die faster with p_T at fixed s.

In Fig. 18 $\tilde{R}_{\gamma\gamma}$ is plotted for the TASSO single tag data as a function of the p_T-cut. Also shown are the predictions from the naive ICQ model (10/3) and the FCQ model (34/27). Figures 19(a) and 19(b) show $\tilde{R}_{\gamma\gamma}$ for the PLUTO low and high Q^2 data respectively as a function of p_T (not p_T^{min}!). In all three figures the data approach the FCQ model expectation at high p_T where $\tilde{R}_{\gamma\gamma}$ is expected to be closer to $R_{\gamma\gamma}$. The naive ICQ model seems to be ruled out although the errors are large. It is remarkable that for the high Q^2 events $\tilde{R}_{\gamma\gamma}$ is almost flat and is always close to the FCQ expectation, again reflecting the fact that competing subprocesses are Q^2-suppressed.

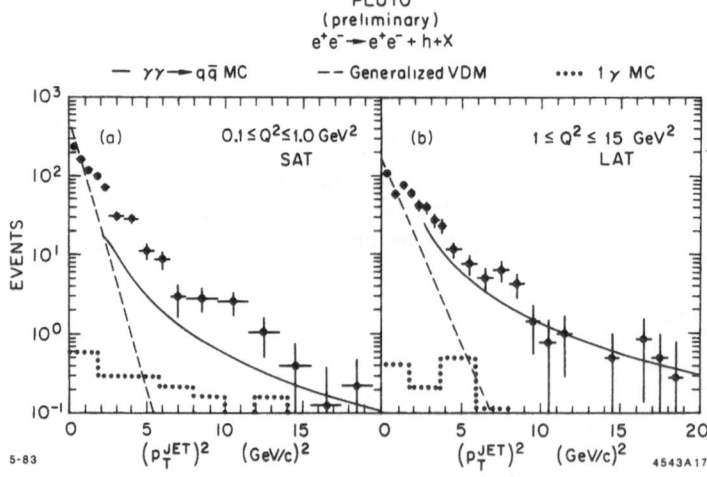

Fig. 17. Jet-p_T distributions for different Q^2 ranges of the tagged photon made by PLUTO; (a) $0.1 \leq Q^2 \leq 1\ GeV^2$ and (b) $1 \leq Q^2 \leq 15\ GeV^2$. The solid line represents the $\gamma\gamma \to q\bar{q}$ calculation. The histogram represents the amount of 1γ background.

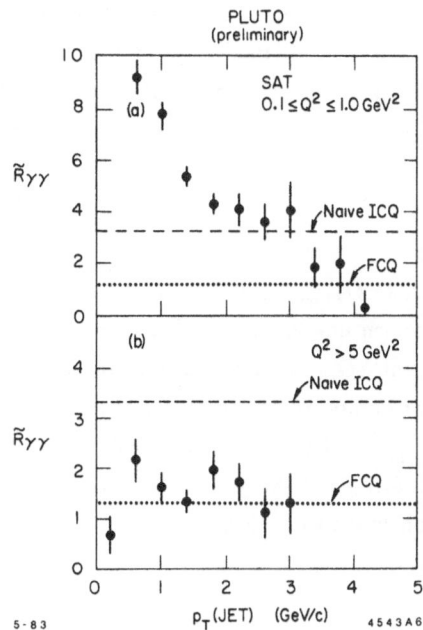

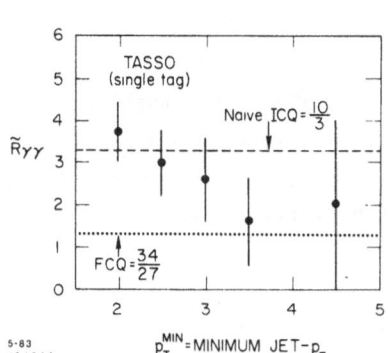

Fig. 18. $\tilde{R}_{\gamma\gamma} = [d\sigma/dt(\gamma\gamma \to jet + X)]/$ $[d\sigma/dt(\gamma\gamma \to \mu\mu)]$ plotted as a function of the p_T^{jet}-threshold p_T^{min} (TASSO). The expectations from FCQ and ICQ models are shown as horizontal lines.

Fig. 19. $\tilde{R}_{\gamma\gamma} = [d\sigma/dt(\gamma\gamma \to jet + X)]/$ $[d\sigma/dt(\gamma\gamma \to \mu\mu)]$ for the PLUTO data; (a) $0.1 \leq Q^2 \leq 1\ GeV^2$ and (b) $Q^2 > 5$ GeV^2. The horizontal lines represent expectations from FCQ and ICQ models.

4. ANSWERS AND CONCLUSIONS

In this conclusion, I want to assess to what extent we have been able to approach the challenging questions phrased in the beginning.

1. I think we can say that hard scattering processes in $\gamma\gamma$-reactions probably do exist. The evidence is most convincing from inclusive particle $d\sigma/dp_T^2$ distributions, particularly when comparing directly to hadron-hadron data, and from jet searches. The observation of high p_T particles cannot be explained by VDM model calculations nor by extrapolating hadron-like behaviour of the photon from hadron-hadron reactions. Substantial progress has been made to establish the Born term process $\gamma\gamma \to q\bar{q}$ by suppressing competing processes through a selection of high Q^2 events.

2. At first glance the measured values of $\tilde{R}_{\gamma\gamma}$, particularly at high p_T^{jet} seem to rule out the ICQ model values of quark charges. But this is only true for non-gauge ICQ models of the naive Han-Nambu-type. For so-called gauge-ICQ models[31] this statement no longer

holds.[32] We have seen that the $\{8\}_C$ part of the photon is responsible for a larger value of $R_{\gamma\gamma}$ in ICQ models. But this allows direct photon-gluon coupling to occur in $\gamma\gamma \rightarrow q\bar{q}$ scattering (Fig. 20) which modifies the effective quark charge seen by the photon by introducing a propagator-type suppression

$$Q_{eff}(q^2) = Q_{\{1\}_C} + \frac{m_g^2}{m_g^2 - q^2} Q_{\{8\}_C} \qquad (11)$$

where $Q_{\{1\}_C}$ = colour singlet charge, $Q_{\{8\}_C}$ = colour octet charge, q^2 = photon 4-momentum squared and m_g = gluon mass. This means that for $q^2 = 0$, i.e. notag events, Q_{eff} is equal to the sum of $Q_{\{1\}_C}$ and $Q_{\{8\}_C}$. For $q^2 \neq 0$, i.e. single or double tag events, the colour-octet charge is suppressed. In the limit of $q^2 \rightarrow \infty$ Q_{eff} becomes $Q_{\{1\}_C}$ and therefore $R_{\gamma\gamma}^{ICQ}$ will approach $R_{\gamma\gamma}^{FCQ}$ when q^2 increases. The authors of Ref. 33 estimated the gluon mass m_g using the JADE and TASSO single tag data.[12,13] They concluded from these data that

$$m_g \lesssim 200 - 300 \ MeV/c^2 \quad .$$

Untagged data ($Q^2 = 0$) will be required to decide whether the ICQ models are ruled out or not.

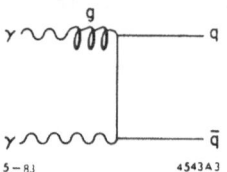

Fig. 20. Feynman diagram for direct photon-gluon coupling in $\gamma\gamma \rightarrow q\bar{q}$ reactions. This diagram requires the photon to carry colour-charge.

3. No real test has been performed so far on the magnitude of the quark propagator in Fig. 10. All differences between data and model predictions were attributed to $R_{\gamma\gamma}$. On the other hand, no dramatic deviations from the standard QED propagator have been observed which would imply that QED is not valid for photon-quark interactions. Most of the experiments are insensitive to whether current or constituent masses are the right quark masses in the propagator. At high p_T the quark mass is only a small correction. Experiments with forward angle coverage (PLUTO and PEP9 at PEP) should be able to attack this question looking for small angle jets.

4. The predicted scale invariance of pointlike scattering processes is strongly indicated in inclusive particle $d\sigma/dp_T^2$ distributes at high p_T. The direct comparison to pp and $p\bar{p}$ data shows very nicely that scaling is achieved at much lower energies than in hadron-hadron reactions. The errors are still too big for a firm conclusion on this issue for the jet-trigger $d\sigma/dp_T^2(\text{jet})$ cross sections.

5. Data at different Q^2 are very important and helpful to disentangle the different contributions of the subprocesses to the $\gamma\gamma \to$ jet $+ X$ reaction. Encouraging preliminary results from PLUTO have been shown in this talk. I think it is the theorists' turn now to tell us what impact the PLUTO data have on the understanding of how the different underlying subprocesses contribute.

6. A positive identification of CIM-subprocesses is still lacking. There is also only little hope that this will be achieved very soon because our understanding of the relative importance of these processes has been modified quite a lot. This is, on the other hand, very fortunate if one is concerned about processes confusing a clean measurement of the Born term process.

7. A big challenge for the dedicated two-photon experiments PLUTO and PEP9 is to look for direct γq- and $q\bar{q}$-scattering subprocesses like the one in Fig. 3. A major warning has always been that it is extremely difficult to identify beam-pipe jets. But a very nice indirect approach would be to perform an "anti-beam-pipe-jet" cut, i.e. requiring that the forward spectrometers did <u>not</u> detect hadrons. Just comparing the $\gamma\gamma \to$ jet $+ X$ cross sections with and without this cut would already provide a first result for the relative importance of these 3- and 4-jet processes.

Let me finish with a word of caution. As first noticed by Berends et al.[34] and recently worked out in more detail by Kang[35] using the Sterman-Weinberg definition[36] of jets, the QCD corrections to the lowest order process $\gamma\gamma \to q\bar{q}$ are expected to be large in the p_T and $\sqrt{s}$ range of the present experimental data. This would not modify the experimental observations presented here but would change the interpretation of the data in terms of $R_{\gamma\gamma}$.

ACKNOWLEDGEMENTS

I am very grateful for private communications and discussions with S. Cartwright and A. Tylka (PLUTO), D. Cords (JADE), H. Lierl (CELLO) and E. Duchovni (TASSO) on the special aspects of the different experimental data and for providing preliminary and unpublished results for this report. I would like to thank S. Brodsky and A. Janah for valuable discussions. I very much appreciate the support I have received from my colleagues at DESY, BONN and SLAC while preparing the talk and finishing the manuscript. Finally I want to thank Prof. C. Berger and his staff for the hospitality extended to me at this interesting conference.

REFERENCES

1. For instance, J. F. Gunion, talk given at the 3rd International Colloquium on $\gamma\gamma$ Collisions, Amiens, 1980, Lecture notes in Physics No. 134 (1980); S. Brodsky, talk presented at the 4th International Colloquium on Photon-Photon Interactions, Paris, 1981, World Scientific, Singapore (1981).

2. S. M. Berman, J. D. Bjorken and J. B. Kogut, Phys. Rev. D $\underline{4}$, 3388 (1971).

3. See e.g., F. E. Close, An Introduction to Quarks and Photons, Academic Press, London, 1979.

4. Y. Ne'eman, Nucl. Phys. $\underline{26}$, 222 (1961); M. Gell-Mann, Phys. Rev. $\underline{125}$, 1067 (1962); O. W. Greenberg, D. Zwanziger, Phys. Rev. $\underline{150}$, 1177 (1966).

5. M. Y. Han and Y. Nambu, Phys. Rev. $\underline{139B}$, 1006 (1965).

6. M. Chanowitz, Proceedings of the 12th Rencontre de Moriond, 1977, ed. Tran Thanh Van; S. J. Brodsky, talk given at the J. Weis Memorial, Seattle, 1978, SLAC-PUB 2240; P. Landshoff, Proc. Lep Summer Study, Les Houches, 1978, CERN 79-01, P. 555; H. K. Lee, ICTP Preprint, IC/78/95, Trieste, Italy.

7. R. Blankenbecler, S. Brodsky and J. F. Gunion, Phys. Rev. D $\underline{18}$, 900 (1978).

8. S. Brodsky, T. A. DeGrand, J. F. Gunion and J. H. Weis, Phys. Rev. D $\underline{19}$, 1418 (1979); and Phys. Rev. Lett. $\underline{41}$, 672 (1978).

9. C. H. Llewellyn Smith, Phys. Lett. $\underline{79B}$, 83 (1978).

10. K. Kajantie, Proceedings of the 4th International Colloquium on $\gamma\gamma$ Collisions, Paris, 1981, World Scientific, Singapore (1981); K. Kajantie and R. Raitio, Nucl. Phys. $\underline{B159}$, 528 (1979).

11. W. Wagner, Proceedings of the International Conference on High Energy Physics, Madison, 1980, p. 576.

12. W. Bartel et al., Phys. Lett. $\underline{107B}$, 163 (1981).

13. R. Brandelik et al., Phys. Lett. $\underline{107B}$, 290 (1981).

14. See also: H. Spitzer, summary of discussion sessions, this conference.

15. H. Kolanoski, talk given at this conference.

16. W. Caswell, R. Horgen and S. Brodsky, Phys. Rev D $\underline{18}$, 2415 (1978).

17. MARK J Collaboration, B. Aderva et al., Phys. Rev. Lett. $\underline{48}$, 721 (1982).

18. M. Pohl, talk given at this conference.

19. S. Cartwright, PLUTO Collaboration, private communication.

20. S. Brodsky and G. Farrar, Phys. Rev. D $\underline{11}$, 1309 (1975).

21. B. Alper et al., Nucl. Phys. $\underline{B100}$, 237 (1975).

22. G. Arnison et al., Phys. Lett. 118B, 167 (1982).

23. M. Yvert, talk given at the Intern. Symp. on High Energy Physics, Vanderbilt (1982).

24. S. Brandt et al., Phys. Lett. 12, 57 (1964); E. Fahri et al., Phys. Rev. Lett. 39, 1587 (1977).

25. K. Lanius, preprint DESY 80/36 (1980); H. J. Daum, H. Meyer and J. Buerger, Z. Phys. C8, 167 (1981).

26. J. A. Bagger and J. F. Gunion, University of California, Davis, Preprint UCD-83/1.

27. J. Stirling, talk given at this conference.

28. J. A. M. Vermaseren, Proceedings of the International Colloquium on $\gamma\gamma$ Collisions, Amiens, 1980, Lecture notes in Physics No. 134, Springer, 1980.

29. R. D. Field and R. P. Feynman, Nucl. Phys. B136, 1 (1978).

30. G. Hansen et al., Phys. Rev. Lett. 35, 1609 (1975).

31. J. C. Pati, A. Salam, Phys. Rev. D 8, 1240 (1973), Phys. Rev. D 10, 275 (1974); J. C. Pati and A. Salam, Phys. Rev. Lett. 36, 11 (1976); G. Rajasekaran and P. Roy, Phys. Rev. Lett. 36, 355 (1976).

32. A. Janah and M. Özer, University of Maryland, PUB-81-221; A. Janah, Doctoral dissertation, University of Maryland (1982).

33. T. Jayaraman, G. Rajasekaran and S. D. Rindani, Phys. Lett. 119B, 215 (1982); K. H. Cho, S. H. Han and J. K. Kim, Phys. Rev. D 27, 684 (1983).

34. F. A. Berends et al., DESY 80/89 (1980); F. A. Berends et al., Phys. Lett. 92B, 186 (1980).

35. I. Kang, Preprint Oxford TP 50/82; see also J. Stirling, talk given at this conference.

36. G. Sterman and S. Weinberg, Phys. Rev. Lett. 39, 1436 (1977).

Q: C. Berger: Are the notag CELLO and TASSO data of similar p_T^{jet}?
(Aachen)   Is there an explanation for the discrepancy?

A: N. Wermes: The CELLO data extend to higher p_T^{jet} due to a higher W_{vis}-cut.
The different ratios between data and $\gamma\gamma \to q\bar{q}$ model for the two
experiments are probably due to systematic effects when subtracting
the amount of 1γ background. In accordance with the PLUTO
measurements at larger Q^2 one would expect the ratio between
data and MC to be high for notag data.

Q: R. D. Field: When you parameterize the $q\bar{q}$-jets in $\gamma\gamma \to q\bar{q}$ do you use the same
(Univ. of Florida)   parametrization that fits the $e^+e^- \to q\bar{q}$ data at the corresponding
W? Is data on $\gamma\gamma \to q\bar{q}$ sensitive enough to say if they are the same?

A: N. W.   The experiments use the Field-Feynman fragmentation procedure
developed for e^+e^- jets with the parameters determined at higher
energies than accessible to $\gamma\gamma$ reactions. The present statistical
significance of the data does not allow to determine the parameters
from the measurements. The TASSO single tag data are compared
to models using different types of fragmentation including a
longitudinal phase space model which might be more appropriate
at low W.

Q: S. Brodsky: (1) Since $\gamma\gamma \to c\bar{c}$ gives a substantial fraction of the jet cross
(SLAC)   section, has allowance been made in the acceptance models to the
features of charm-jet fragmentation, specifically the Bjorken-Suzuki
effect?
(2) Although "direct" higher twist processes such as $\gamma q \to \pi q$ are
undoubtedly negligible for jet triggers, such processes may be
accessible in single particle triggers ($\gamma\gamma \to hX$), where h is
produced unaccompanied.

A: N. W.   So far the experiments employed a flat fragmentation function
for the c-quark. At present PLUTO is investigating
systematic effects due to c-fragmentation using the Lund-model.

Q: H. Kolanoski: Is it true that $\tilde{R}_{\gamma\gamma} \to R_{\gamma\gamma}$ for $s \to \infty$ if the 3- and 4-jet cross
(Bonn)   sections also scale?

A: N. W.   The 3- and 4-jet processes die faster with x_T than the Born
process $\gamma\gamma \to q\bar{q}$. This is mainly a phase space effect. Therefore
at fixed s, $\tilde{R}_{\gamma\gamma}$ approaches $R_{\gamma\gamma}$ with increasing p_T^{jet}. If $s \to \infty$,
$\tilde{R}_{\gamma\gamma}$ approaches $R_{\gamma\gamma}$ for $x_T \to 1$.

C: A. Janah: I want to make three comments on $R_{\gamma\gamma}$ in ICQ models.

(Univ. of Calif., Irvine)   (1) One expects $R_{\gamma\gamma}$ to decrease with p_T for notag and presumably also for low Q^2 in ICQ theory.

(2) One expects $R_{\gamma\gamma}^{ICQ}$ to approach $R_{\gamma\gamma}^{FCQ}$ as Q^2 increases.

(3) One expects $R_{\gamma\gamma}^{ICQ}$ to increase with W, so one should plot $\tilde{R}_{\gamma\gamma}$ versus W (with p_T-cutoff).

This variation of $R_{\gamma\gamma}^{ICQ}$ with p_T, Q^2 and W is due to the Q^2 dependent colour suppression factors and due to the charged gluon contribution which are not present in FCQ theory.

Q: M. Gorn: Presumably you do the jet analysis in the laboratory system (LAB).

(Univ. München)   What about the change of the jet shape if you go to the $\gamma\gamma$-CMS? Is it possible that a narrow jet-like event in one of the systems does not look like this in the other system?

A: N. W.   Only TASSO performs the jet analysis in the LAB frame. PLUTO, CELLO and JADE evaluate their jets in the $\gamma\gamma$-CMS which is appropriate if the final state is almost completely detected. Figure 11(a) proves that quark axis and jet axis are strongly correlated for high p_T jet events even when using the LAB frame. The influence on the jet-shape is surprisingly small if one requires high p_T of the jets.

HARD HADRONIC FINAL STATES IN TWO PHOTON PROCESSES

W.J. Stirling

Department of Applied Mathematics and Theoretical Physics,
University of Cambridge, Silver Street, Cambridge CB3 9EW
England.

Abstract

The predictions of perturbative Quantum Chromodynamics for hard hadronic final states in photon-photon collisions are reviewed. Particular attention is paid to perturbative corrections to inclusive jet and single hadron cross sections and to the predictions for wide angle exclusive final states.

I.   INTRODUCTION

Two photon collisions at high energy can provide many important tests of perturbative Quantum Chromodynamics. The principal theoretical advantage over hadronic collisions is the simplicity of the initial state, which allows processes such as large transverse momentum hadronic jet production to be calculated exactly to lowest order in perturbation theory. With the advent of high quality experimental data, theoretical analyses have begun to focus on higher order corrections to the basic processes and here again can be found interesting tests of the theory. In addition to inclusive hard scattering cross sections, $\gamma\gamma$ processes can also play an important role in testing the recently developed application of QCD to exclusive processes. Although such phenomenology is in its infancy at present, the prospects look encouraging.

Detailed comparisons of theoretical predictions with experimental data can be found elsewhere in these Proceedings. This review concentrates instead on the underlying theoretical framework of hard scattering processes. Inclusive jet and single hadron cross sections are derived from first principles and higher order corrections — both leading and higher twist — are discussed. The application of perturbative QCD to exclusive processes is reviewed and the relevant results for $\gamma\gamma$ processes are derived.

It will become clear that most theoretical calculations refer to on-shell initial state photons collinear with the e^+e^- beams. In practice, of course, this is only an approximation — the Weizsäcker-Williams or equivalent photon approximation (EPA) — since the virtual photons can have $Q^2 \neq 0$. Originating from bremsstrahlung, however, most photons do have very low Q^2 with a $1/E$ energy spectrum and limited transverse momentum relative to the beam. The $\gamma\gamma$ system is therefore characterised by an energy $s_{\gamma\gamma} \ll s_{e^+e^-}$ and is in general moving with large momentum in the e^+e^- centre-of-mass frame. More precisely, hard scattering cross sections have the form

$$d\sigma^{e^+e^- \to e^+e^-X} = \int_0^1 dx_1 dx_2 \ G(x_1)G(x_2) \ d\hat\sigma^{\gamma\gamma \to X} \tag{1.1}$$

where the photon/electron "structure function" is

$$G(x) = \frac{\alpha}{2\pi} \ln\eta \ \frac{1 + (1-x)^2}{2}$$

$$\eta = \frac{s}{4m_e^2} \quad \text{[no tag]} \tag{1.2}$$

$$\eta = \left(\frac{\theta_{max}}{\theta_{min}} \right)^2 \quad \text{[tag]}$$

and $\hat{s} = x_1 x_2 s$, $\hat{t} = x_1 t$, ... etc. in $d\hat\sigma$. Note also that if $d\hat\sigma$ is a function of s and (x_1x_2) only then

$$d\sigma^{e^+e^- \to e^+e^-_\_X} = \int_0^1 dz\ L(z)\ d\hat{\sigma}^{\gamma\gamma \to X}\Big|_{x_1x_2 = z} \tag{1.3}$$

where the $\gamma\gamma$ "luminosity" function is

$$L(z) = (\frac{\alpha}{2\pi})^2\ \ell n\ \eta_1\ \ell n\ \eta_2\ \frac{1}{z}\ [(2+z)^2\ \ell n \frac{1}{z} - 2(1-z)(3+z)] \tag{1.4}$$

The validity of the EPA can be checked for some simple processes (e.g. $e^+e^- \to e^+e^-\mu^+\mu^-$) by comparing the results from using eqns.(1.1) and (1.2) with an exact multi-dimensional phase space integral. As expected, the approximation improves with increasing energy and momentum transfer [1,2].

Other important ingredients in $\gamma\gamma$ cross section calculations are the probability distributions for partons in photons. These arise in processes where the hard scattering is mediated by quarks and gluons rather than the photons themselves. The photon structure functions are important objects for theoretical study [3], but in the present context only the specific forms are required. There are essentially three levels of approximation: at lowest order in α_s, the box diagram of Fig 1.2 gives

$$G_{q_i/\gamma}\ (x,Q^2) = \frac{\alpha}{2\pi}\ e_i^2\ \ell n\ \frac{Q^2}{m_i^2}\ [x^2 + (1-x)^2] \tag{1.5}$$

for a quark of charge e_i and mass m_i. Note that at this order $G_{g/\gamma} = 0$. In QCD, the box diagram is dressed with gluons, Fig 1.2, and in the Leading Logarithm Approximation the all-orders result has the form

$$G_{q_i/\gamma}\ (x,Q^2) = \frac{\alpha}{2\pi}\ \frac{1}{\alpha_s(Q^2)}\ \begin{cases} f_u(x) & |e_i| = \frac{2}{3} \\ \\ f_d(x) & |e_i| = \frac{1}{3} \end{cases} \tag{1.6a}$$

$$G_{g/\gamma}\ (x,Q^2) = \frac{\alpha}{2\pi}\ \frac{1}{\alpha_s(Q^2)}\ f_g(x) \tag{1.6b}$$

The moments of the functions $f_a(x)$ can be calculated directly and the x-dependence obtained numerically by inverting the Mellin transform. In this review, the simple polynomial approximations derived in [4] will be used for numerical calculations, i.e.

$$f_u(x) = \frac{1}{b_o} \frac{0.32}{x} (\frac{1}{2} + x)$$

$$f_d(x) = \frac{1}{b_o} \frac{0.06}{x} (1 + x)$$

$$f_g(x) = \frac{1}{b_o} 0.348 \ x^{-1.6} (1-x)$$

$$[b_o = \frac{33 - 2N_f}{12\pi}]$$

(1.7)

These forms are adequate for $x \geq 0.1$. For smaller values of x the leading logarithm results are unstable with respect to higher order QCD corrections and, in fact, the precise behaviour cannot be determined [3]. A related problem concerns the non-perturbative hadronic structure of the photon. Although QCD predicts asymptotic scale free couplings of on-shell photons to quarks in large momentum transfer processes, at low momentum transfer many features of photon induced reactions can be interpreted in terms of vector meson dominance models. Although the ambiguities in such models preclude precise theoretical tests in the low p_T domain, the non-pointlike contributions are expected to fall off rapidly with increasing p_T. In this review, therefore, it will always be assumed that the momentum transfers are sufficiently large that vector dominance contributions can be ignored [5].

2.   LARGE p_T INCLUSIVE CROSS SECTIONS

2.1 Large p_T Jets — Lowest Order

The simplest way to produce hadrons with large transverse momentum in $\gamma\gamma$ collisions is via quark-antiquark pair production, $\gamma\gamma \to q\bar{q}$ [6-9]. The Born diagrams are shown in Fig 2.1 and a straightforward calculation yields

$$\frac{d\sigma^{\gamma\gamma \to q\bar{q}}}{dt} = \frac{2\pi\alpha^2}{s^2} \cdot 3e_q^4 \cdot (\frac{t}{u} + \frac{u}{t})$$

(2.1)

for a single quark flavour summed over colour degrees of freedom. In the "ideal" approximation where partons fragment collinearly and all the resulting hadrons are detected, the jet cross section is trivially related to the parton cross section:

$$\frac{d\sigma^{\gamma\gamma \to JX}}{dt} = \frac{2\pi\alpha^2}{s^2} \cdot 2R_{\gamma\gamma} \cdot (\frac{t}{u} + \frac{u}{t})$$

(2.2)

where $R_{\gamma\gamma} = \sum_q 3e_q^4$ and the factor 2 arises from the sum over both quark and antiquark jets. For a jet of total energy E_J and momentum $\vec{p}_J$, the single jet inclusive cross section is

$$E_J \frac{d\sigma^{\gamma\gamma\rightarrow JX}}{d^3 p_J} = \frac{s}{\pi} \delta(s+t+u) \frac{d\sigma^{\gamma\gamma\rightarrow JX}}{dt} \tag{2.3}$$

The corresponding cross section for the process $e^+e^- \rightarrow e^+e^-JX$ is obtained simply by convoluting this with the structure functions $G(x)$ introduced in Section 1:

$$E_J \frac{d\sigma^{e^+e^-\rightarrow e^+e^-JX}}{d^3 p_J} = \int_0^1 dx_a dx_b \; G(x_a) \; G(x_b) \; . \; \frac{\hat{s}}{\pi} \delta(\hat{s}+\hat{t}+\hat{u}) \; . \; \frac{d\sigma^{\gamma\gamma\rightarrow JX}}{d\hat{t}} \tag{2.4}$$

where the subprocess invariants are related to physical e^+e^- centre-of-mass variables by

$$\hat{s} = x_a x_b s$$

$$\hat{t} = -x_a \; p_T \; \sqrt{s} \; \tan\frac{\theta_{cm}}{2} = -x_a \; p_T \; \sqrt{s} \; e^y$$

$$\hat{u} = -x_b \; p_T \; \sqrt{s} \; \cot\frac{\theta_{cm}}{2} = -x_b \; p_T \; \sqrt{s} \; e^{-y} \tag{2.5}$$

$$x_T = 2p_T/\sqrt{s}$$

Of more practical interest are cross sections at fixed p_T and rapidity, y. With the explicit (EPA) forms for the $G(x)$ given in the preceding section,

$$\left. \frac{d\sigma}{dp_T^2 dy} \right|_{y=0} = (\frac{\alpha}{2\pi})^2 \; \ell n \, \eta_1 \; \ell n \, \eta_2 \; R_{\gamma\gamma} \frac{\pi\alpha^2}{p_T^4} \; F^o(x_T) \tag{2.6a}$$

$$\frac{d\sigma}{dp_T^2} = (\frac{\alpha}{2\pi})^2 \; \ell n \, \eta_1 \; \ell n \, \eta_2 \; R_{\gamma\gamma} \frac{\pi\alpha^2}{p_T^4} \; F(x_T) \tag{2.6b}$$

The scaling functions F^o and F are shown in Fig 2.2. Analytic, though rather cumbersome, expressions can also be derived. It is important to note that away from $x_T = 0,1$ the curves decrease only slowly with increasing x_T. This means that at fixed centre-of-mass energy the cross sections in (2.6) do scale approximately as p_T^{-4} for p_T values which restrict the corresponding x_T values to the range $.1 \lesssim x_T \lesssim .6$.

In principle, these jet cross sections provide simple direct tests of the underlying field theory embodied in the Born diagrams of Fig 2.1. The point-like quark-photon coupling gives p_T^{-4} scaling, the structure functions G and the hard scattering amplitude combine to give the scaling functions F and F^o and the

overall normalization is determined by $R_{\gamma\gamma}$ which measures the sum of the fourth
power of the quark charges. In practice, however, the situation is less clear:
(a) real jets are far from "ideal", (b) the equivalent photon approximation becomes
invalid near the edges of phase space, and (c) there are higher order contributions
— for example, from gluon corrections to the Born diagrams and from power-
suppressed higher twist contributions. It is important therefore to investigate
the domain of validity, if any, of the approximations which lead to the simple
expressions in eqn.(2.6). The remainder of this section will focus in particular
on the corrections which arise in QCD perturbation theory and on higher twist
contributions to single hadron production.

2.2 Quark, Gluon Subprocess Contributions

An important class of corrections arises from subprocess hard scatterings
mediated by the parton "constituents" of the photon. In Section 1 the structure
functions for quarks and gluons in a photon were introduced. Convoluting these
with the quark, gluon subprocess cross sections familiar from large p_T hadronic
physics gives additional contributions to large p_T jet production in $\gamma\gamma$
collisions [4,7,8,9].

Examples of such contributions are illustrated in Fig 2.3. The different
processes can be classified according to the number of distinct jets in the final
state. The Born contributions (Fig 2.3a) are then "2-jet processes", those with
one initial state quark or gluon (Fig 2.3b) are "3-jet processes" — the extra jet
being the spectator fragments of the photon moving along the beam direction — and
those with two initial state quarks are "4-jet processes", with two extra jets in
the beam direction. An important feature of these additional contributions is
that they all have the same asymptotic scaling behaviour since the extra powers
of α_s in the subprocess cross sections are compensated by the corresponding powers
of α_s^{-1} in the structure functions [9]. This is in contrast to the analogous
corrections to the Drell-Yan process $(h_1 h_2 \to \mu^+\mu^- X)$ which are suppressed
asymptotically by powers of α_s [10].

The complete expression for the single jet inclusive cross section is
therefore given by

$$E_J \frac{d^3\sigma}{d^3 p_J} = \sum_{a,b=\gamma,q,g} \int_0^1 dx_a dx_b \, G_{a/e}(x_a) G_{b/e}(x_b) \cdot \frac{\hat{s}}{\pi} \, \delta(\hat{s}+\hat{t}+\hat{u}) \, \frac{d\sigma^{ab\to JX}}{d\hat{t}} \qquad (2.7)$$

where the subprocess cross sections can be found, for example, in [11]. Again it
is useful to consider the rapidity integrated cross section, which now has the form

$$\frac{d\sigma}{dp_{T_J}^2} = (\frac{\alpha}{2\pi})^2 \, \ln\eta_1 \, \ln\eta_2 \, R_{\gamma\gamma} \, \frac{\pi\alpha^2}{p_T^4} \sum_{N=\# \text{ of jets}} F^N(x_T) \qquad (2.8)$$

Fig 2.4 shows the ratios F^3/F^2, F^4/F^2 and $(F^2 + F^3 + F^4)/F^2$ as functions of x_T. It is clear that for $x_T \gtrsim 0.1$ the 3-, 4-jet contributions are small and the cross section is well approximated by the Born (2-jet) contribution of Section 2.1.

Since the existence of 3- and 4-jet processes is a rigorous prediction of the theory, it is important to find tests which can distinguish the different contributions. Some (theoretical) possibilities are listed below:

(i) Detect and identify gluon jets. Fig 2.5 shows the probability that the detected single large p_T jet is a gluon jet and also the probability that the initial state contained at least one gluon constituent. Again these are sizeable only at small x_T where, as discussed in Section 1, there is considerable uncertainty concerning the precise form of the photon structure functions and also non-perturbative vector-dominance contributions.

(ii) Detect additional beam direction jets. This is, unfortunately, very difficult in practice because of the finite size of the beam pipe.

(iii) Measure away side rapidity distributions at small x_T. In particular, the double inclusive jet cross section can be written as [4,8]

$$\frac{d\sigma^{e^+e^- \to e^+e^- J_1 J_2 X}}{dp_T^2 dy_1 dy_2} = (\frac{\alpha}{2\pi})^2 \, \ell n \, \eta_1 \, \ell n \, \eta_2 \, \pi\alpha^2 \, R_{\gamma\gamma} \, \frac{1}{p_T^4} \, H(x_T, y_1, y_2) \qquad (2.9)$$

Fig 2.6 shows that the distribution in y_2 of the away side jet at fixed y_1, x_T is rather different for the different N-jet processes.

(iv) Measure opposite side charge correlations. In particular, for a pair of pions with equal and opposite large p_T, $\sigma(\pi^+\pi^-) > \sigma(\pi^+\pi^+)$ for the 2-jet contribution whereas $\sigma(\pi^+\pi^-) \approx \sigma(\pi^+\pi^+)$ for the 3- and 4-jet contributions.

Finally, it should be noted that gluon jets can also be produced in a 2-jet configuration, by the mechanism shown in Fig 2.7 [12,13]. A detailed calculation shows that the ratio $\sigma(\gamma\gamma \to gg)/\sigma(\gamma\gamma \to q\bar{q})$ at $y = 0$ is approximately 10% over the whole x_T range.

2.3 Single Hadron Cross Sections

Large p_T single hadron cross sections are readily obtained from jet cross sections by convoluting with the appropriate fragmentation functions. The general result for a hadron H is

$$E \frac{d\sigma}{d^3p} (e^+e^- \rightarrow e^+e^- HX) = \sum_{\substack{a,b=\gamma,q,g \\ c=q,g}} \int_0^1 dx_a dx_b \frac{dz}{z^2}$$

$$\cdot\ G_{a/e}(x_a,Q^2)\ G_{b/e}(x_b,Q^2)\ G_{H/c}(z,Q^2) \cdot \frac{\hat{s}}{\pi} \delta(\hat{s}+\hat{t}+\hat{u})$$

$$\cdot \frac{d\sigma^{ab \rightarrow cX}}{d\hat{t}} \tag{2.10}$$

For example if H is any charged hadron then, approximately,

$$z\ G_{H/q}(z) = \frac{4}{3}(1-z) \tag{2.11}$$

and the contribution from the $\gamma\gamma \rightarrow q\bar{q}$ subprocess can be written (cf. eqn.(2.6b))

$$\frac{d\sigma^H}{dp_T^2} = (\frac{\alpha}{2\pi})^2 \ln \eta_1 \ln \eta_2\ R_{\gamma\gamma}\ \frac{1}{p_T^4} F^{2,H}(x_T) \tag{2.12}$$

where

$$F^{2,H}(x_T) = \int_{x_T}^1 dz\ z^2\ G_{H/q}(z)\ F^2(\frac{x_T}{z}) \tag{2.13}$$

Fig 2.2 shows $F^{2,H}(x_T)$ as a function of x_T for the above fragmentation function. Compared with F^2 the curve goes faster to zero as $x_T \rightarrow 1$ ($0(1-x_T)^2$ in this case) but p_T^{-4} scaling should again be approximately valid for small x_T values.

2.4 Higher Order QCD Corrections

The conclusion of Section 2.2 was that at large p_T the jet cross sections are dominated by the $\gamma\gamma \rightarrow q\bar{q}$ contribution. Another important class of corrections is the finite perturbative corrections to the hard scattering subprocesses — the so-called K-factors. With additional gluons in the final state, the definition of a "jet" requires more precision. Before considering jet cross section corrections, it is conceptually simpler to examine the corrections to the single hadron cross section.

According to QCD factorization theorems, the single hadron cross section is a convolution of a single parton cross section, calculated perturbatively to all orders, with a non-perturbative (parton → hadron) fragmentation function. In the previous section, the lowest order contribution ($0(\alpha_s^0)$) to the former was considered. The $0(\alpha_s)$ contribution to the inclusive quark cross section is obtained from diagrams with an additional gluon — either real or virtual — some of which are shown in

Fig 2.8. In the cross section calculation various infra-red and ultra-violet singularities arise. These can be regulated (diagram by diagram) either by introducing small parton masses and an ultra-violet cut-off, or by performing the calculation in $N \neq 4$ dimensions. The singularities then appear as large logarithms or poles at $N = 4$ respectively. When all diagrams are summed, all the ultra-violet and soft gluon infra-red singularities cancel, but there remain single logarithmic/ pole singularities from the collinear configurations shown in Fig 2.9 (and others obtained by crossing the photons.) However the collinear branching of the photon in Fig 2.9. corresponds to the lowest order contribution to the photon structure function and this piece is, by definition, already included in the 3-jet contribution to the cross section. Likewise the collinear branching of the final state quark is already included in the non-scaling fragmentation function. The <u>finite</u> $0(\alpha_s)$ correction is therefore obtained by first calculating the full $0(\alpha_s)$ cross section for $\gamma\gamma \to q\bar{q}g$ and then subtracting off the $0(\alpha_s)$ piece of the quark fragmentation function, as measured say in $e^+e^- \to$ hadrons, and the $0(\alpha_s)$ piece of the 3-jet contribution. This leads to a result of the form

$$E \frac{d\sigma^{\gamma\gamma \to HX}}{d^3p} = \int_0^1 \frac{dz}{z^2} G_{H/q}(z,Q^2) \{\delta(\hat{s}+\hat{t}+\hat{u})\sigma_o^{\gamma\gamma}(\hat{s},\hat{t},\hat{u}) + \frac{\alpha_s}{\pi} \sigma_1^{\gamma\gamma}(\hat{s},\hat{t},\hat{u}) + 0(\alpha_s^2)\}$$

$$+ \int_0^1 dx_a \frac{dz}{z^2} G_{q/\gamma}(x_a,Q^2)G_{H/q}(z,Q^2)\sigma^{\gamma q}(\hat{s},\hat{t},\hat{u}) + \text{(other 3-,4-jet contributions)} \quad (2.14)$$

The K-factor is then defined as the $\sigma_o^{\gamma\gamma} + \sigma_1^{\gamma\gamma}$ term divided by the $\sigma_o^{\gamma\gamma}$ term. A similar analysis of the Drell-Yan cross section reveals an $0(\alpha_s)$ correction which is rather large [10]. The calculation of the $\sigma_1^{\gamma\gamma}$ correction in eqn.(2.14) is under way [14] and preliminary results suggest that the correction is very small. The single hadron inclusive cross section at large transverse momentum appears, therefore, to be stable with respect to higher order perturbative corrections and additional subprocess contributions.

The simplest way to calculate perturbative corrections to <u>jet</u> cross sections is to define an "infra-red safe" jet measure which cannot resolve final state collinear configurations and is therefore less singular than the inclusive parton cross section. Several such jet definitions have been proposed [15,16]. The simplest method is to choose, in analogy with $e^+e^- \to$ hadrons, the Sterman-Weinberg Jet Criterion. This defines a 2-jet event as one with all but at most $2\epsilon\sqrt{s}$ of final state hadronic energy in oppositely directed cones of semi-angle $\delta(\epsilon,\delta \ll 1)$ [17]. Thus final states with a gluon which is either virtual, real and soft, or real and collinear with the quark or antiquark are all degenerate and <u>final state</u> collinear and soft singularities cancel by the Bloch-Nordsieck, KLN theorems.

Through $0(\alpha_s)$ the Sterman-Weinberg $\gamma\gamma \rightarrow$ 2-jet cross section is [15,16]

$$\frac{d\sigma}{d\cos\theta} = \frac{d\sigma^{(o)}}{d\cos\theta} [1 + \frac{\alpha_s(s)}{\pi} R(\epsilon,\delta,\theta)] \qquad (2.15)$$

where

$$\frac{d\sigma^{(o)}}{d\cos\theta} = \frac{2\pi\alpha^2}{s} R_{\gamma\gamma} \frac{1 + \cos^2\theta}{1 - \cos^2\theta} \qquad (2.16)$$

$$R\Big|_{\epsilon,\delta \ll 1} = \frac{4}{3} [- \ln\delta(3 + 4\ln(2\epsilon)) - \frac{\pi^2}{3} + 3$$

$$+ \frac{1}{4(1+\cos^2\theta)}\{ (5 - 4\cos\theta + \cos^2\theta) \ln^2 (\frac{2}{1-\cos\theta})$$

$$- (1+\cos\theta) (5+\cos\theta) \ln (\frac{2}{1-\cos\theta})$$

$$+ (\cos\theta \leftrightarrow - \cos\theta) \}]$$

$$(2.17)$$

and θ is the angle between the cone axis and the photon beam direction in the two photon centre-of-mass. It would at first sight appear that the above cross section is also free of the initial state collinear singularities of Fig 2.9. However these singularities are still present, but in terms of $0(\epsilon,\delta)$ which have been neglected in eqn.(2.17). [Strictly, therefore, the quantity R defined in eqn.(2.15) should also depend on the regulator.] Once again, these singularities are correctly accounted for by including 3,4-jet subprocesses.

Kang has recently presented a quantitative analysis of the above corrections [16]. He finds, for $\epsilon,\delta \ll 1$,

$$\frac{d\sigma}{d\cos\theta}\Big|_{90°} = \frac{d\sigma^{(o)}}{d\cos\theta}\Big|_{90°} [1 + \frac{\alpha_s}{\pi} R(\epsilon,\delta,90°) + 1.2 \min (\epsilon,\delta)] \qquad (2.18)$$

where the last term in the square brackets is the 3-jet contribution. The 4-jet contribution is $0(\epsilon\delta)$ and is negligible. In particular, with $\epsilon = .15$, $\delta = .17$ the corrections are $\frac{\alpha_s}{\pi} R = - 0.34$ and $1.2 \min(\epsilon,\delta) = 0.18$. Although the net effect is therefore reasonably small, Kang's calculation does not include the finite $0(\epsilon,\delta)$ pieces not associated with the multi-jet contributions. Suppressing these terms by taking the limit $\epsilon,\delta \rightarrow 0$ causes an apparent breakdown in the convergence of the perturbation series from terms of order $(\alpha_s \ln\delta\ln\epsilon)^n$. In addition, other

definitions of jet cross sections lead to quantitatively different corrections [15].
Thus for reasons both experimental and theoretical, it seems preferable to use
single hadron inclusive cross sections to investigate higher order perturbative
corrections [14].

2.5 Higher Twist Contributions

The cross sections considered so far have all been "leading twist", i.e. the
inclusive cross section scales as p_T^{-4} at fixed x_T . QCD also predicts "higher
twist" contributions which scale as p_T^{-n} with $n \geq 6$ [18]. The motivation for
such contributions is that constituents can sometimes scatter collectively, as
mesons, diquarks etc. In the case of single meson production at large p_T , the
hard scattering subprocess $\gamma q \to Mq$ enables the meson M to be produced directly,
avoiding the suppression from jet fragmentation [Fig 2.2]. On the other hand the
meson "size" or form factor gives additional powers of p_T^{-1} . It is possible that
such contributions actually dominate in certain kinematic regions — the scaling
behaviour observed at FNAL/ISR energies for $p_T \lesssim 8$ GeV/c in meson production
$[Ed\sigma/d^3p \sim p_T^{-8} f(x_T,\theta_{cm})]$ can be interpreted as evidence for such higher twist
contributions [19]. It is therefore important to investigate similar contributions
in $\gamma\gamma$ collisions.

The original higher twist calculations were in the framework of the
Constituent Interchange Model where the power-law scaling behaviour, but not in
general the normalisation, of higher twist contributions could be determined [19].
These early calculations have recently been refined using the QCD exclusive process
formalism developed by Brodsky and Lepage and others [20]. In particular, Bagger
and Gunion have calculated the dominant higher twist contribution to $\gamma\gamma \to \pi X$
at large p_T [21]. A typical diagram is shown in Fig 2.10. [Note that details of
the pion wave function will be discussed more fully in Section 3.] The higher twist
contribution to the cross section can be written as

$$E \left.\frac{d\sigma}{d^3p}\right|^{\gamma\gamma\to\pi X}_{HT} = \sum_q \int_0^1 dx \; G_{q/\gamma}(x,Q^2) \; \frac{\hat{s}}{\pi} \; \delta(\hat{s}+\hat{t}+\hat{u}) \cdot \{ \frac{d\sigma}{d\hat{t}}^{\gamma q\to\pi q'} + (\hat{t} \leftrightarrow \hat{u}) \} \qquad (2.19)$$

where

$$\frac{d\sigma^{\gamma q\to\pi q'}}{d\hat{t}} = \frac{8\pi^2\alpha}{3} \; C_F \; [\Delta(\hat{s},\hat{u},e_q,e_{q'})]^2 \cdot \hat{s}^{-2} \; (-\hat{t})^{-1} \; (\hat{s}^{-2} + \hat{u}^{-2}) \qquad (2.20)$$

The quantity Δ is related to the pion form factor $F_\pi(Q^2)$, and therefore the
normalisation can be completely fixed in principle. Effectively there is a quark-
antiquark-meson vertex of strength g_M/p_T where the dimensionful coupling g_M^2 is
proportional to $Q^2 F_M(Q^2)$. Fig 2.11 shows the leading and higher twist contri-

butions to large p_T π^+ production, calculated in [21]. The p_T^{-6} contribution is very small, due mainly to the fact that $g_\pi^2/4\pi \approx 0.04$ GeV^2. Although dominated by the leading twist contribution in the single hadron inclusive cross section, the higher twist contribution does have a very distinctive topology in that the large p_T hadron is unaccompanied by collinear jet fragments. In the language of the previous section, there is one beam-line jet $(\gamma \rightarrow qX)$ but only one large p_T jet, balancing the direct hadron trigger. Clearly the relative higher twist contribution will be enhanced by selecting events with this structure.

3. HARD EXCLUSIVE PROCESSES

3.1 General Framework of QCD Exclusive Process Calculations

Until recently, detailed quantitative tests of perturbative QCD were confined to hard <u>inclusive</u> processes. Despite the inaccessibility of real parton initial and final states, non-perturbative confinement dynamics can be isolated in "universal" process-independent probability distributions for partons in hadrons. For inclusive cross sections, the separation of short and long distance dynamics is embodied in factorization theorems [22]:

$$\begin{bmatrix} \text{cross} \\ \text{section} \end{bmatrix} = \begin{bmatrix} \text{hard scattering} \\ \text{parton subprocesses} \end{bmatrix} * \begin{bmatrix} \text{distribution functions} \\ \text{hadrons} \leftrightarrow \text{partons} \end{bmatrix} , \qquad (3.1)$$

which permit detailed predictions, once the appropriate distribution functions are measured.

A similar theoretical framework for large momentum transfer <u>exclusive</u> processes has recently been developed [23-25], extending the testing ground of QCD to include elastic scattering cross sections and hadronic form factors. The basic principle is the representation of hadronic bound states in terms of Fock State wave functions for quark, gluon constituents, e.g.

$$|\pi\rangle = |q\bar{q}\rangle \, \psi_\pi^{(2)} + |q\bar{q}g\rangle \, \psi_\pi^{(3)} + \ldots \qquad (3.2)$$

where the wave functions $\psi_H^{(n)}(x_i, \vec{k}_{T_i})$ represent the probability to find n on-mass-shell quarks and gluons with fractional light-cone momenta x_i and transverse momenta $\vec{k}_{T_i}$ $(i = 1,..n)$ with

$$\sum_{i=1}^{n} x_i = 1 , \qquad \sum_{i=1}^{n} \vec{k}_{T_i} = \vec{0} \qquad (3.3)$$

by momentum conservation [Fig 3.1]. A general lepton-photon-hadron exclusive amplitude is then built up by convoluting these wave functions with hard scattering amplitudes T_h involving their parton constituents [Fig 3.2].

In practice, VALENCE or minimal Fock States ($|q\bar{q}\rangle$ for mesons, $|qqq\rangle$ for baryons) dominate the hard scattering, higher states being suppressed by powers of

$1/Q^2 \ll 1$. Thus for an exclusive amplitude with L leptons/photons, M mesons and B baryons, the number of participating partons in T_h is

$$N = L + 2M + 3B ,\qquad (3.4)$$

to leading power. If N is large the calculations become intractable due to the profusion of diagrams, but testable predictions can be made for many processes especially those with B = 0 , M small and vice versa. Furthermore, an exclusive amplitude involving a single large momentum scale Q does not in general resolve the transverse momenta $\vec{k}^2_{T_i}$ up to scales of order Q^2 , and it is the k_T-integrated wave functions

$$\phi_H^{(n)}(x_i,Q^2) = \int^{Q^2} [d^2\vec{k}_{T_i}] \ \psi_H^{(n)}(x_i,\vec{k}_{T_i}) ,\qquad (3.5)$$

for finding n partons collinear up to scale Q and with momentum fractions x_i , which appear in calculations. To leading order, therefore, elastic amplitudes are obtained by convoluting valence wave functions ϕ_H with hard scattering parton subprocess amplitudes T_h . This leads to the following general features:

(i) The power law behaviour is generally consistent, up to calculable logarithmic corrections, with familiar dimensional counting rules, i.e. for hadronic form factors

$$F_H(Q^2) \sim (Q^2)^{1-m}\qquad (3.6)$$

where m is the number of valence constituents, and for elastic scattering cross sections

$$\frac{d\sigma}{d\cos\theta} \sim s^{3-n} f(\theta)\qquad (3.7)$$

where n is the number of fundamental constituents participating in the hard scattering.

(ii) The angular dependence of elastic cross sections probes both the x_i dependence of the wave functions and the form of the subprocess amplitude.

(iii) The normalisation of elastic amplitudes is directly related to the wave function normalisation. Although this cannot be calculated at sub-asymptotic energies, the relative magnitude of different cross sections can be predicted.

These features are illustrated in the following sections where two simple examples — the pion form factor and the elastic cross section $\gamma\gamma \to \pi\bar\pi$ — are analysed in detail.

3.2 The Pion Form Factor

The form factor of the pion is related to the elastic amplitude for $\gamma^* \pi \to \pi$ by

$$M[\gamma^*(Q) + \pi(p) \to \pi(p' = Q+p)]$$

$$= ie_\pi \, <p' \, |J_\mu^{em}|\, p> \tag{3.8}$$

$$= ie_\pi \, F_\pi(Q^2) \, (p + p')_\mu \quad .$$

At large Q^2 the form factor can be calculated by the methods described above [23,25]. To leading power in $1/Q^2$, the factorized structure illustrated in Fig 3.3 is obtained, and the result is

$$F_\pi(Q^2) = \int_0^1 dx\, dy \, \phi_\pi^*(x,Q^2) \, \phi_\pi^*(y,Q^2) \, T_h(x,y,Q^2) \quad . \tag{3.9}$$

The hard scattering amplitude can be expanded in a power series in the strong coupling α_s :

$$T_h(x,y,Q^2) = \frac{\alpha_s(Q^2)}{\pi} \, t^{(1)}(x,y,Q^2) + [\frac{\alpha_s(Q^2)}{\pi}]^2 \, t^{(2)}(x,y,Q^2) + \dots \tag{3.10}$$

The Born diagrams are shown in Fig 3.4 and their calculation yields [23]

$$t^{(1)}(x,y,Q^2) = \frac{16\pi^2 C_F}{Q^2} \cdot \frac{1}{(1-x)(1-y)} \quad . \tag{3.11}$$

The Q^2 dependence of the wave function $\phi_\pi(x,Q^2)$ can be obtained either from an operator product expansion on the light-cone or by direct evaluation of Feynman diagrams [23,26]. In a physical gauge, the leading behaviour is obtained from the ladder diagrams (together with vertex and self-energy insertions) shown in Fig 3.5. This ladder structure leads to an evolution equation

$$Q^2 \frac{\partial}{\partial Q^2} \phi_\pi(x,Q^2) = \frac{\alpha_s(Q^2)}{\pi} \int_0^1 dy \, V(x,y) \, \phi_\pi(y,Q^2) \tag{3.12}$$

with the kernel $V(x,y)$ given by

$$V(x,y) = \frac{C_F}{2} \left[\frac{1-y}{1-x} (1 + \frac{1}{y-x})_+ \theta(y-x) + \frac{y}{x} (1 + \frac{1}{x-y})_+ \theta(x-y) \right] \quad . \tag{3.13}$$

The solution of (3.12) is

$$\phi_\pi(x,Q^2) = x(1-x) \sum_{n=0}^{\infty} a_n \left[\frac{\alpha_s(Q^2)}{\alpha_s(Q_o^2)} \right]^{d_n} C_n^{(\frac{3}{2})}(2x-1) \tag{3.14}$$

where

$$d_n = \frac{8}{33-2N_f} \left[\frac{1}{2} - \frac{1}{(n+1)(n+2)} + 2 \sum_{i=2}^{n+1} \frac{1}{i} \right] \tag{3.15}$$

is the familiar non-singlet anomalous dimension and the $G_n^{(\frac{3}{2})}(x)$ are Gegenbauer polynomials. The coefficients a_n are determined once the wave function is known at some Q_o^2 :

$$a_n = \frac{4(2n+3)}{(n+1)(n+2)} \int_0^1 dx \, C_n^{(\frac{3}{2})} (2x-1) \, \phi_\pi(x,Q_o^2) \ . \tag{3.16}$$

Two important constraints on the wave function follow from the fact that $d_o = 0$, $d_n > 0 \ (n \geq 1)$:

(i) The integrated wave function is independent of Q^2 to leading power, i.e.

$$\int_0^1 dx \, \phi_\pi(x,Q^2) = \frac{a_o}{6} \ . \tag{3.17}$$

In fact it is precisely this quantity which determines the $\pi \rightarrow e\nu$ decay rate (Fig 3.6). Thus

$$a_o = \sqrt{3} \, f_\pi \ , \qquad f_\pi = 93 \text{ MeV} \ . \tag{3.18}$$

(ii) Only the first term in (3.14) survives in the $Q^2 \rightarrow \infty$ limit, i.e.

$$\lim_{Q^2 \rightarrow \infty} \phi_\pi(x,Q^2) = a_o x(1-x) = \sqrt{3} \, f_\pi \, x(1-x) \ . \tag{3.19}$$

This, in turn, fixes the asymptotic behaviour of the pion form factor

$$F_\pi(Q^2) \overset{Q^2 \rightarrow \infty}{\sim} \frac{16\pi\alpha_s(Q^2)}{Q^2} \, f_\pi^2 \ . \tag{3.20}$$

Unfortunately, an analysis of the sub-asymptotic terms reveals that this limit is approached only slowly in practice. The experimental measurement [27] comes from analysing data on $ep \rightarrow e\pi n$ at small momentum transfer. The interpretation of these data in terms of a pion form factor is, however, far from obvious. There is perhaps some suggestion of a $1/Q^2$ power behaviour

$$Q^2 F_\pi(Q^2) \approx 0.4 \qquad (Q^2 \leq 3 \text{ GeV}^2) \tag{3.21}$$

but this cannot be reconciled with the asymptotic form — equating (3.20) and (3.21) gives $\alpha_s = 0.87$. In practical applications at finite Q^2, therefore, the choice of $\phi_\pi(x,Q^2)$ is somewhat arbitrary. In some cases, for example the higher twist direct meson production discussed in Section 2.5, the hard scattering amplitude is such that the cross section is proportional to $F_\pi(Q^2)$ and the experimental measurement can be used. Otherwise there is some ambiguity in the predictions.

Finally, note that the QCD result for $F_\pi(Q^2)$ is consistent with the dimensional counting prediction (3.6) and that the "scale" for the inverse power of Q^2 is set by $\alpha_s f_\pi^2 \sim O(10^{-3} GeV^2)$. This accounts for the small size of higher twist contributions to direct pion production in QCD.

3.3 End Point, Pinch Singularities

The power behaviour of exclusive cross sections is determined by the structure of the hard scattering amplitude T_h . In general, all internal quark and gluon lines in the Born diagrams for T_h are far off-shell:

$$k_r^2 = f_r(x_i,y_i,\ldots)Q^2 + O(m^2) \tag{3.22}$$

and the power behaviour can be obtained simply by counting the internal hard propagators. However problems arise if one or more of the f_r can vanish in the allowed integration range and the typical outcome is anomalous scaling behaviour.

In general there are two cases to consider:

(i)   end point singularities, i.e. $f_r \to 0$ as one or more of the $\{x_i,y_i,..\} \to 0,1$. An example is the baryon form factor [23,28].

(ii) pinch (or Landshoff) singularities, when $f_r \to 0$ in the middle of the $\{x_i,y_i,\ldots\}$ range. An example is the elastic cross section $BB \to BB$ [29].

In such cases a more careful analysis is required. The "dangerous" regions often appear to be suppressed by Sudakov-like form factors $\exp[-\kappa \ln Q^2 \ln \ln Q^2]$ which vanish as $Q^2 \to \infty$ and change the naive dimensional counting behaviour [30]. However it is difficult to estimate the precise subasymptotic behaviour and the theoretical analysis is still incomplete. Fortunately, there are many processes which do not suffer from these ambiguities, including F_π and $d\sigma/d\cos\theta(\gamma\gamma \to \pi\pi)$, and for these rigorous predictions can be obtained.

3.4 Exclusive Wide Angle Scattering: $\gamma\gamma \to H\bar{H}$

The production at wide angle of a hadron pair in $\gamma\gamma$ collisions provides a particularly interesting and important test of the foregoing theoretical ideas. The leading order predictions are free of the end point and pinch singularity ambiguities discussed in the preceding section and also relatively insensitive to the choice of wave function. Furthermore, preliminary comparisons with experimental data, presented elsewhere in these Proceedings, are encouraging and suggest that these processes will indeed provide fundamental tests of the theory.

The theoretical analyses can be classified according to the nature of the final state hadron H . Of particular interest is the production of (i) light scalar mesons (π) (ii) light vector mesons (ρ) (iii) heavy quark mesons (ψ) and (iv) baryons (p) . The latter two calculations are discussed in detail elsewhere in these Proceedings, and so will only be treated briefly here.

3.4.1 Light Helicity Zero Mesons: $\gamma\gamma \rightarrow \pi^+\pi^-$, $\pi^0\pi^0$

Consider first the production of $\pi\bar{\pi}$ at angle θ in the $\gamma\gamma$ centre of mass. The cross section can be written as [31]

$$\frac{d\sigma}{d\cos\theta} = \frac{1}{32\pi s} \sum_{\substack{\text{photon helicities} \\ r,r' = \pm}} | M_{rr'}(s,\theta)|^2 \ .$$
(3.23)

The lowest order perturbative QCD contribution to the amplitude M has the factorized structure shown in Fig 3.7a, which implies

$$M_{rr'}(s,\theta) = \int_0^1 dx\, dy\, \phi_\pi^*(x,s)\, \phi_\pi^*(y,s)\, T_h^{rr'}(x,y,s,\theta) \ .$$
(3.24)

The wave function ϕ_π is the same as that appearing in the pion form factor calculation, eqn. (3.9). A sampling of the lowest order hard scattering subprocess diagrams is shown in Fig 3.7b and a straightforward calculation yields [31]

$$T_h^{++} = T_h^{--} = \frac{128\pi^2\alpha\alpha_s C_F}{s} \frac{(e_1+e_2)^2}{1-\cos^2\theta}\left[\frac{1}{xy} + \frac{1}{(1-x)(1-y)}\right]$$

$$T_h^{+-} = T_h^{-+} = \frac{128\pi^2\alpha\alpha_s C_F}{s}\left\{ \frac{(e_1+e_2)^2}{1-\cos^2\theta}\left[\frac{1}{x(1-y)} + \frac{1}{y(1-x)}\right]\right.$$

$$- \frac{e_1 e_2}{4xy(1-x)(1-y) + (1-x-y)^2(1-\cos^2\theta)}\left[\frac{1-y}{x} + \frac{1-x}{y} + \frac{x}{1-y} + \frac{y}{1-x}\right]$$

$$\left. + (e_1^2 - e_2^2)\frac{(x-y)}{2xy(1-x)(1-y)}\right\}$$
(3.25)

where e_1 and e_2 are the quark charges ($e_\pi = e_1 + e_2$) . Using the symmetry of the wave functions, $\phi_\pi(x,Q^2) = \phi_\pi(1-x,Q^2)$, these expressions reduce to

$$T_h^{++} = T_h^{--} = \frac{16\pi\alpha(e_1+e_2)^2}{1-\cos^2\theta} \cdot \frac{16\pi\alpha_s C_F}{(1-x)(1-y)s}$$

$$T_h^{+-} = T_h^{-+} \rightleftharpoons T_h^{++} - \frac{256\pi^2\alpha\alpha_s C_F}{s} \cdot e_1 e_2 \cdot \left(\frac{1-y}{x} - \frac{x}{1-y}\right)$$

$$\cdot \frac{1}{4xy(1-x)(1-y) + (1-x-y)^2(1-\cos^2\theta)}$$
(3.26)

A comparison of eqns.(3.26) and (3.11) shows that the amplitudes T_h^{++} and T_h^{--} are proportional to the corresponding amplitude for the pion form factor. Thus

$$M_{++} = M_{--} = \frac{16\pi\alpha e_\pi^2}{1 - \cos^2\theta} \cdot F_\pi(s)$$

$$M_{+-} = M_{-+} = M_{++} - 32\pi\alpha \, F_\pi(s) \, e_1 e_2 \, g(s,\theta)$$

(3.27)

where

$$g(s,\theta) = \frac{< (\frac{1}{x(1-x)} + \frac{1}{y(1-y)}) \frac{(1-x)(1-y) + xy}{4xy(1-x)(1-y) + (1-x-y)^2(1-\cos^2\theta)} >}{< \frac{1}{xy(1-x)(1-y)} >}$$

(3.28)

and the notation

$$< f(x,y) > = \int_0^1 dx \, dy \, \phi_\pi^*(x,s) \, \phi_\pi^*(y,s) \, f(x,y)$$

(3.29)

has been used. Note that the only explicit dependence on the wave function is in the residual term $g(s,\theta)$. Substituting the amplitudes in (3.27) into eqn.(3.23) gives the cross section

$$\frac{d\sigma^{\gamma\gamma \to \pi\pi}}{d\cos\theta} = \frac{8\pi\alpha^2}{s} F_\pi^2(s) \left\{ \frac{e_\pi^4}{(1-\cos^2\theta)^2} - \frac{2e_\pi^2 \overline{(e_1 e_2)} g(s,\theta)}{1-\cos^2\theta} + 2\overline{(e_1 e_2)}^2 \, g^2(s,\theta) \right\}.$$

(3.30)

Here $\overline{e_1 e_2}$ denotes a charge averaging, appropriate for $\gamma\gamma \to \pi^0\pi^0$. Substituting specific values for the quark charges, the result is

$$\frac{d\sigma^{\gamma\gamma \to \pi^+\pi^-}}{d\cos\theta} = \frac{8\pi\alpha^2}{s} F_\pi^2(s) \left\{ \frac{1}{(1-\cos^2\theta)^2} - \frac{4}{9} \frac{g}{(1-\cos^2\theta)} + \frac{8}{81} g^2 \right\}$$

(3.31a)

$$\frac{d\sigma^{\gamma\gamma \to \pi^0\pi^0}}{d\cos\theta} = \frac{8\pi\alpha^2}{s} F_\pi^2(s) \left\{ \frac{25}{162} g^2 \right\}.$$

(3.31b)

The dependence on the wave function of these cross sections is illustrated in Fig 3.8, which shows the θ dependence at fixed s for the two choices

$$\phi_\pi(x,s) \sim \delta(x - \tfrac{1}{2})$$

$$\phi_\pi(x,s) \sim x(1-x)$$

(3.32)

For $\gamma\gamma \to \pi^+\pi^-$, the first term in eqn.(3.31a) dominates and the dependence on ϕ_π is weak. There is more variation in the prediction for $\gamma\gamma \to \pi^0\pi^0$ — in general, the more peaked ϕ_π is about $x = \frac{1}{2}$, the weaker the dependence of g on θ . In principle, therefore, the angular variation of $d\sigma/d\cos\theta$ $(\gamma\gamma \to \pi^0\pi^0)$ provides a measure of the shape of the pion wave function. Note also that g can be eliminated from eqns.(3.31a,b) to obtain a relation between the two cross sections and F_π .

The above results are seen to be consistent with the dimensional counting prediction, eqn.(3.7) with $n = 6$, the form factor supplying the extra powers of s^{-1} . Normalizing to the QED cross section $\gamma\gamma \to \mu^+\mu^-$ at the same energy gives

$$\frac{\dfrac{d\sigma^{\gamma\gamma\to\pi^+\pi^-}}{d\cos\theta}}{\dfrac{d\sigma^{\gamma\gamma\to\mu^+\mu^-}}{d\cos\theta}} = 4F_\pi^2(s)\left[\frac{1}{1-\cos^4\theta} - \frac{4}{9}\frac{g}{1+\cos^2\theta} + \frac{8}{81}g^2\frac{1-\cos^2\theta}{1+\cos^2\theta}\right] \qquad (3.33)$$

i.e. the pion cross section is suppressed by a factor s^{-2} . That these exclusive cross sections are therefore very small is illustrated by Fig 3.9 which shows the e^+e^- cross section predictions for $e^+e^- \to e^+e^- + (\mu^+\mu^-, \pi^+\pi^-, \pi^0\pi^0)$,

$$\frac{d\sigma^{e^+e^-\to e^+e^- + (a\bar{a})}}{dp_T^2} = \int_0^1 dz\, L(z)\left[\frac{4}{\hat{s}\cos\theta}\frac{d\sigma^{\gamma\gamma\to a\bar{a}}}{d\cos\theta}\right]$$

$$\cos\theta = (1 - \frac{4p_T^2}{\hat{s}})^{1/2}$$

$$\hat{s} = zs \qquad (3.34)$$

$$a = \mu^+, \pi^+, \pi^0$$

as functions of p_T^2 at fixed s .

Assuming that the wave functions are the same in each case, the predictions for other helicity zero mesons are obtained by substituting the appropriate decay constant f_M in the form factor. This gives the ratios [31]:

$$
\begin{array}{cccccc}
\pi^+\pi^- & : & K^+K^- & : & \rho^+\rho^- & \\
1 & : & 2 & : & 7.5 &
\end{array}
\qquad (3.35a)
$$

$$
\begin{array}{cccccccccc}
\pi^0\pi^0 & : & K_L K_S & : & \rho^0\rho^0 & : & \omega\omega & : & \phi\phi \\
1 & : & .3 & : & 7.5 & : & 8 & : & 1.4
\end{array}
\qquad (3.35b)
$$

where the zero helicity projection of the vector mesons is understood.

3.4.2 Helicity One Mesons

In a similar way, predictions are obtained for light helicity ± 1 ("transverse") mesons $(M_T \bar{M}_T = \rho_T^+ \rho_T^- , \rho_T^0 \rho_T^0 , \ldots)$. The result is [31]

$$
\frac{d\sigma^{\gamma\gamma \to M_T \bar{M}_T}}{d\cos\theta} = \frac{8\pi\alpha^2}{s} F_{M_T}^2 (s) \left\{ e_M^4 + 4\overline{(e_1 e_2)} e_M^2 \cos^2\theta \ g_T + 2\overline{(e_1 e_2)}^2 \cos^2\theta (1+\cos^2\theta) g_T^2 \right\},
$$

(3.36)

where

$$
F_{M_T} = \frac{16\pi\alpha_s C_F}{s} < \frac{1}{(1-x)(1-y)} >
$$

$$
g_T(s,\theta) = \frac{< \dfrac{1}{(xy(1-x)(1-y)} \cdot \dfrac{(1-x-y)^2}{4xy(1-x)(1-y) + (1-x-y)^2(1-\cos^2\theta)} >}{< \dfrac{1}{xy(1-x)(1-y)} >}
$$

(3.37)

and the expectation values are with respect to the transverse projection of the wave function. The problem here is that, because of the helicity conservation embodied in the QCD structure, the transverse form factors cannot be measured directly. At present therefore, detailed predictions require extra assumptions about the transverse wave function.

3.4.3 Other Exclusive Final States

The above analysis can be extended to include the production of heavy quark mesons, in particular, the quarkonia states $\psi\psi$, TT , etc. For these a non-relativistic wave function approximation can be used — $\phi_{Q\bar{Q}} \sim \delta(x - \frac{1}{2})$ — and the normalisation can again be fixed by an appropriate decay constant. Fig 3.10 shows the results of a calculation of the total $e^+e^- \to e^+e^-\psi\psi$ cross section [32]. It has also been shown that certain higher order contributions, shown in Fig 3.11, can be important especially at forward angles [33].

Exclusive baryon production has also been considered [34]. Although the analysis is in principle the same, in that the exclusive amplitude again has the form

$$
M = \int_0^1 [dx][dy] \ \phi_B^*([x],s)\phi_B^*([y],s) \ T_h([x],[y],s,\theta)
$$

(3.38)

there are several additional complications. First, there are over 200 diagrams to be considered and second, the corresponding baryon form factor calculation suffers from end point singularity ambiguities. However there are other processes, e.g.

$\psi \rightarrow B\bar{B}$, which are suitable for comparison. In this way relatively unambiguous predictions of the form

$$\frac{d\sigma^{\gamma\gamma \rightarrow B\bar{B}}}{d\cos\theta} = S^{-5} f(\theta) \tag{3.39}$$

have been obtained [34], and a preliminary comparison with data appears encouraging [35].

Acknowledgements

It is a pleasure to thank the organizers for such an excellent Workshop. In preparing this review I have benefitted enormously from discussions with Peter Landshoff and David Scott. I am also grateful to the SERC of Great Britain and to Peterhouse, Cambridge for financial support.

References and Footnotes

[1] See, for example, J.H. Field, Nucl. Phys. B168, 477 (1980), erratum B176, 545 (1980).

[2] G. Missonier, College de France preprint.

[3] D.W. Duke, these proceedings.

[4] K. Kajantie and R. Raitio, Nucl. Phys. B159, 528 (1979).

[5] For quantitative estimates of vector dominance contributions see, for example, S.J. Brodsky et al. reference 7.

[6] J.D. Bjorken, S. Berman and J. Kogut, Phys. Rev. D4, 3388 (1971).

[7] S.J. Brodsky, T. DeGrand, J. Gunion and J. Weis, Phys. Rev. Lett. 41, 672 (1978). Phys. Rev. D19, 1418 (1979).

[8] K. Kajantie, Phys. Scripta 29, 230 (1979).

[9] C.H. Llewellyn Smith, Phys. Lett. 79B, 83 (1978).

[10] See, for example, F. Khalafi and W.J. Stirling, Cambridge University preprint DAMTP 83/2 (1983).

[11] B.L. Combridge, J. Kripfganz and J. Ranft, Phys. Lett. 70B, 234 (1977).

[12] R. Cahn and J. Gunion, Phys. Rev. D20, 2253 (1979).

[13] K. Kajantie and R. Raitio, Phys. Lett. 87B, 133 (1979).

[14] F. Khalafi, P.V. Landshoff and W.J. Stirling, Cambridge University preprint in preparation.

[15] F.A. Berends, Z. Kunszt and R. Gastmans, Nucl. Phys. B182, 397 (1981).

[16] I. Kang, Oxford University preprint 50/82 (1982).

[17] G. Sterman and S. Weinberg, Phys. Rev. Lett. 39, 1436 (1977).

[18] See, for example, S.J. Brodsky, E.L. Berger and G.P. Lepage, Proceedings of the Drell-Yan Workshop, Fermilab, October 1982, page 187.

[19] For a review see M. Jacob and P.V. Landshoff, Phys. Reports 48, 285 (1978).

[20] See Section 3 and references therein.

[21] J.A. Bagger and J.F. Gunion, U.C. Davis preprint UCD 83/1 (1983).

[22] For a recent review see, for example, E. Reya, Phys. Reports 69, 195 (1981).

[23] G.P. Lepage and S.J. Brodsky, Phys. Rev. D22, 2157 (1980); Phys. Lett. 87B, 359 (1979), and references therein.

[24] A.V. Efremov and A.V. Radyushkin, Rev. Nuovo Cimento 3, 1 (1980); Phys. Lett. 94B, 245 (1980).

[25] A. Duncan and A. Mueller, Phys. Rev. D21, 1636 (1980); Phys. Lett. 98B, 159 (1980).

[26] S.J. Brodsky, Y. Frishman, G.P. Lepage and C. Sachrajda, Phys. Lett. 91B, 239 (1980). See also A. Duncan and A. Mueller, Reference 25, and M.K. Chase, Nucl. Phys. B167, 125 (1980).

[27] C. Bebek et al., Phys. Rev. D17, 1693 (1978).

[28] S.J. Brodsky, G.P. Lepage and S.A.A. Zaidi, Phys. Rev. D23, 1152 (1981); S.J. Brodsky and G.P. Lepage, Phys. Rev. Lett. 43, 545 (1979); Erratum ibid. 43, 1625 (1979).

[29] P.V. Landshoff, Phys. Rev. D10, 1024 (1974); P. Cvitanovic, Phys. Rev. D10, 338 (1974); S.J. Brodsky and G. Farrar, Phys. Rev. D11, 1309 (1975).

[30] P.V. Landshoff and D.J. Pritchard, Z. Phys. C6, 69 (1980).

[31] S.J. Brodsky and G.P. Lepage, Phys. Rev. D22, 2157 (1980).

[32] R.E. Ecclestone and D.M. Scott, Cambridge University preprint, DAMTP 82/31 (1982).

[33] S.L. Grayson, R.R. Horgan and P.V. Landshoff, Cambridge University preprint DAMTP 83/8 (1983).

[34] P.H. Damgaard, Nucl. Phys. B211, 435 (1983).

[35] H. Kolanoski, these proceedings.

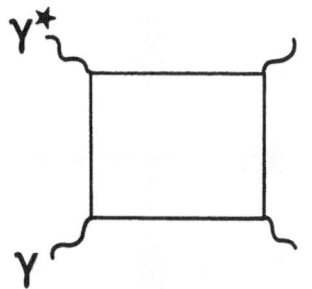

Fig.1.1 Lowest order (box) diagram for the photon structure function.

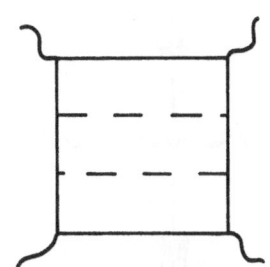

Fig. 1.2 Box diagram dressed with gluons (dashed lines).

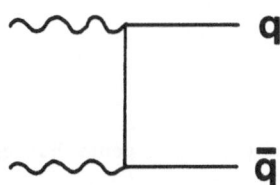

Fig. 2.1 Born diagrams for $\gamma\gamma \to q\bar{q}$.

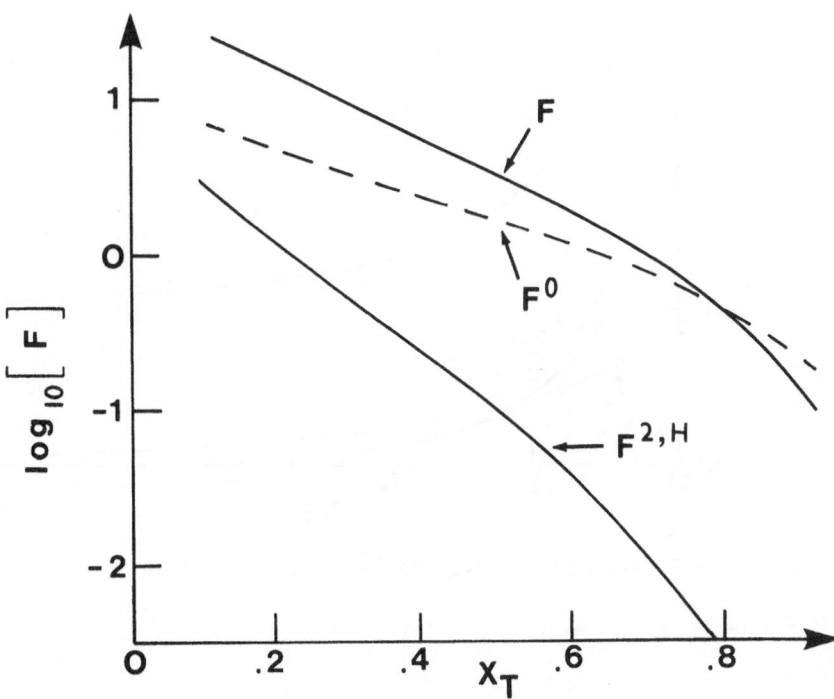

Fig. 2.2 Large p_T scaling functions defined in Section 2.

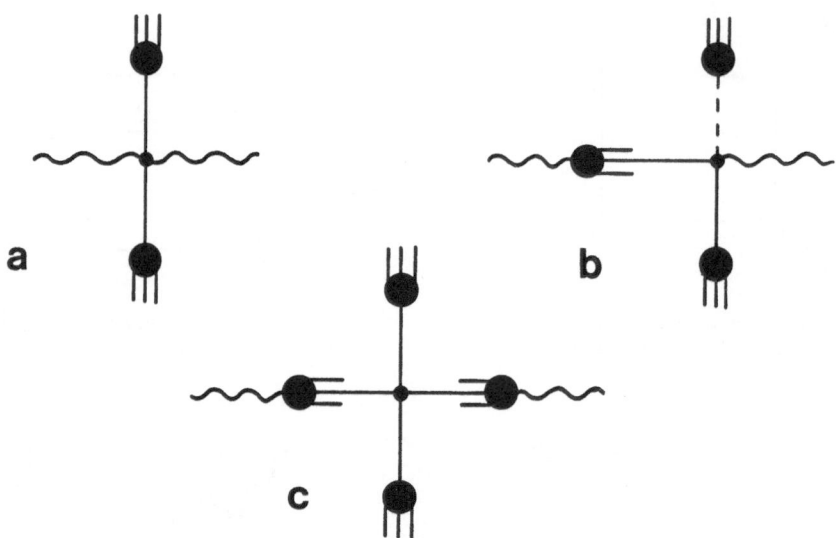

Fig. 2.3 Examples of (a) 2-jet (b) 3-jet and (c) 4-jet contributions to large p_T jet production.

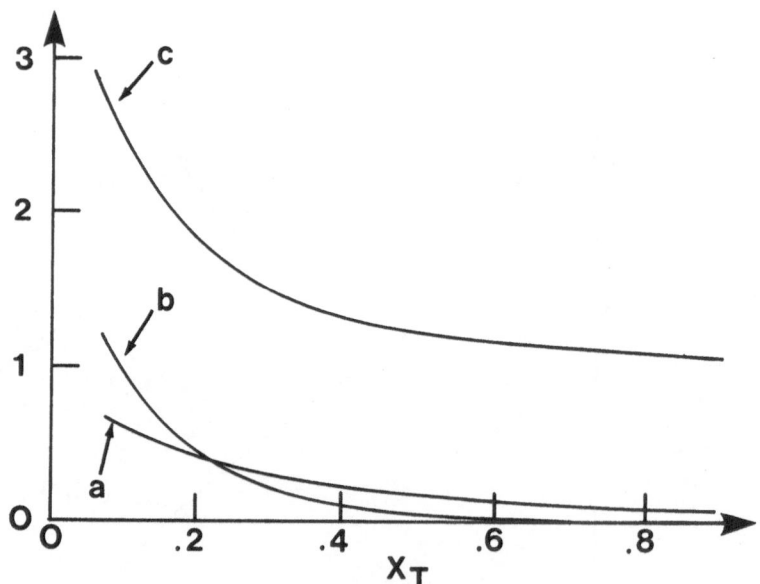

Fig. 2.4 Relative contributions to $\dfrac{d\sigma}{dp_T^2}$ ($e^+e^- \rightarrow e^+e^- JX$):

(a) 3-jet/2-jet (b) 4-jet/2-jet (c) (2+3+4)-jet/2-jet.

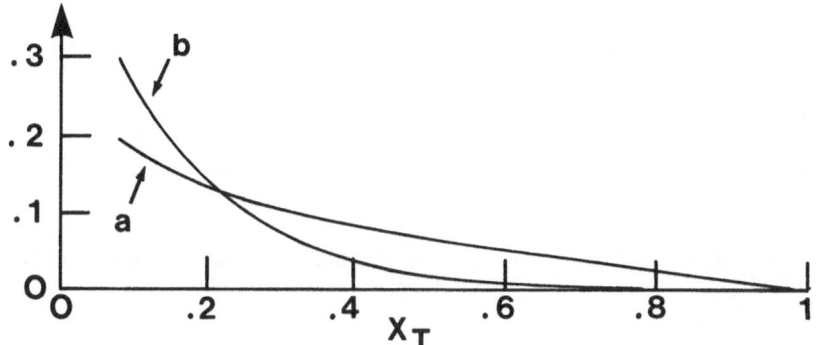

Fig. 2.5 Relative contributions to $\dfrac{d\sigma}{dp_T^2}$ $(e^+e^- \to e^+e^-JX)$ from subprocesses involving gluons: (a) J = gluon jet (b) contributions with $G_{g/\gamma}$.

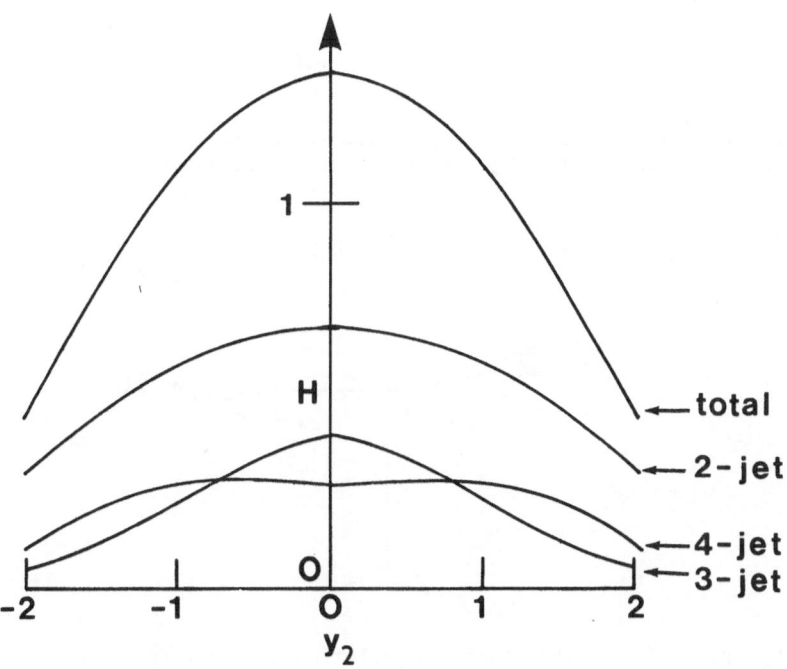

Fig. 2.6 Away side rapidity correlations: $\gamma\gamma \to J_1 J_2 X$. The graph shows the contributions to $H(.2,.1,y_2)$ where H is defined in eqn.(2.9).

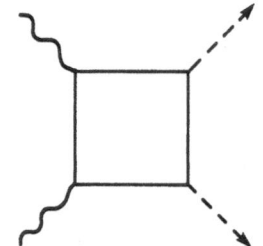

Fig. 2.7 Box diagram for
$\gamma\gamma \rightarrow$ 2 gluon jets.

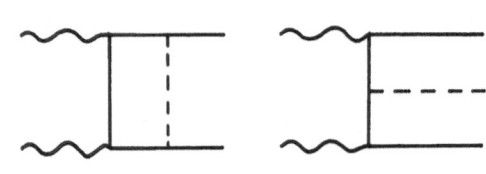

Fig. 2.8 Examples of one gluon corrections to
Fig. 2.1.

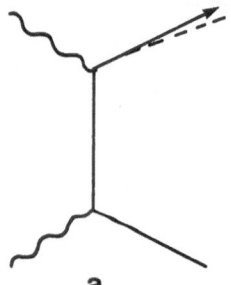

a

b

Fig. 2.9 Singular collinear configurations which are included in (a) the
fragmentation function and (b) the photon structure function.

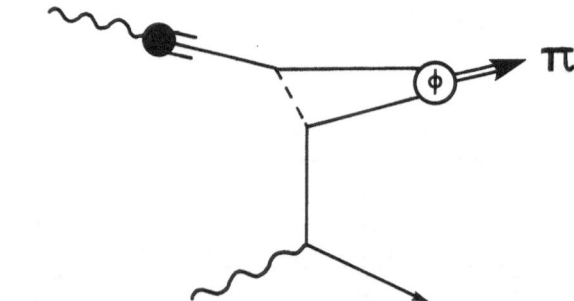

Fig. 2.10 A higher twist contribution to the large p_T pion cross
section.

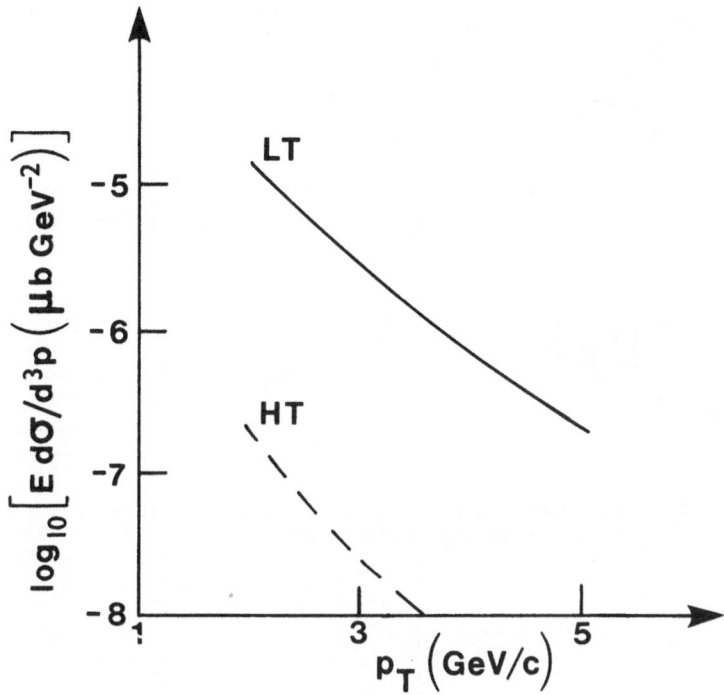

Fig. 2.11 Leading twist (LT) and higher twist (HT) contributions to the $\gamma\gamma \to \pi^+X$ inclusive cross section at $\theta_{cm} = 90°$, $\sqrt{s} = 10$ GeV, from [21].

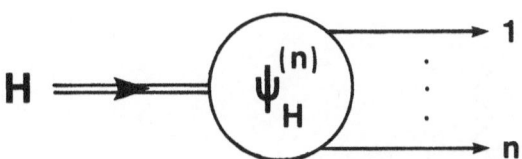

Fig. 3.1 The n parton Fock state wave function for a hadron H .

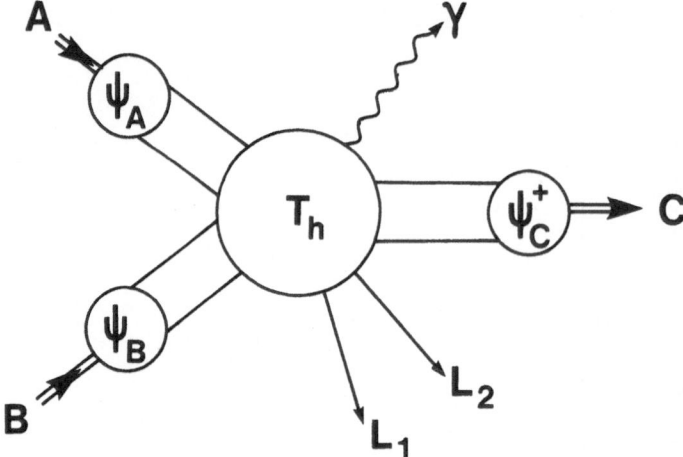

Fig. 3.2 An example illustrating the general structure of a hadron, lepton exclusive amplitude in QCD.

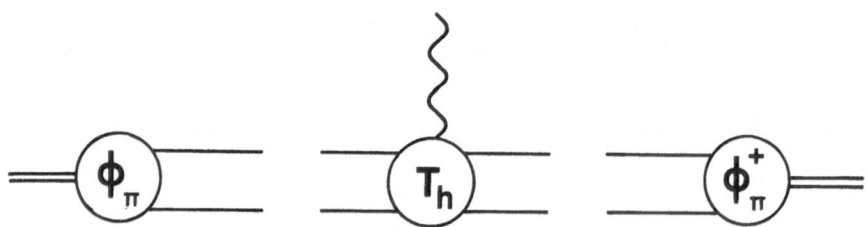

Fig. 3.3 The factorized structure of the pion form factor.

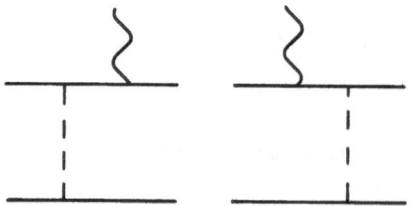

Fig. 3.4 Lowest order contributions to T_h.

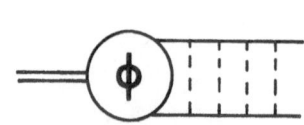

Fig. 3.5 Ladder diagram for wave function Q^2 evolution.

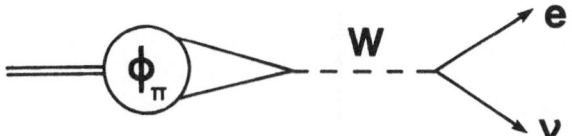

Fig. 3.6 Diagram for the decay $\pi \to e\nu$.

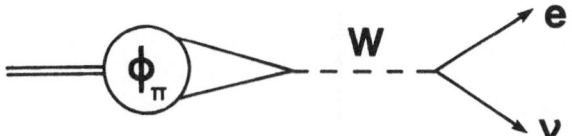

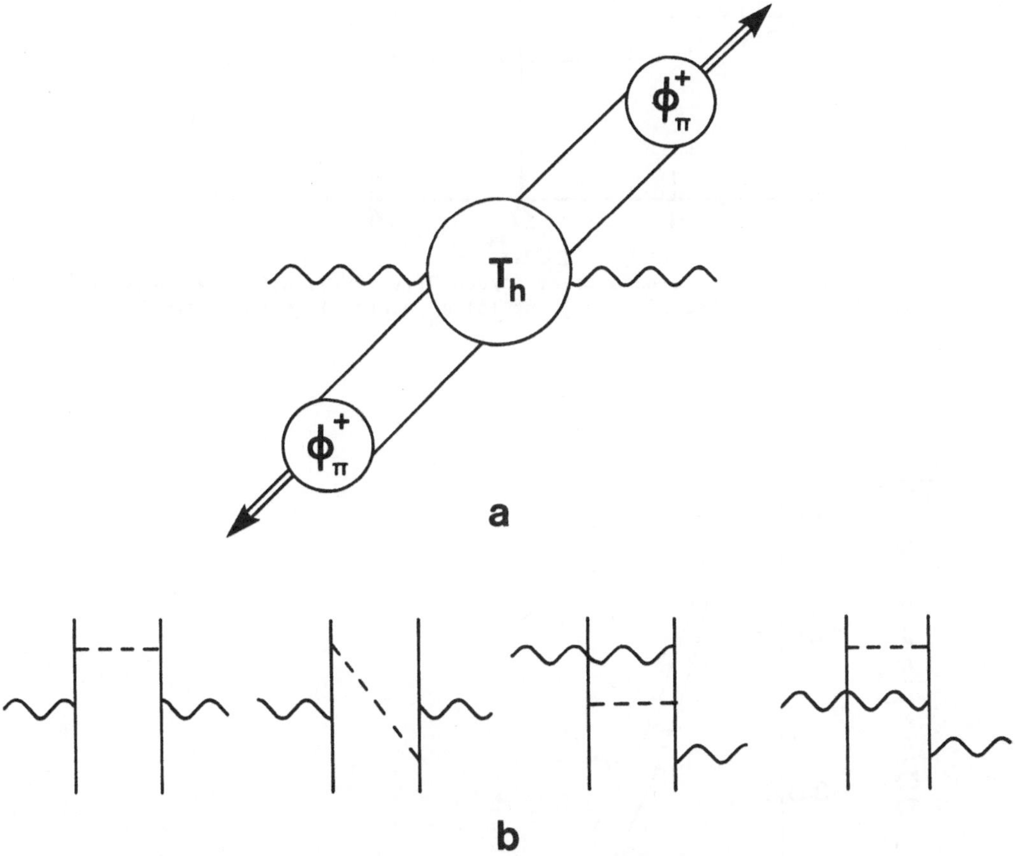

a

b

Fig. 3.7 (a) Factorized structure for the amplitude $\gamma\gamma \to \pi\pi$. (b) Lowest order contributions to T_h.

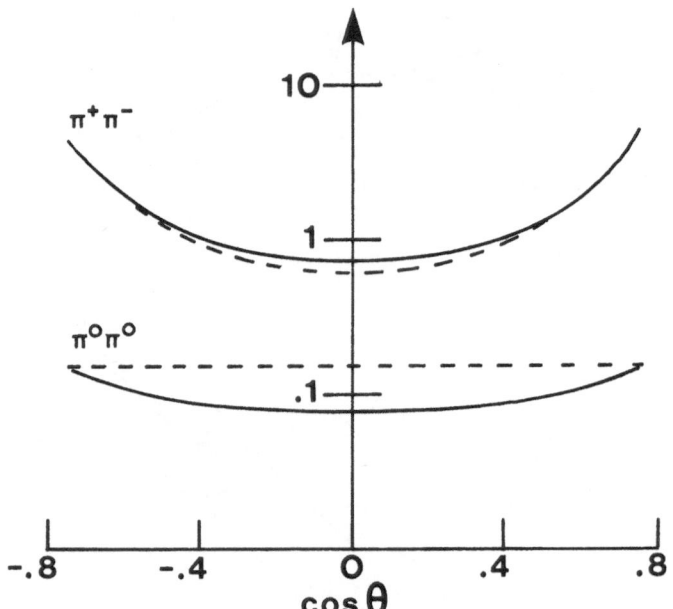

Fig. 3.8 Angular dependence of $d\sigma/d\cos\theta(\gamma\gamma \to \pi\bar{\pi})$, with wavefunctions
(a) $\phi_\pi \propto x(1-x)$ [solid lines] and (b) $\phi_\pi \propto \delta(x-\frac{1}{2})$ [dashed lines].

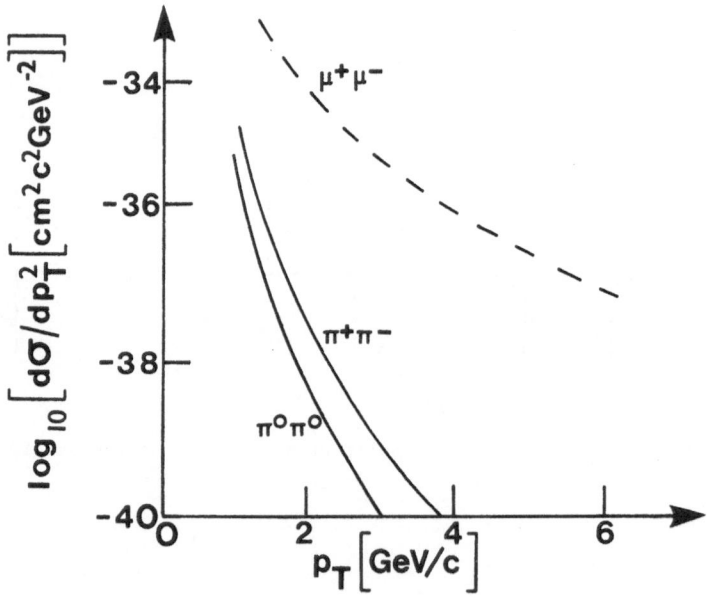

Fig. 3.9 Cross sections for $e^+e^- \to e^+e^-a\bar{a}$.

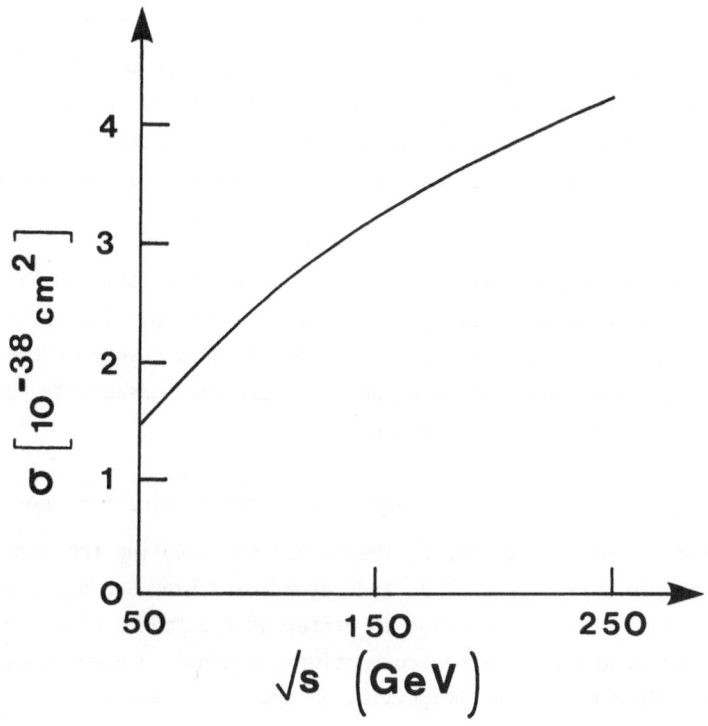

Fig. 3.10 Total cross section for $e^+e^- \rightarrow e^+e^-\psi\psi$ from [32].

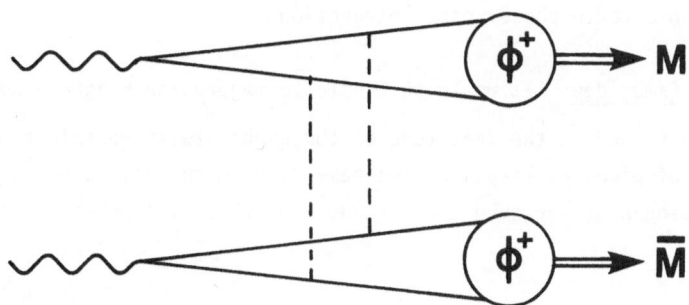

Fig. 3.11 Higher order QCD contributions which dominate at forward angles [33].

N. Wermes, SLAC: I have two questions concerning the 'inclusive' part of your talk.

1) If I recall correctly, the number given by Berends et al. for the QCD - corrections using Sterman-Weinberg parameters is in the order of ~ + 40 % dependent on cut-off masses.   How can I understand the number you quoted given by Kang calculating - 34 % + 18 % = - 16 %? Is this only because the beam-pipe jets are treated differently?

2) As you have seen in the previous talk there are now data available on the Q^2 - dependence of jet-trigger cross sections. It would be nice to calculate on a theoretical basis what that means for the presence or non-presence of the different underlying subprocesses. What can you say about the feasability for making definite calculations on this subject?

J. Stirling, Cambridge: 1) Corrections to jet cross sections depend on two things:

a) the precise definition of a 'jet' and b) the method for handling the beam-pipe jets. It is true that the corrections of Berends et al. and Kang are quantitatively rather different and this appears to be due to different treatments of a) and b). I believe it is better to analyse instead corrections to single hadron cross sections, which do not suffer from such ambiguities to the same extent.

2) The experimentalists are clearly one step ahead of the theorists in having data on the Q^2 - dependence of jet cross sections. I agree that it would be very interesting to use this to investigate the various subprocesses - I know of no theoretical calculation which looks at this. There might, however, be a problem in that such an analysis would presumably require a much better treatment of the kinematics than is embodied in the equivalent photon approximation. This would mean Monte Carlo phase space integration.

P.V. Landshoff, Cambridge: It would be unsafe to regard the Bagger - Gunion calculation as the last word on the magnitude of the higher-twist contribution to inclusive production of pions at large p_T. The same diagram can give production of ρ or many other resonances, or even a continuum, and this will enhance the pion production.

J. Stirling, Cambridge: I agree. That is a good comment.

HADRONIC FINAL STATES IN SOFT PHOTON–PHOTON SCATTERING

Hermann Kolanoski

Physikalisches Institut der Universität Bonn
Federal Republic of Germany

Abstract

In this talk experimental results on the non-resonant production of hadrons in soft $\gamma\gamma$ scattering are reviewed:
The measured cross sections for the production of hadron pairs by two photons are compared to theoretical predictions (QCD calculations above about 2 GeV).
Analyses of four pion final states are reported, including an angular correlation analysis of $\rho^\circ\rho^\circ$, first results on $\rho^+\rho^-$ production and the observation of a narrow structure in the four pion mass spectrum near 2.1 GeV.
The experimental results on the total hadronic cross section for photon-photon scattering are critically discussed.

1.0 INTRODUCTION

The title of this talk raises the very general question: How do the photons turn into hadrons? However, I will not be all that general. Since this is a review of the experimental status, the framework for the talk is given by the results available on hadronic final states in soft $\gamma\gamma$ scattering. "Soft" is meant to distinguish this talk from the talk given by N.Wermes at this workshop on hard scattering phenomena such as jet production by two photons and from the talk given by W.Wagner on deep inelastic electron-photon scattering involving high Q^2 values of the photons. The experimental aspects of resonance production by two photons are reviewed by J.Olsson at this workshop.

This talk will cover three topics:

- Two-photon production of hadron pairs.

- Two-photon production of four pions.

- The total cross section for two-photon production of hadrons (concentrating on the low energy, low Q^2 region).

2.0 HADRON PAIR PRODUCTION BY TWO PHOTONS

I will begin with summarizing our knowledge about the continuum production of hadron pairs at low energies. The understanding of these reactions is important for the analysis of two-photon production of resonances decaying into hadron pairs. One cannot always get rid of the background from the continuum by a simple subtraction, but interferences between resonances and continuum may also change the resonance shapes.

In the second part of this chapter new results on hadron pair production for $\gamma\gamma$ invariant masses ($W_{\gamma\gamma}$) above 2 GeV are presented. The experimental results will be compared to QCD calculations.

2.1 HADRON PAIR PRODUCTION AT LOW ENERGIES

In the resonance region most experimental information is available on the $\pi^+\pi^-$ continuum which will be discussed in more detail below.

The $\pi^\circ\pi^\circ$ continuum below ~1 GeV has not yet been analyzed, though data are available from the JADE experiment. This channel may be best suited to look for broad $\pi\pi$ resonances below ~1 GeV.

The reaction $\gamma\gamma\rightarrow K\bar{K}$ has been measured by the TASSO group for the K^+K^- and $K^0_sK^0_s$ final states /1/. The $K\bar{K}$ mass spectra have been used to determine the $\gamma\gamma$-width of the f' (see J.Olsson's talk). A more general analysis of the spectra

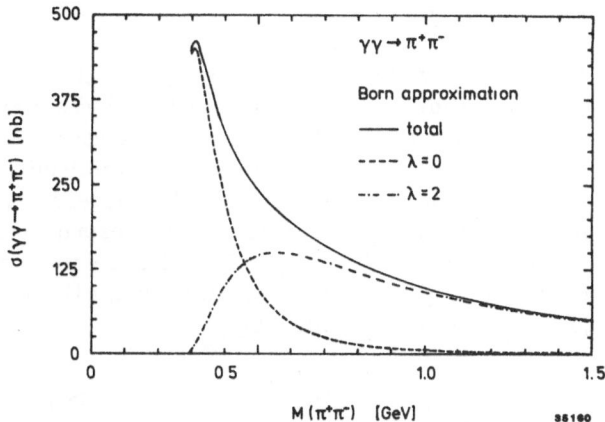

Figure 1. $\sigma(\gamma\gamma\to\pi^+\pi^-)$ (Born): curves for $\gamma\gamma$-helicities 0, 2 and sum of both

including both resonances and continuum is desireable because it would reduce the systematic error on the $\gamma\gamma$-width. A model which describes continuum and resonances by unitarizing the combined scattering amplitude is given in /2/.

2.1.1 The Reaction $\gamma\gamma\to\pi^+\pi^-$ Near Threshold

At low energies the measured cross section for $\gamma\gamma\to\pi^+\pi^-$ may be compared to the Born approximation for this process given by the following graphs:

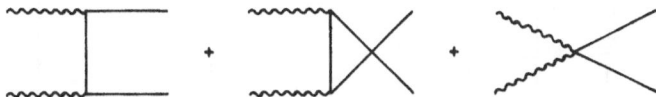

The resulting cross section as well as its decomposition into $\gamma\gamma$-helicity components is shown in Figure 1. When searching for a broad resonance below ~1 GeV, as for the expected scalar meson $\varepsilon(\sim 700)$, a first step in the analysis should be a quantitative comparison of the results with the expectations from a pure Born cross section.

The process $\gamma\gamma\to\pi^+\pi^-$ has been measured at the DCI storage ring in the $\pi\pi$ mass range from threshold to about 0.7 GeV /3/. The pion pairs have been separated from the electron and muon pair background by determining the particle mass from a kinematical reconstruction of the event. With such a method for $\pi\pi$ detection near threshold this experiment is superior to experiments at PETRA or PEP. Unfortunately the experiment suffers from lack of statistics. The measured cross section (within the acceptance) of $\sigma(\gamma\gamma\to\pi^+\pi^-)$ = 69±15 pb has to be compared to 34 pb for the Born approximation. The difference is only 2.3 standard deviations, not enough to establish a resonance in this region. At present, more data are being collected /4/.

2.1.2   The Reaction $\gamma\gamma \to \pi^+\pi^-$ in the f^0-Region

In Figure 1 we see that the $\gamma\gamma$-helicity 0 contribution to the Born cross section dies out very fast above threshold. In the region of the f^0 resonance helicity 2 is virtually the only contribution. Below the f^0-resonance strong interactions are expected to have only a small effect on the helicity 2 amplitude because the $\pi\pi$ phase shifts are small for $\pi\pi$ angular momenta $J \geqq 2$. Therefore it has been suggested that the Born approximation should be a good description of $\gamma\gamma \to \pi^+\pi^-$ as far above threshold as $W_{\gamma\gamma} \sim 1$ GeV (see e.g. /5/). Hence we may compare the data in the f^0-region with an approximation obtained from the Born terms plus a resonance contribution as given by the following graphs:

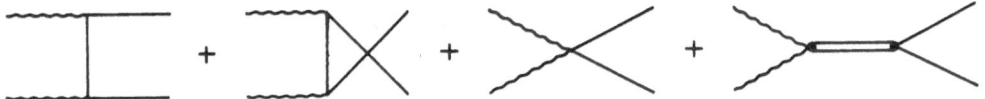

If the resonance is also dominantly produced via helicity 2 a strong interference between the Born graphs and the resonance is expected. The interference is completely determined by the Breit-Wigner phase (the Born amplitude is purely real) except for a relative sign between the Born and the resonance amplitudes. Curves for the two choices of sign are shown in Figure 2.

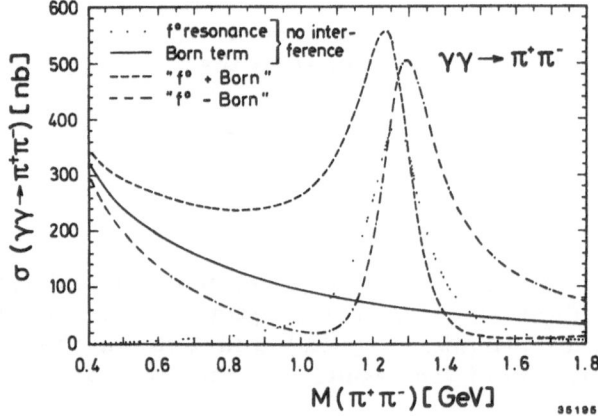

Figure 2.   Cross section for the two-photon production of the f^0 (assuming $\lambda=2$) and $\pi^+\pi^-$-continuum : without interference (dotted and full curve), resonance and Born amplitudes interfering with a positive relative sign (dashed curve) and with a negative relative sign (dotted curve).

The justification for such a simple approach is given by the data. Figure 3 shows the $\pi^+\pi^-$-mass spectrum obtained by the Mark II collaboration at SPEAR /6/. The data are compared to calculations for the Born term and the f^0-resonance. Com-

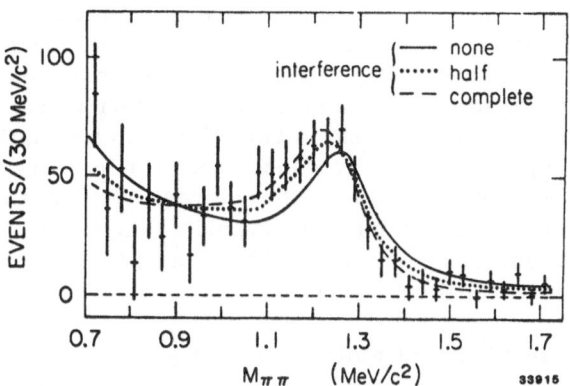

Figure 3. Mark II: Invariant mass of events with two charged tracks assigning pion masses to all tracks, QED background subtracted.

plete interference, which is expected if the f⁰ is produced mainly via helicity 2, gives a good description of the data.

New data from the CELLO group exhibit the same structure in the $\pi^+\pi^-$-mass spectrum (Figure 4) /7/. This spectrum was obtained selecting two oppositely charged tracks with $|cos\theta|<0.8$ and transverse momenta w.r.t. the beam $p_T>0.35$ GeV/c. The resulting transverse momentum of the two particles was required to be smaller than 0.1 GeV/c. Since the pions have not been identified, the electron and muon pairs from the QED processes had to be subtracted.

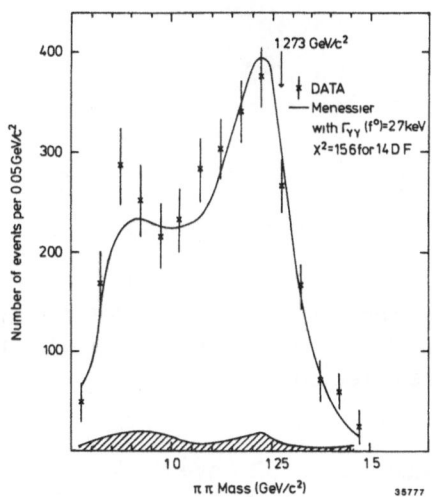

Figure 4. CELLO: Invariant mass of events with two charged tracks assigning pion masses to all tracks, QED background subtracted. The hatched area is the estimated K⁺K⁻ background. The solid curve is explained in the text.

The CELLO results on $\pi^+\pi^-$ production between 0.8 and 1.5 GeV are well described by a model of Mennessier /2/ (solid curve in Figure 4). In this model the amplitudes for a given partial wave and given isospin are composed of a unitarized Born term, a unitarized term for vector exchange and terms describing the direct couplings of the photons to resonances. In the CELLO analysis only the unitarized Born term and the coupling to the f^0-resonance were necessary. In particular, a contribution of a scalar resonance was not required by the data. The modification of the Born term obtained from the unitarization procedure using strong interaction data turns out to be small /7/. This may be due to the smallness of the D-wave phase shifts below the f^0-mass.

The status of the analysis of the $\gamma\gamma$-width of the f^0 was reviewed by J.Olsson at this workshop.

2.2 PRODUCTION OF PROTON-ANTIPROTON PAIRS

The TASSO collaboration published an analysis of the reaction $\gamma\gamma\to p\bar{p}$ based on 8 observed events /8/. This analysis has now been repeated with much higher integrated luminosity ($\sim$74pb^{-1}) increasing the number of events by an order of magnitude /9/.

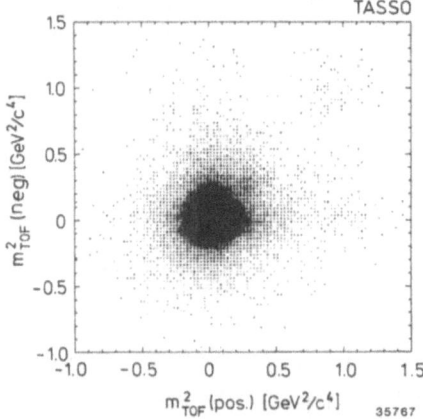

Figure 5.   Square of the mass calculated from TOF for the negative track vs. the same quantity for the positive track   in an event with 2 charged tracks detected

Candidate events for the reaction $\gamma\gamma\to p\bar{p}$ have been selected from events with two oppositely charged tracks in a polar angular range $|\cos\theta|<0.8$ and in a momentum range $0.35\leq p\leq 1.6$ GeV/c. The $p\bar{p}$ pairs have been identified by means of the time-of-flight (TOF) information of the counters surrounding the TASSO central drift chamber. In Figure 5 the square of the mass calculated from TOF for the positive track is plotted versus the same quantity for the negative track. The cluster of $p\bar{p}$ events is clearly separated. The final sample contains 72 $p\bar{p}$ events

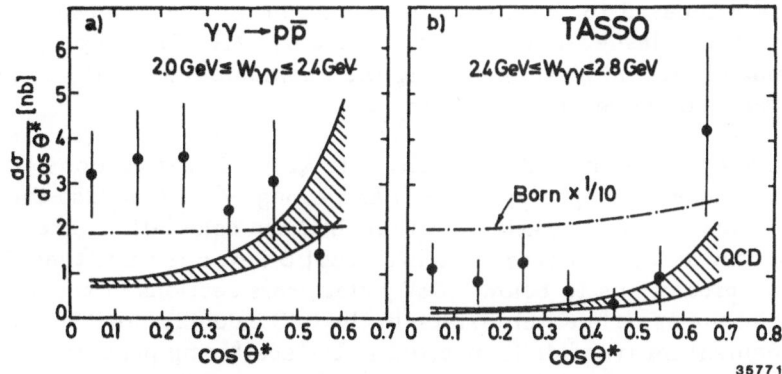

Figure 6. $d\sigma/d\cos\Theta^*$ for $\gamma\gamma\to p\bar{p}$:   a) $2.0 < W_{\gamma\gamma} < 2.4$ GeV and b) $2.4 < W_{\gamma\gamma} < 3.0$ GeV.   Data are compared to the Born approximation for a Dirac proton   (dashed-dotted   curve,   scaled   by   1/10)   and   to   QCD calculations /11/ (hatched   area).   The   band   for   the   QCD calculations   corresponds   to   different   choices   for   the   parton distribution in the protons.

with invariant masses between 2.0 and 3.1 GeV and with $\left|\sum\vec{p}_T\right| < 0.1$ GeV/c ($\left|\sum\vec{p}_T\right|$ is the transverse momentum of the detected system w.r.t. the beam). The background contributes less than 5 events.

In   Figure 6 the differential cross section for $\gamma\gamma\to p\bar{p}$ as a function of $|\cos\Theta^*|$ ($\Theta^* =$ polar angle of the p in the $\gamma\gamma$ rest system) is given for two $p\bar{p}$-mass intervals.   In both   intervals   the   differential   cross   section   is   consistent   with   a   flat $\cos\Theta^*$ dependence. The differential cross section averaged over the accepted Θ^* range is plotted in Figure 7 as a function of the $p\bar{p}$-mass.   The measured differential cross

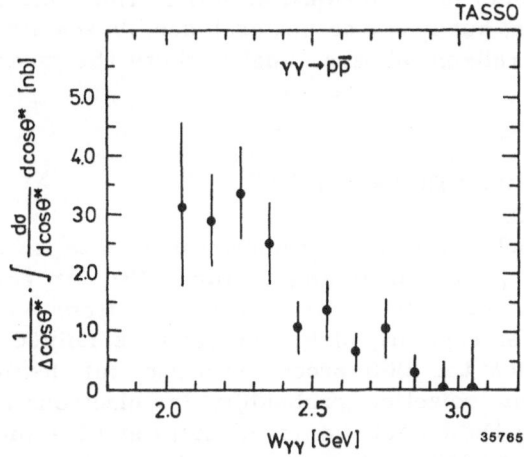

Figure 7. $d\sigma/d\cos\Theta^*$ for $\gamma\gamma\to p\bar{p}$, averaged over the accepted angular range, as a function of the $p\bar{p}$ mass.

sections are compared to the Born approximation for the process $\gamma\gamma\to p\bar{p}$ assuming Dirac protons (dashed-dotted curve, scaled by 1/10). This approximation is an order of magnitude too large. Including the anomalous magnetic moment of the proton yields even higher cross sections /10/.

The results are also compared to an absolute QCD calculation done by Damgaard /11/ with methods developed by Brodsky and Lepage /12/ (see also the discussion in Stirling's talk at this workshop). Although these calculations are meant to be meaningful only at higher energies, above about 3 GeV, one finds that the extrapolation of the predictions to below 3 GeV yields cross sections of the same order of magnitude as the experimental results. In the QCD calculations it is assumed that, if a high momentum transfer is involved in the scattering process, the scattering amplitude factorizes into a hard scattering amplitude T_H and a parton distribution function Φ:

$$M \sim \Phi^* \cdot T_H \cdot \Phi$$

The soft part, the parton distribution function Φ, cannot be calculated. For the form of parton distribution one has to make reasonable assumptions. The absolute normalisation of Φ is then obtained by comparing with other processes containing the same distribution function (in this case $J/\Psi\to p\bar{p}$). The hard scattering amplitude T_H can be calculated perturbatively by evaluating QCD diagrams. In /11/ the differential cross section for $p\bar{p}$ production was calculated for $W_{\gamma\gamma} = M_{J/\Psi}$. An extension to other energies is obtained from the dimensional counting rule according to:

$$\frac{d\sigma}{dt} = \frac{1}{s^6} f(t)$$

Here s,t are the usual Mandelstam variables for the process $\gamma\gamma\to p\bar{p}$. The function f(t) strongly rises towards the forward direction. This behaviour has not yet been observed in the data, but it is also not excluded. On the other hand, the QCD calculations are most reliable at large angles where the measurements have been done.

2.3   HADRON PAIR PRODUCTION ABOVE 2 GEV

The PLUTO group looked for charged hadron pairs with invariant masses above 2 GeV not separating pions, kaons and protons /13/. An earlier analysis yielded upper limits /14/. The new analysis uses data from ~40 pb^{-1}. The charged hadrons have not been explicitly identified, but were defined as not being electron or muon pairs. Since the QED processes are by far dominant, this procedure requires an excellent rejection probability for electrons and muons. That has been achieved using the barrel shower counters and the muon detection system of the PLUTO detector. Events with two oppositely charged tracks with $|\cos\Theta|<0.6$ and momenta larger than 0.9 GeV/c were selected. The photons were restricted to have low Q^2 values by the requirement that there was no tag in the small and large angle taggers. Exclusively produced pairs were selected by demanding

$|\sum \vec{p}_T| < 0.5$ GeV/c and that the tracks were coplanar with the beam within 6°. The process $\gamma\gamma \to e^+e^-$ has been rejected by a cut in the shower energies leading to about 20% loss of hadron pairs but virtually no remaining background from electrons. The remaining 987 events are mainly events from the reaction $\gamma\gamma \to \mu^+\mu^-$ with a small contribution from hadron pair production.

For the separation of the hadron pairs from the muons it was demanded that the particles should have momenta and angles such that if they were μ's they could be identified with a probability greater than 98%. 651 events fulfilled this requirement. Only 16 of them were not identified as $\mu^+\mu^-$. After subtraction of the estimated $\mu^+\mu^-$ background 15.1 hadron pair events were left yielding the ratio of hadron pair production to muon pair production:

$$\frac{\sigma(e^+e^- \to e^+e^-h^+h^-)}{\sigma(e^+e^- \to e^+e^-\mu^+\mu^-)} \ (W_{\pi\pi} > 2.0 \ \text{GeV}) = 0.042 \pm 0.013(\text{stat.}) \pm 0.008(\text{syst.})$$

This result is more than an order of magnitude smaller than expected for a pointlike production of hadrons, but it is in reasonable agreement with the QCD calculations of /12/ which predict for this ratio 0.020 ± 0.001. In Figure 8 the ratio of the two-photon produced hadron pairs to muon pairs is plotted versus the mass of the hadron pair assuming pion masses for all hadrons. The lowest $\pi\pi$ mass of 2 GeV corresponds to a KK-mass of 2.21 GeV and a p$\bar{p}$ mass of 2.73 GeV. The QCD calculation suggests that the hadron composition within the acceptance is 48% $\pi^+\pi^-$, 48% K^+K^- and 4% p$\bar{p}$. The QCD curve in Figure 8 was obtained with this composition assuming pion masses for all particles as in the data.

In conclusion, in this analysis and in the TASSO analysis of $\gamma\gamma \to p\bar{p}$ the cross sections predicted from perturbative QCD calculations have the correct order of magnitude.

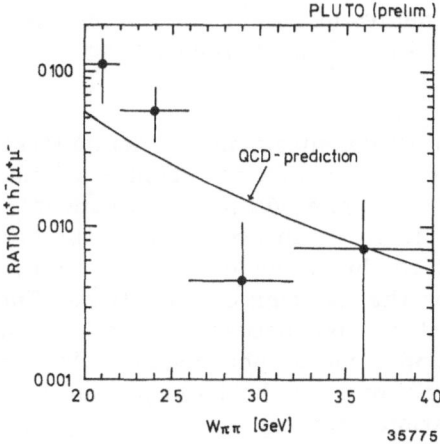

Figure 8. Ratio of two-photon production of charged hadron pairs to two-photon production of muon pairs as a function of the hadron pair mass assuming all hadrons to be pions.

3.0 TWO-PHOTON PRODUCTION OF FOUR PIONS

3.1 THE FOUR CHARGED PION FINAL STATE

The TASSO collaboration /15/ and the Mark II collaboration /16/ reported the observation of a strong enhancement in

$$\gamma\gamma \rightarrow \rho^\circ\rho^\circ \rightarrow \pi^+\pi^-\pi^+\pi^-$$

near the $\rho^\circ\rho^\circ$ threshold. The CELLO collaboration contributed an analysis of this channel to this workshop (see below).

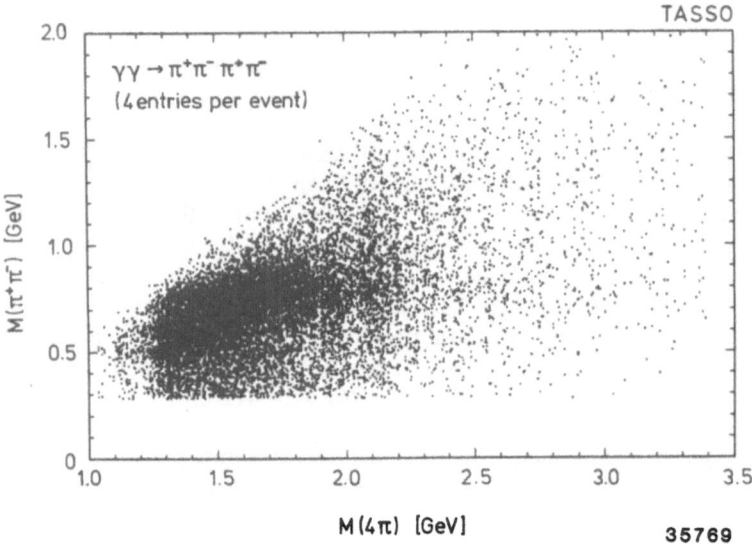

Figure 9.   Plot of the masses of all $\pi^+\pi^-$ combinations versus the four pion mass in events with four charged tracks.

The evidence for the $\rho^\circ\rho^\circ$ enhancement is demonstrated in Figure 9 and Figure 10. In Figure 9 the masses of all $\pi^+\pi^-$ combinations are plotted versus the four pion mass ($M_{4\pi}$). The ρ° signal seems to be concentrated between $M_{4\pi}$ ~1.2 GeV and ~2.2 GeV. Figure 10 shows in two $M_{4\pi}$-bins the correlations between the corresponding pairs of $\pi\pi$ combinations, on the left for the unlike sign combinations and on the right for the like sign combinations. The $\rho^\circ\rho^\circ$ signal is clearly visible even in the lowest $M_{4\pi}$-bin between 1.3 and 1.4 GeV, although the production is considerably suppressed by phase space effects below the nominal $\rho^\circ\rho^\circ$ threshold. Possible explanations for the large $\rho^\circ\rho^\circ$ cross section near threshold have been offered by various authors in the past /17, 18, 19, 20, 21/. If the enhancement is due to resonance production, the final state would have well defined quantum numbers, like isospin, spin and parity. A check on the isospin quantum number can be made by measuring the related channel $\gamma\gamma\rightarrow\rho^+\rho^-$ as will be discussed in the next section. A spin-parity analysis has been carried out by the TASSO group by studying the angular correlations in the four pion final state.

3.1.1   Angular Correlation Analysis (TASSO)

In the no-tag case the four pion final state is defined by 7 variables. One can choose two masses (m_{12}, m_{34}), the ρ-production angle $\vartheta_\rho{}^{12}$ and two angles for each decaying ρ ($\vartheta_\pi{}^{12}$, $\varphi_\pi{}^{12}$, $\vartheta_\pi{}^{34}$, $\varphi_\pi{}^{34}$). The indices 12 and 34 refer to a pion pair as defined if one numbers the pions as follows: $\pi^+{}_1$, $\pi^-{}_2$, $\pi^+{}_3$, $\pi^-{}_4$. The rotational properties of the $\rho^\circ\rho^\circ$ system with spin-parity J^P and helicity J_z are then given by:

$$\psi^{JP,Jz} \propto \sum a_{L,Lz,S12,S34}{}^{JP,Jz}\ Y_L{}^{Lz}(\vartheta_\rho{}^{12},)\ Y_1{}^{S12}(\vartheta_\pi{}^{12},\varphi_\pi{}^{12})\ Y_1{}^{S34}(\vartheta_\pi{}^{34},\varphi_\pi{}^{34})$$

The spherical harmonics describe the ρ-production and the decay of each ρ. The coefficients $a_{L,Lz,S12,S34}{}^{JP,Jz}$ (see table in /22/) are unambiguously defined if one assumes that only the lowest multipole in the $\gamma\gamma$-system and the lowest orbital angular momentum in the $\rho^\circ\rho^\circ$-system contribute. The matrix elements for $\rho^\circ\rho^\circ$-production (without the $W_{\gamma\gamma}$-dependence) were then defined by:

$$g_{\rho\rho}{}^{JP,Jz} = 1/\sqrt{2}\ \Bigg(\ BW(m_{12})\ BW(m_{34})\ \psi^{JP,Jz}(\vartheta_\rho{}^{12},\vartheta_\pi{}^{12},\varphi_\pi{}^{12},\vartheta_\pi{}^{34},\varphi_\pi{}^{34})$$

$$+\ BW(m_{14})\ BW(m_{32})\ \psi^{JP,Jz}(\vartheta_\rho{}^{14},\vartheta_\pi{}^{14},\varphi_\pi{}^{14},\vartheta_\pi{}^{32},\varphi_\pi{}^{32})\ \Bigg)\ .$$

BW denotes the ρ Breit-Wigner amplitude. This form is symmetric with respect to the interchange of identical bosons in the final state.

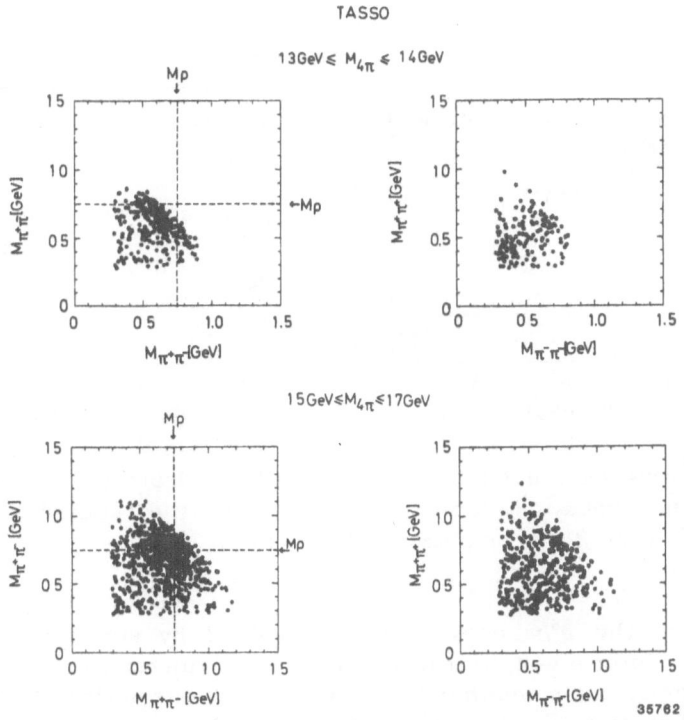

Figure 10.   Plot of one $\pi\pi$ mass combination versus the other one for $\pi^+\pi^-\pi^+\pi^-$ events in two different $M_{4\pi}$ intervals; left side: $\pi^+\pi^-$ vs. $\pi^+\pi^-$; right side $\pi^-\pi^-$ vs. $\pi^+\pi^+$.

The full matrix element, i.e. including all mass and angular correlations, has been used in maximum likelihood fits for the determination of the spin-parity contributions $J^P = 0^+$, 0^-, 2^+, 2^-. The fits have been carried out in 100 MeV wide $M_{4\pi}$-bins in the range $1.2 \leq M_{4\pi} \leq 2.0$ GeV. In the fits the four charged pion yields were described by a sum of the different spin-parity states for $\rho^\circ\rho^\circ$ plus additional contributions from $\rho^\circ\pi^+\pi^-$ and 4π phase space. In addition to this fit with 6 parameters also a fit with 3 parameters was tried where the $\rho^\circ\rho^\circ$ contribution was described by isotropic production and decay of the ρ's.

The results of the fits presented in the following have been obtained with the full integrated luminosity collected by the TASSO experiment until summer 82 (~80 pb^{-1}). For the published data /22/ about 40 pb^{-1} were used. The new analysis was done in the same way as the published one, except for a more stringent cut in $|\sum \vec{p}_T|$ (new analysis: $|\sum \vec{p}_T| < 70$ MeV/c; old analysis: $|\sum \vec{p}_T| < 150$ MeV/c). This cut reduces the background from events with undetected particles to about 7% (before: 17%).

Figure 11a shows the $\rho^\circ\rho^\circ$ cross section obtained from the fits with isotropic $\rho^\circ\rho^\circ$ production. The data are well described by such a production mechanism. The most striking feature of the cross section is that it stays high below the nominal $\rho^\circ\rho^\circ$ threshold, although the phase space decreases drastically. That means that the matrix element has to increase steeply below threshold (see discussion in /22/).

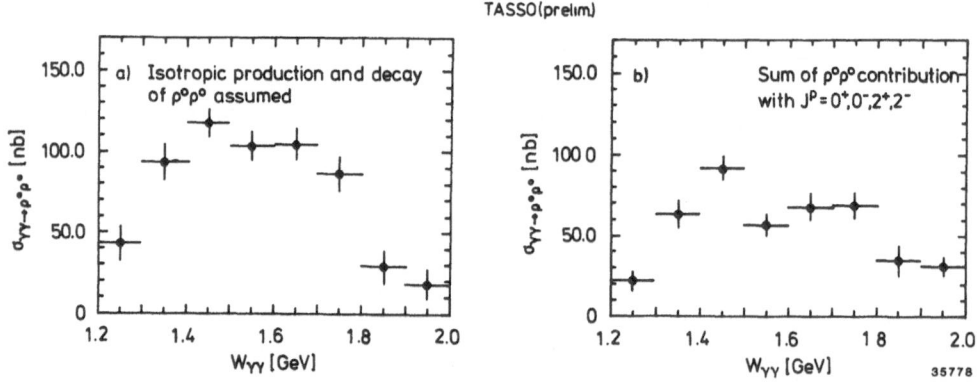

Figure 11.   Cross section for $\gamma\gamma \to \rho^\circ\rho^\circ$ (TASSO):   a) Isotropic $\rho^\circ\rho^\circ$ production and decay assumed;   b) Sum of the contributions from $J^P = 0^+, 0^-, 2^+, 2^-$ as obtained from the fit.

Figure 11b shows the $\rho^\circ\rho^\circ$ cross section obtained by summing the different J^P-contributions determined in the fits with 6 parameters. On the average the cross section comes out somewhat lower than in the fits with isotropic $\rho^\circ\rho^\circ$ production. That has two reasons: this fit assigns less events to $\rho^\circ\rho^\circ$ (more to $\rho^\circ\pi^+\pi^-$) and the detection efficiency for the sum of J^P-states is different than for the $\rho^\circ\rho^\circ$ phase space. The spin-parity decomposition of the cross section is shown in Figure 12. The following conclusions can be drawn from these plots: The spin-parity states $J^P = 0^-$ and 2^- are not dominant in the investigated $M_{4\pi}$ range. The $J^P = 0^+$ contribution is large except for the two highest $M_{4\pi}$ bins, whereas the contribution of $J^P = 2^+$ increases with $M_{4\pi}$ and is dominant in the two highest $M_{4\pi}$ bins.

Some words of caution have to be added concerning especially the region below the nominal $\rho^\circ\rho^\circ$ threshold. The results have been obtained with the special choice of the matrix elements as described above. There is no guarantee that the real world is like that. In fact, it has been reported in a parallel session at this workshop that there are indications of final state interactions at low $\rho^\circ\rho^\circ$ masses /23/. That was found by studying correlations in the $\pi\pi$-masses using TASSO data. Such final state interactions are not included in the matrix elements used in the fits.

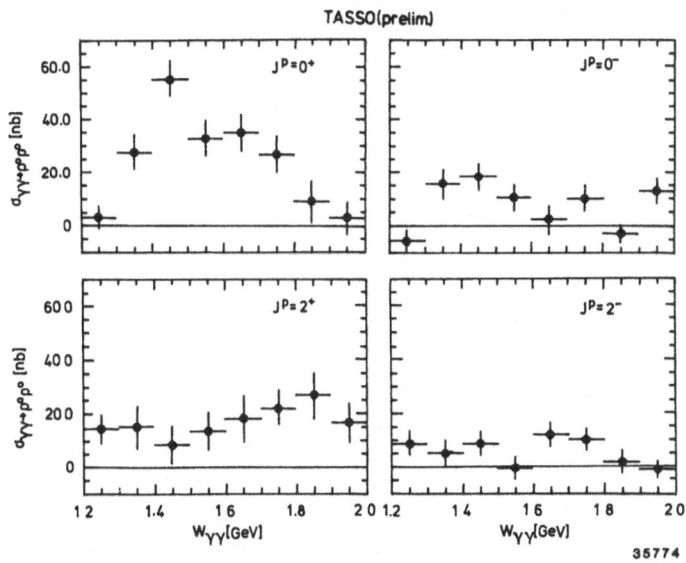

Figure 12.   Decomposition of the cross section for $\gamma\gamma\to\rho^\circ\rho^\circ$ into the different spin-parity contributions (TASSO) : $J^P = 0^+$, $J^P = 0^-$, $J^P = 2^+$ and $J^P = 2^-$

3.1.2 The CELLO Analysis of $\gamma\gamma\to\pi^+\pi^-\pi^+\pi^-$

An analysis of the four charged pion final state has been presented at this workshop by the CELLO collaboration /24/. They used data from an integrated luminosity of 11.3 pb^{-1}. The event selection can be briefly summarized as follows: The selected events had to have four charged tracks with momenta p≥120 MeV/c detected in a polar angular range IcosϑI≤0 95 (TASSO: IcosϑI≤0.84). The total transverse momentum of the four pion system was required to be $|\sum\vec{p}_T|$<120 MeV/c. After these cuts 835 events with an estimated background of 9% were found in the range 1.1 ≤ $M_{4\pi}$ ≤ 2.5 GeV.

In a first step the CELLO group determined the cross section for the production of four charged pions including possibleintermediate resonance states. To account for differences in the acceptance depending on whether resonances are formed or not, the acceptance was calculated as a function of the masses m_{12}, m_{34}, m_{14}, m_{32} of $\pi\pi$-combinations (the indices of the masses have been explained above). Since for a full determination of the system 7 variables are needed additional information from the measured angular distributions was included in the acceptance calculations. The ρ-production was assumed to be isotropic, as justified by the distribution in Figure 13a. Different decay angular distributions of the ρ's in the ρ-helicity system (Θ_π^H) were used depending on the ρ-production angle. Figure 13b shows that the decay angular distribution is flat for $|\cos\Theta_\rho| < 0.8$ and $\sim\sin^2\Theta_\pi^H$ (indicative for ρ's with helicity ±1) for $|\cos\Theta_\rho| > 0.8$. Figure 14a shows the four charged pion cross section obtained with these acceptance corrections.

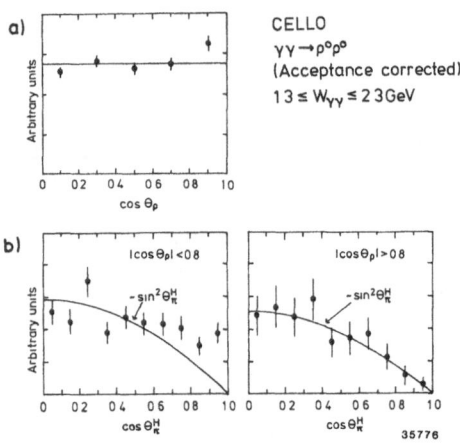

Figure 13. CELLO: Acceptance corrected angular distributions in the reaction $\gamma\gamma\rightarrow\rho^\circ\rho^\circ$ ($\rho^\circ\rho^\circ$ is defined by cuts in the $\pi^+\pi^-$ masses, $\Delta M = \pm100$ MeV) : a) Polar angular distribution of the produced ρ's w.r.t. the γ-direction in the $\gamma\gamma$ c.m.s.. b) Decay angular distribution of the ρ's in the ρ-helicity system for two $\cos\Theta_\rho$-intervals.

In a second step the CELLO group separated the three contributions $\rho^\circ\rho^\circ$, $\rho^\circ\pi^+\pi^-$ and 4π phase space in the range $1.1 \leq M_{4\pi} \leq 2.5$ GeV. For each contribution the number of events in 200 MeV wide $M_{4\pi}$-bins was determined by fitting the sum of the three contributions to the two density distributions $m(\pi^+\pi^-)$ vs. $m(\pi^+\pi^-)$ and $m(\pi^+\pi^+)$ vs. $m(\pi^-\pi^-)$.

The $\rho^\circ\rho^\circ$ cross section is shown in Figure 14a. Shape and height of the cross section are in good agreement with the TASSO result. The $\rho^\circ\rho^\circ$ channel accounts for about half the four charged pion cross section. The cross sections for $\rho^\circ\pi^+\pi^-$ and 4π phase space production are shown in Figure 14b. Note that the $\rho^\circ\pi^+\pi^-$ cross section, though of similar size, does not exhibit a threshold behaviour as dramatic as observed for $\rho^\circ\rho^\circ$, since the $\rho^\circ\pi^+\pi^-$ threshold is much lower.

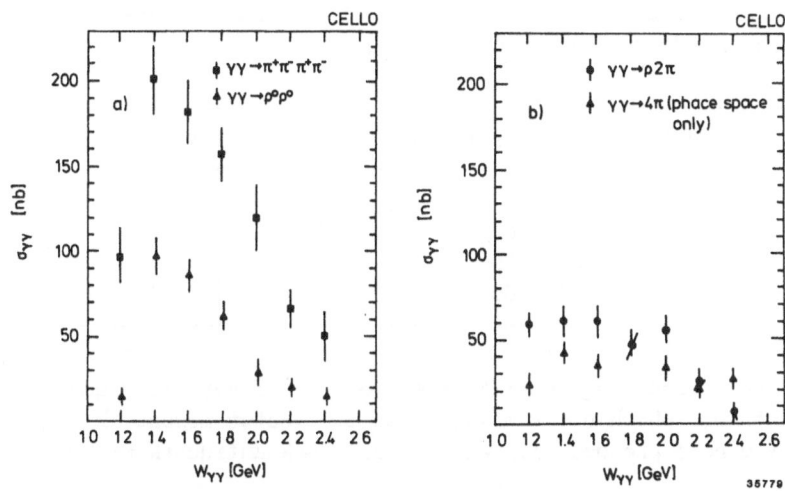

Figure 14. CELLO: a) Cross section for $\gamma\gamma\to\pi^+\pi^-\pi^+\pi^-$ (all 4 pion final states) and $\gamma\gamma\to\rho^\circ\rho^\circ$ b) Cross section for $\gamma\gamma\to\pi^+\pi^-\pi^+\pi^-$ (phase space only) and $\gamma\gamma\to\rho^\circ\pi^+\pi^-$.

3.1.3 Summary of the Results on the Four Charged Pion Final State

The results of the measurements of the four charged pion final state are: The large cross section for $\rho^\circ\rho^\circ$ production extends below the nominal $\rho^\circ\rho^\circ$ threshold. A sizeable $\rho^\circ\pi^+\pi^-$ cross section is observed. The partial wave analysis for $\rho^\circ\rho^\circ$ yields: $J^P = 0^-$ and 2^- are not dominant; $J^P = 0^+$ dominates below ~1.7 GeV and $J^P = 2^+$ above ~1.7 GeV.

What could be a possible explanation of a $\rho^\circ\rho^\circ$ enhancement with spin-parity 0^+ and 2^+ ? One interesting possibility, suggested in /20, 21/, is to explain the effect with the production of four quark bound states /25/:

$$\gamma\gamma \;\to\; qq\bar{q}\bar{q} \;\to\; \rho^\circ\rho^\circ$$

This interpretation accounts for both the threshold enhancement and for the measured spin-parity structure. The four quark states are ordered in multiplets. The lowest multiplet is a $J^{PC} = 0^{++}$ nonet with masses around 1 GeV. It has been suggested that the scalar states $\varepsilon(700)$, $S^*(980)$ and $\delta(980)$ belong into this multiplet /26, 27/. These states couple dominantly to a pair of pseudoscalar mesons (PP) and the $\gamma\gamma$-width should be suppressed (up to now in agreement with experiment). According to VMD we expect the largest couplings to $\gamma\gamma$ for the multiplets which have a dominant coupling to pairs of vector mesons (VV). Table 1 gives a summary of some expected multiplets with the average masses and the relative amplitudes for the couplings to pairs of pseudoscalar and vector mesons ('recoupling coefficients'). In general, a four quark state should be broad because it can easily decay into pairs of $q\bar{q}$ mesons. An exception could be, for example, the 0^{++}-state at about 1.45 GeV (Table 1), which couples dominantly to $\rho^\circ\rho^\circ$. This state is predicted to have a relatively narrow width (i.e. typical hadronic width) because it lies below the threshold for its fall apart pieces ($\rho°\rho°$) /26/. Above threshold the widths become gradually larger.

Multiplet	$J^{PC}(I)$	Mass [GeV]	PP	VV
9*	$0^{++}(0)$	1.45	−0.177	0.644
36*	$0^{++}(0,2)$	1.80	0.041	0.743
9	$2^{++}(0)$	1.65	0	0.816
36	$2^{++}(0,2)$	1.65	0	0.577

Table 1. The recoupling coefficients of the S-wave four quark
states which decay mainly into two vector mesons. P and
V denote pseudoscalar and vector quark-antiquark states.

Many other explanations for the large $\rho°\rho°$ production have been discussed. A clue for the solution of this question is expected from measurements of the production of other vector meson pairs. Table 2 shows that different models can differ largely in their predictions for the relative yields of vector meson pair production. For the following discussion I want to emphazise that a resonance with isospin 0 decays twice as frequently into $\rho^+\rho^-$ than into $\rho°\rho°$.

Model	$\rho°\rho°$	$\rho^+\rho^-$	$\omega\omega$	$\rho°\omega$
VMD	1	0	1/81	1/9
VMD /18/	1	–	5.8	0.5
Quark model	1	2	1	36/25
Quark model /19/	1	4/25	allowed	allowed
Resonance (I=0)	1	2	allowed	0
$qq\bar{q}\bar{q}$ /20/	1	~0	–	~1
$qq\bar{q}\bar{q}$ /21/	1	~0	0.1	0.05

Table 2. Ratios of the cross sections for $\gamma\gamma\to VV'$ to $\gamma\gamma\to\rho°\rho°$
as predicted from different models.

3.2 THE JADE ANALYSIS OF $\gamma\gamma\rightarrow\pi^+\pi^-\pi^0\pi^0$

The JADE group analyzed $\gamma\gamma\rightarrow\rho^+\rho^-$ studying the final state $\pi^+\pi^-\pi^0\pi^0$ in a data sample collected for an integrated luminosity of 77 pb^{-1} /28/. Events were selected requiring two charged tracks with momenta p > 100 MeV/c and four photons with shower energies E_γ > 60 MeV each. The evidence for two correlated π^0's in an event is shown in Figure 15, where one $\gamma\gamma$-mass is plotted versus the other. Photon pairs with invariant masses between 60 and 220 MeV were considered to be π^0's. For the subsequent analysis the kinematics of the two photons was fitted with a constraint on the π^0 mass.

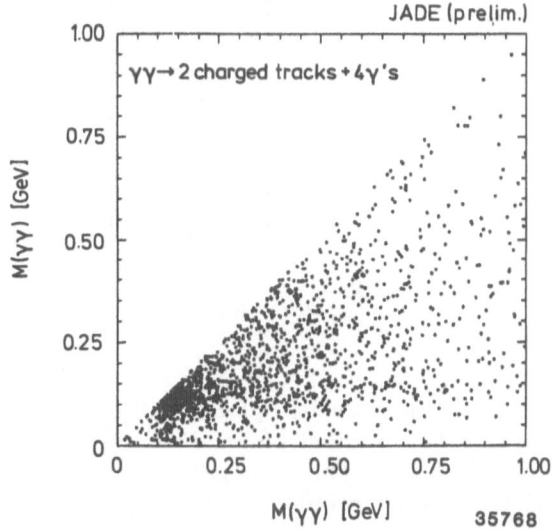

Figure 15. Plot of one $\gamma\gamma$ mass versus the other for events with 4 photons and two charged tracks detected (JADE). There are 3 entries per event.

Plotting the masses of the $\pi^+\pi^0$ and $\pi^-\pi^0$ combinations (Figure 16) one obtains a signal at the ρ-mass, which is not seen in the neutral combinations $\pi^+\pi^-$ and $\pi^0\pi^0$. The absence of a ρ^0 signal in the $\pi^+\pi^-$ combination is easily explained: Since the four pion final state with charge conjugation C=+1 has even isospin I, the presence of a ρ^0 with I=1 requires that the $\pi^0\pi^0$ combination be also in an I=1 state which is not possible for $\pi^0\pi^0$. Although there is a charged ρ signal, the two-dimensional distribution of $\pi^+\pi^0$-masses versus $\pi^-\pi^0$-masses in Figure 17a does not show a clear evidence for correlated $\rho^+\rho^-$ production. For comparison the corresponding plot for the neutral mass combinations is shown in Figure 17b.

The measured event rate as a function of the 4 pion mass is plotted in Figure 18 (open histogram). The hatched histogram is the rate restricted to the $\rho^+\rho^-$-band (0.55 < m_ρ < 0.95 GeV). The observed shift of the distribution after this cut may be just a phase space effect. The cross section derived from the events in the $\rho^+\rho^-$-band is plotted in Figure 19. Since the $\rho^+\rho^-$ rate has not been explicitly determined, this cross section has to be considered as an upper limit for

$\sigma(\gamma\gamma \rightarrow \rho^+\rho^-)$. The errors in the plot are statistical only; the systematic errors are on the order of 30%.

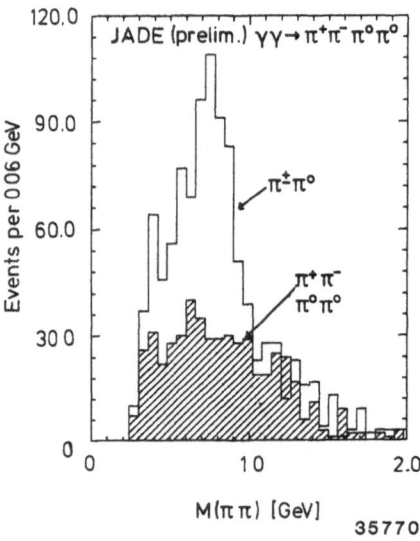

Figure 16. Mass distributions for two pion combinations in events with two charged and two neutral pions (JADE). Open histogram: charged pion pairs (4 entries per event), hatched histogram: neutral pion pairs (2 entries per event).

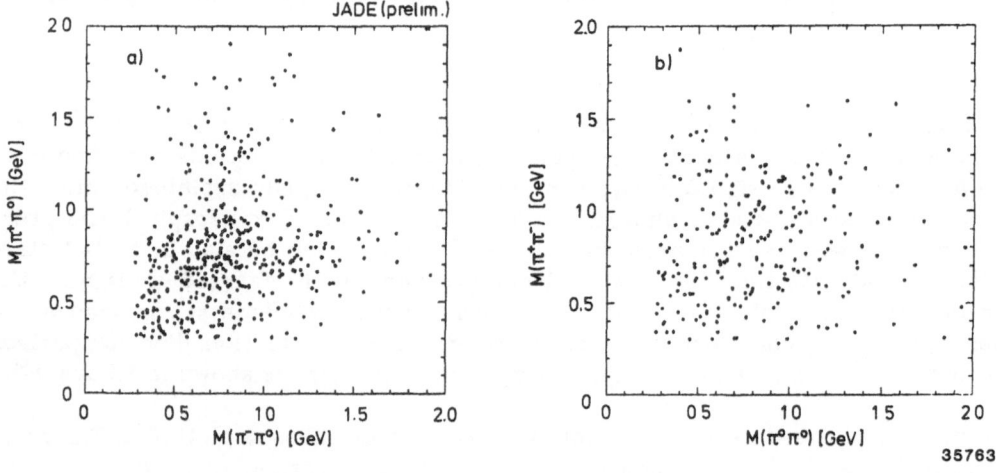

Figure 17. Two-dimensional plot of the $\pi\pi$ masses shown in Figure 16 (JADE). (a: charged pion pairs, 2 entries per event; b: neutral pion pairs, 1 entry per event).

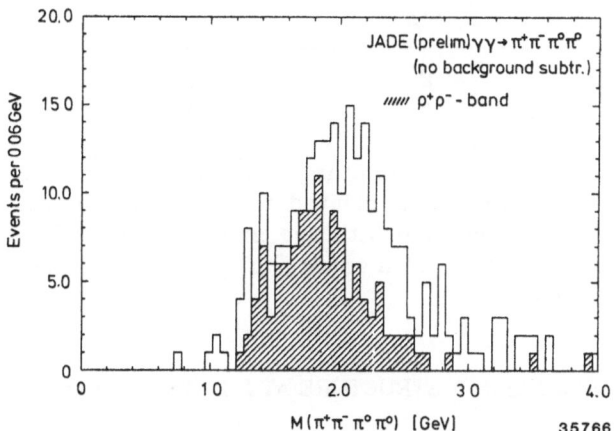

Figure 18. Four pion mass spectrum of the reaction $\gamma\gamma \to \pi^+\pi^-\pi^0\pi^0$ (JADE). The hatched histogram is the same spectrum with mass combinations in the $\rho^+\rho^-$ band.

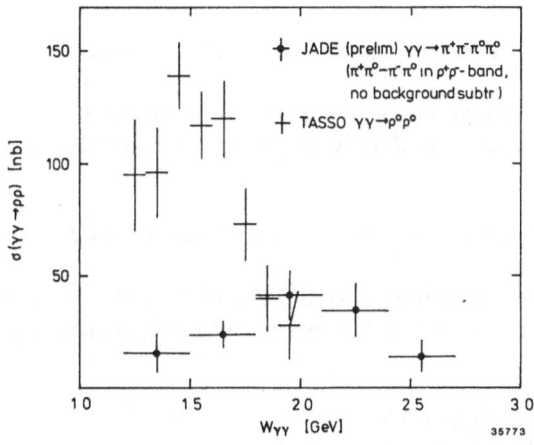

Figure 19. Cross section for $\gamma\gamma \to \rho^+\rho^-$ obtained from the hatched histogram in previous figure (JADE, preliminary). Also shown is the cross section for $\gamma\gamma \to \rho^0\rho^0$ from TASSO /22/.

A comparison of this cross section with the $\rho^0\rho^0$ cross section shows very clearly that both cross sections have a totally different behaviour at threshold. That means in particular that <u>the $\rho^0\rho^0$ enhancement cannot be due to a single resonance</u> with well defined isospin. The relative rates expected for the decay of a resonance with isospin I (I = 0, 2) into pairs of neutral and charged ρ's are:

I	$\rho^+\rho^-$	$\rho^\circ\rho^\circ$
0	2	1
2	1	2

Which possibilities are now remaining for the explanation of the $\rho^\circ\rho^\circ$ enhancement? Models, which are related to VMD, will always prefer $\rho^\circ\rho^\circ$- over $\rho^+\rho^-$-production and thus conform with the measurements. The four-quark models of /20/ and /21/[1] also predict a small $\rho^+\rho^-$ cross section due to interferences between different states.

3.3   OBSERVATION OF A NARROW STRUCTURE AT 2.1 GEV IN $\gamma\gamma\rightarrow\pi^+\pi^-\pi^+\pi^-$

The TASSO collaboration observed a narrow structure around 2.1 GeV in the four charged pion mass spectrum (Figure 20) /29/. A smooth curve was fitted to the mass spectrum between 1.65 and 3.0 GeV including the signal region. For masses between 2.05 and 2.15 GeV the observed event rate is 4.3 standard deviations larger than expected from this fit. Including a Breit-Wigner function in addition to the smooth curve the fit gives a good description of the data and yields for the Breit-Wigner parameters, mass and width, and for the number of events in the peak:

$M=2.103\pm0.01$ GeV, $\Gamma=0.030\pm0.034$ GeV, $N=125.6\pm46$

In this fit the 4 pion mass resolution, which is about 60 MeV (FWHM) near 2 GeV, has been taken into account. If this structure is indeed a resonance its $\gamma\gamma$ coupling is:

$\Gamma_{\gamma\gamma}\cdot(2J+1)\cdot B(4\pi^\pm) = 1.25 \pm 0.5$ (stat.) ± 0.5 (syst.) keV

J is the spin of the resonance and $B(4\pi^\pm)$ the decay branching ratio into four charged pions. Upper limits for the branching ratios into $\rho^\circ\rho^\circ$ and $K_0\bar{K}_0$ are found to be

$B(\rho^\circ\rho^\circ)/B(4\pi^\pm) < 0.6$ (95% c.l.),
$B(K_0\bar{K}_0)/B(4\pi^\pm) < 0.06$ (95% c.l.).

The signal does not seem to be correlated with the large $\rho^\circ\rho^\circ$ enhancement observed between threshold and $M_{4\pi} \sim 2$ GeV. Removing events which have a $\pi^+\pi^--\pi^+\pi^-$ combination in the $\rho^\circ\rho^\circ$-band (cut in the ρ°-mass: ±150 MeV) the $M_{4\pi}$ distribution in Figure 21a was obtained. Although the signal appears more pronounced in this plot a quantitative analysis is more difficult because the effect of the applied cut on the background introduces uncertainties. The signal to back-

---

[1]   At the time of the workshop there was a discrepancy between the predictions for the $\rho^+\rho^-$ cross section from /20/ and /21/. Meanwhile the authors of /21/ corrected their calculations.

ground ratio is improved if one plots only events which appear planar in the laboratory system (Figure 21b). "Planar" is defined by $Q_1 < 0.05$, where Q_1 is the smallest eigenvalue of the sphericity tensor. Monte Carlo studies showed that the latter cut also enhances 4π phase space over $\rho°\rho°$ final states. The effects of these kinematical cuts are still under study by the TASSO group.

If the observed narrow structure will be confirmed as a resonance it cannot easily be associated with a known state with similar resonance parameters. The nearest candidate would be the h(2040), but it does not seem to have the same resonance parameters. Note however, that the authors of /30/ predicted a relatively large $\gamma\gamma$-width for the h(2040), which they called in the paper the $f°$-recurrence f^*.

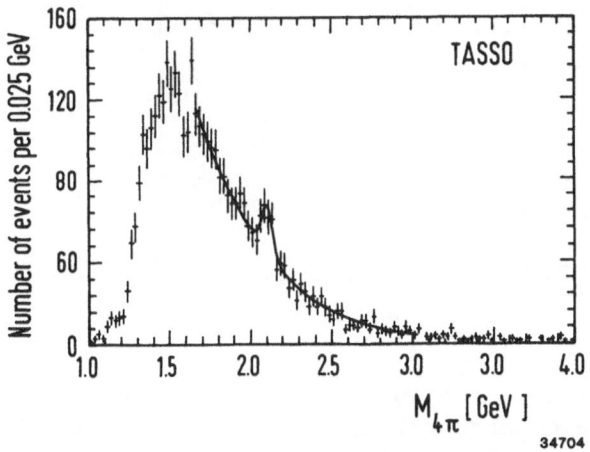

Figure 20.   Four charged pion mass spectrum (TASSO).

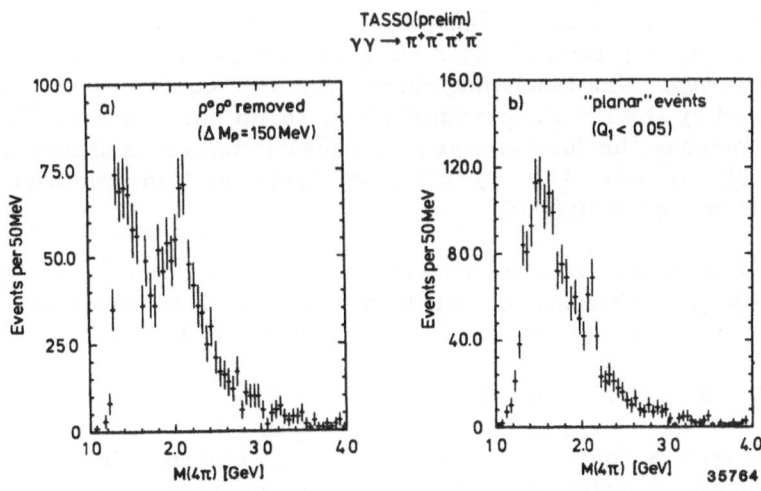

Figure 21.   Four charged pion mass spectrum (TASSO):   a) $\rho°\rho°$ events removed; b) events with $Q_1 < 0.05$

4.0 THE TOTAL CROSS SECTION FOR TWO-PHOTON PRODUCTION OF HADRONS AT LOW Q^2

4.1 EXCLUSIVE FINAL STATES AND THE TOTAL CROSS SECTION

The large amount of experimental results on exclusive final states in two-photon scattering enables us to compare the exclusive measurements with the experimental determinations of the total $\gamma\gamma$ cross section and theoretical expectations.

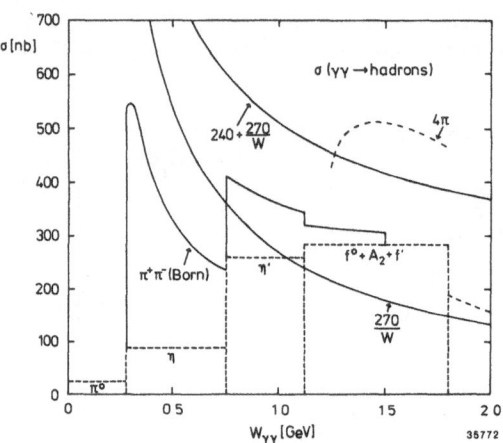

Figure 22. Sum of measured $\gamma\gamma$ cross sections compared to predictions from Regge model calculations /31/. The sum of cross sections includes resonance production, the Born cross section for $\gamma\gamma\to\pi^+\pi^-$ and 4 pion production, $\gamma\gamma\to\pi^+\pi^-\pi^+\pi^-$ and $\gamma\gamma\to\pi^+\pi^-\pi^0\pi^0$ (restricted to $\rho^+\rho^-$-band).

In Figure 22 the measured exclusive channels are summed up. The resonances are schematically represented by rectangular boxes with areas equal to those under the corresponding resonance curve. The $\pi^+\pi^-$ continuum was assumed to be represented by the Born approximation up to 1.5 GeV. The curve for four pion production includes the four charged pion production cross section taken from Figure 14 (CELLO) and the $\gamma\gamma\to\pi^+\pi^-\pi^0\pi^0$ cross section (restricted to the $\rho^+\rho^-$-band) from Figure 19 (JADE).

The measured exclusive cross sections are compared to estimates using the Regge model. Relating $\gamma\gamma$ scattering to γ-nucleon and nucleon-nucleon cross sections at high energies the $\gamma\gamma$ total cross section was predicted to be /31/:

$$\sigma_{\gamma\gamma} = (240 + 270/W) \text{ nb} \quad \text{(W in GeV).}$$

The constant term accounts for pomeron exchange and the $1/W$-term arises from the leading Regge trajectories (f^0, A_2). Contributions from non-leading Regge trajectories (like the f'-trajectory) giving rise to terms with higher powers in $1/W$ were estimated to be small. Via duality the $1/W$-term should reproduce the average of the resonance cross sections. In fact, the measured resonances seem to saturate roughly this predicted cross section (Figure 22).

On the other hand, it has been suggested that in addition to the Regge contributions there may be non-Regge terms arising from the pointlike coupling of the photon. According to /32/ a rough estimate for this contribution could be the simple quark loop box diagram. That would add a term $\sim 1/W^2$ to the total cross section. In /32/ it was argued that such an additional term was needed to account for the resonance contributions, which had been estimated using superconvergent sum rules. Now the measured cross sections for the resonances show that this argument is no longer valid. The superconvergent sum rule discussed in /32/ requires a $\gamma\gamma$-width of the f^0 of ~ 9 keV, which has to be compared to the measured value of ~ 3 keV (see also discussion in /33/). One may wonder if this too large estimate is caused by the assumption that below ~ 2 GeV the total cross section is saturated by resonances (i.e. the pseudoscalar, scalar and tensor $q\bar{q}$-resonances). Does the discrepancy between the predicted and the measured $\gamma\gamma$-width of the f^0 mean that there must be additional production of 2^+-states? For example, what is the role of four-quark states, in the estimates of resonance cross sections via duality arguments as done in /31/? Since a lot of experimental results on $\gamma\gamma$ resonance production and other exclusive final states are now available, these questions should be reinvestigated.

4.2 THE EXPERIMENTAL DETERMINATION OF THE TOTAL HADRONIC CROSS SECTION

Measurements of the $\gamma\gamma$ total cross section have a major disadvantage compared to the measurements of the one-photon annihilation cross section: The center-of-mass energy, $W_{\gamma\gamma}$, of an event is not known because the photons provide a continuous spectrum of $W_{\gamma\gamma}$-values. Hence $W_{\gamma\gamma}$ has to be measured. In a double-tag experiment $W_{\gamma\gamma}$ is in principle determined by the measured kinematics of the scattered leptons. But it is difficult to obtain sufficient resolutions in the tagging devices. For typical forward detectors at PETRA and PEP the resolution is not good enough for measuring the total cross section in the resonance region. In the single-tag or no-tag case $W_{\gamma\gamma}$ has to be reconstructed from the measured hadrons. Because of the imperfect detection the measured $W_{\gamma\gamma}$ (W_{vis}) will in general be smaller than the real $W_{\gamma\gamma}$. The determination of the $W_{\gamma\gamma}$ dependence of the cross section therefore requires the use of a model, which realistically describes the hadronic final states. The ingredients of such models are for example:

- a multibody phase space with limited transverse momenta of the particles w.r.t. the direction of the photons (p_T), $d\sigma/dp_T^2 \sim \exp\text{-}ap_T^2$ or $\sim \exp\text{-}bp_T$;

- the average multiplicities as a function of $W_{\gamma\gamma}$ for different particle species (charged and neutral π's, K's, ρ's,...);

- multiplicity distributions;

- the Q^2-dependence of the cross section.

The Q^2-parametrization is also needed for the extrapolation to the cross section for real photons. In Figure 23 the dependence of the cross section on Q^2 obtained from simple ρ-pole dominance is compared to the expectation from a generalized vector meson dominance model (GVMD) /34/. For the Q^2 range covered by typical small angle taggers the change in the extrapolated cross section is about 30 to 50% in the single-tag case.

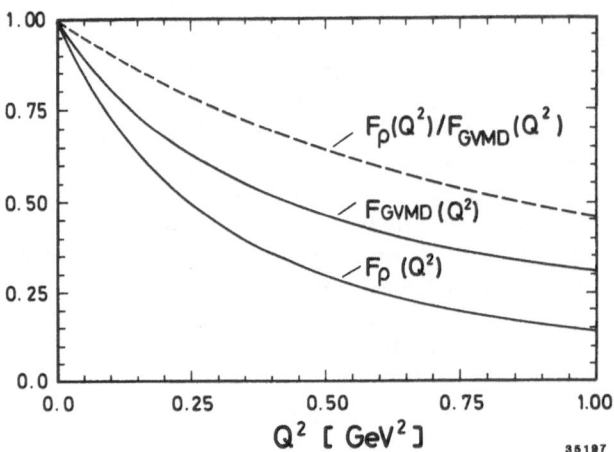

Figure 23. Q^2 dependence of the two-photon cross section for ρ-pole dominance and GVMD. The dashed curve is the ratio of both (F_ρ/F_{GVMD}).

There have been two early analyses of the total cross section using the single tag method: the published analysis of the PLUTO group /35/ and a preliminary analysis of the TASSO group /36/. The discrepancies between both analyses, especially in the low $W_{\gamma\gamma}$ region, have been often discussed in the past. At low $W_{\gamma\gamma}$ the PLUTO group found a steep increase of the cross section described by a large coefficient of the $1/W^2$-term in a parametrization of the cross section according to:

$$\sigma_{\gamma\gamma} = A + B/W + C/W^2$$

In the preliminary TASSO analysis the inclusion of a $1/W^2$-term was not necessary.

To understand the reason for this discrepancy the TASSO group carried out a detailed study of the model dependence of the total cross section determination /37/. This study was done with data corresponding to an integrated luminosity of $\sim$9 pb^{-1}, which is about 3 times the amount available for the previous analysis. Events with at least 3 charged tracks were selected for the analysis; neutral energy measurements were not available for these data.

With these data an attempt was made to fit simultaneously the model parameters, such as the p_T-slope, the average charged multiplicity and the different cross section terms, by comparing experimental distributions to the Monte Carlo generated distributions. For this purpose a method was developed which supplied the contents of the Monte Carlo distributions as continuous functions of the parameters. This procedure allows one to study the correlations between the parameters.

The result of the fits was rather disappointing: The correlations between the parameters turned out to be so large that a reliable unfolding of the $W_{\gamma\gamma}$ distribution was not possible. The stability of the fit could be improved if some of the parameters, like the p_T-slope and the multiplicity, were fixed. But this approach is of questionable validity since these parameters are in principle unknown.

This study was done with a specific detector. It is clear that a higher detection efficiency for both charged and neutral particles will improve the results of the unfolding procedure. However, some conclusions are of more general validity: The study has shown that the determination of the total cross section is more complicated than assumed in the first analyses. Only a <u>simultaneous fit of all model parameters</u> can reveal correlations and allows to prove the significance of a result for the W-dependence of the cross section. The significance of the result is considerably increased if one can show that all distributions, especially the p_T and the multiplicity distributions, are equally well described for all W_{vis} values. In particular, the determination of the p_T-slope before the proper unfolding, without regarding correlations with other parameters and averaging over all W_{vis} values, was found to be dangerous.

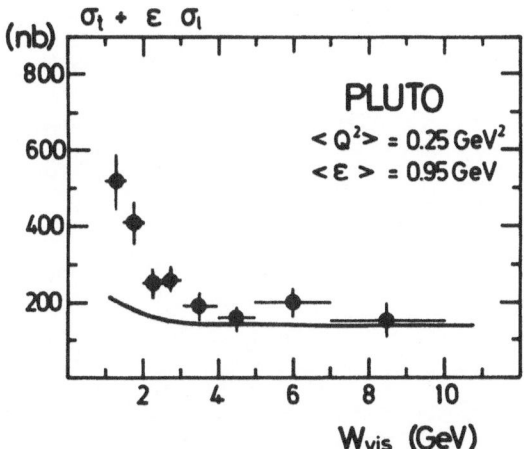

Figure 24.   The total cross section versus the visible invariant mass (Pluto): $\sigma_{TT} + \varepsilon\sigma_{LT}$ at $\langle Q^2 \rangle = 0.25$ GeV^2. The prediction of the Regge estimate is given by the solid curve.

The determination of the cross section below $W_{\gamma\gamma} \sim 2$ GeV is particularly difficult (see also the discussion in /38/). Figure 24 demonstrates that the large $1/W^2$-term in the cross section found by the PLUTO group is required only by the data points below $W_{\gamma\gamma} \sim 2$ GeV. However, in the resonance region one has to expect special problems: The acceptance for events below 2 GeV is only ~8% /38/. On the other hand, about 85% of the events observed at $W_{vis} < 2$ GeV come from higher $W_{\gamma\gamma}$ values due to smearing /39/. In such a situation the cross section cannot be obtained applying simple acceptance corrections as it was done in Figure 24, where the cross section is given as a function of W_{vis}. A proper unfolding procedure is necessary as described e.g. in /37/. Concerning the applicability of the used model to the resonance region, one has to question whether a multi-pion phase space model with limited p_T and some statistical multiplicity distribution works well for resonances. Another uncertainty has been mentioned in /38/: At low $W_{\gamma\gamma}$ the Q^2 extrapolation to $Q^2=0$ is especially large, about a factor 3 in the PLUTO analysis.

My personal conclusion is that the total hadronic cross section for two real photons below $W_{\gamma\gamma} \approx 2$ GeV is not yet known. A similar conclusion was drawn by Ch.Berger at the Paris meeting /38/.

New results are expected from JADE, the improved PLUTO detector and the PEP4/PEP9 experiment. Recently, the ARGUS collaboration proposed the installation of a 0°-tagger at DORIS /40/. This device is supposed to achieve a resolution good enough to measure the total cross section down to the resonance region in a double-tag experiment.

5.0 SUMMARY

HADRON PAIR PRODUCTION. The $\pi^+\pi^-$ continuum seems to be well approximated by the Born term up to the f^0 region. To search for scalar resonances below 1 GeV one should look for deviations from the Born cross section. The charged hadron pair production above 2 GeV is roughly in agreement with absolute QCD calculations for pion, kaon and proton pairs. Differential cross sections for $\gamma\gamma\to p\bar{p}$ for $2.0 < M(p\bar{p}) < 3.1$ GeV have been presented.

FOUR PION PRODUCTION. The most interesting new result on this reaction: The strong $\rho^0\rho^0$ threshold enhancement has no counterpart in $\rho^+\rho^-$. That excludes an interpretation of the $\rho^0\rho^0$ enhancement as one single resonance. A partial wave analysis of $\rho^0\rho^0$ yielded large $J^P = 0^+$ intensities below ~1.7 GeV and $J^P = 2^+$ intensities above ~1.7 GeV. A four-quark model predicts roughly this structure. The narrow structure observed in the four charged pion mass spectrum around 2.1 GeV waits for confirmation by other experiments.

THE TOTAL TWO-PHOTON CROSS SECTION. The measured resonances are approximately in agreement with the estimated contributions from leading Regge trajectories. Some inconsistencies between sum rule results for resonance production by two photons and the now available measurements have been pointed out. Future analyses of the total cross section will show whether the experimental problems, which have been encountered in the past with unfolding the W_{vis} distribution, can be overcome.

I am indebted to many collegues for valuable discussions and support in preparing this talk. In particular, I want to mention here: H.J.Behrend, Ch.Berger, S.Cooper, E.Hilger, R.Kellogg, L.Köpke, H.Kück, J.Olsson, W.Wagner, R.Wedemeyer, N.Wermes and M.Wollstadt. I would like to thank the organizers of the workshop for an extremely pleasent atmosphere in and around the workshop.

REFERENCES

/1/   TASSO Collaboration, M.Althoff et al., Phys.Lett. 121B (1983) 216;
/2/   G.Mennessier, Z.Phys. C16 (1983) 241
/3/   A.Courau et al., Phys.Lett. 96B (1980) 402;
       R.Wedemeyer, Proc. of the Intern. Symposium on Lepton and Photon Inter-
       actions at High Energies, Bonn, 1981 and paper No.48 submitted to the con-
       ference
/4/   A.Courau, private communication
/5/   V.M.Budnev et al., Phys.Rep. 15 (1975) 181
/6/   A.Roussarie et.al., Phys. Lett. 105B (1981) 304
/7/   F.Kovacs (CELLO), presented at the experimental parallel session at this
       workshop
/8/   TASSO Collaboration, R.Brandelik et al., Phys. Lett. 108B (1982) 67
/9/   H.Kück (TASSO), presented at the experimental parallel session at this
       workshop
/10/   N.Arteaga-Romero, Seminar on $\gamma\gamma$-Physics, LPC/82-14, Paris (1982)
/11/   P.H.Damgaard, Nucl.Phys. B211 (1983) 435.
/12/   G.P.Lepage and S.J.Brodsky, Phys.Rev. D21 (1980) 2157
/13/   R.Kellogg (PLUTO), presented at the experimental parallel session at this
       workshop
/14/   PLUTO Collaboration, Ch.Berger et al., Nucl.Phys. B202 (1982) 189
/15/   TASSO Collaboration, R.Brandelik et al., Phys.Lett. 97B (1980) 448
       TASSO Collaboration, M.Althoff et al., Z.Phys. C16 (1982) 13
/16/   D.L.Burke et.al., Phys. Lett. 103B (1981) 153
/17/   J.Layssac and F.M.Renard, Montpelier preprint PM/80/11 (1980)
       H.Goldberg and T.Weiler, Phys. Lett. 102B (1981) 63;
       R.M.Godbole and K.V.L.Sarma, Phys. Lett. 109B (1982) 504;
       S.Minami, Lett. Nuov. Cim. 34 (1982) 125;
/18/   G.Alexander, U.Maor and P.Williams, Phys.Rev. D26 (1982) 1198
/19/   K.Biswal and S.P.Misra, Phys.Rev. D26 (1982) 3020
/20/   N.N.Achasov, S.A.Devyanin, and G.N.Shestakov, Phys. Lett. 108B (1982) 134;
       Z.Phys. C16 (1982) 55;
/21/   Bing An Li and K.F.Liu, Phys.Lett. 118B (1982) 435 and Erratum, Phys.Lett.
       124B (1983) 550.
/22/   TASSO Collaboration, M.Althoff et al., Z.Phys. C16 (1982) 13
/23/   A.Shapira (TASSO), presented at the experimental parallel session at this
       workshop
/24/   H.J.Behrend (CELLO), presented at the experimental parallel session at this
       workshop
/25/   R.L.Jaffe, Phys.Rev. D15 (1977) 267 and 281;
       R.L.Jaffe and K.Johnson, Phys.Lett. 60B (1976) 201
/26/   M.S.Chanowitz, Lectures given at the SLAC Summer Institute 1981
/27/   L.Montanet, Rep.Prog.Phys. 46 (1983) 337
/28/   J.Olsson (JADE), presented at the experimental parallel session at this
       workshop
/29/   D.Lüke, XXI International Conference on High Energy Physics, Paris(1982);
       H.Kolanoski, Proc. of the Seminar on $\gamma\gamma$ Physics, Montpellier, Dec. 9-10,
       1982
/30/   B.Schrempp et al., Phys.Lett. 36B (1971) 463.

/31/   J.L.Rosner, Brookhaven report CRISP 71 26 (1971)

/32/   M.Greco and Y.Srivastava, Nuovo Cim. 43A (1978) 88;
       M.Greco, Talk given at the International Workshop on $\gamma\gamma$-Collisions
       (Amiens), Lecture Notes in Physics No 134, Springer (1980)

/33/   J.Field, Proc. of the Seminar on $\gamma\gamma$ Physics, Montpellier, Dec. 9-10, 1982

/34/   I.F.Ginzburg and V.G.Serbo, Phys.Lett. 109B (1982) 231

/35/   PLUTO Collaboration, Ch.Berger et al., Phys.Lett. 89B (1981) 287
       F.A.Raupach, Thesis, DESY PLUTO 81/10 (1981)

/36/   E.Hilger, Talk given at the International Workshop on $\gamma\gamma$-Collisions
       (Amiens), Lecture Notes in Physics No 134, Springer (1980)
       W.Hillen, Thesis Bonn 1981, BONN-IR-81-7

/37/   N.Wermes, Thesis Bonn 1982, BONN-IR-82-27;
       N.Wermes (TASSO), presented at the experimental parallel session at this
       workshop

/38/   Ch.Berger, Talk given at the 4th Internat. Coll. on $\gamma\gamma$-Interactions, Paris
       1981

/39/   W.Wagner, private communication

/40/   ARGUS Collaboration, Proposal to DESY-PRC 1983/06

DISCUSSION

Q: Ch.Berger (Aachen): You said, σ_{TOT} is not known yet, i.e. the PLUTO experiment is wrong. Does this statement come from a new analysis or new experimental facts ?

A: H.K.: I concluded that the cross section <u>below 2 GeV</u> is not yet known. I cannot prove that the PLUTO experiment is wrong, but I think PLUTO cannot prove that their result is right.

C: Ch.Berger (Aachen): One can get rid of false experiments in two ways: a) finding a bug in the analysis, b) presenting a new and better experiment but <u>not</u> by digging out an old experiment which is in every respect worse than the PLUTO experiment.

A: H.K.: One should also be allowed to criticize an experiment if one finds problems in the analysis. We want to learn to do better.

Q: J.Kovacs (Paris): On Greco's superconvergent sum rules which I discussed in a parallel session: The expression assumes resonance saturation and this is obviously not true if you look at the CELLO results of dipions in the f^0 region. But is it possible to theorists to formulate another expression relating cross sections with helicity 2 and 0 and taking into account the continuum?

C: P.Singer (Technion): I would like to make two remarks: 1) From the small ratio of the $\gamma\gamma\to\rho^+\rho^-$ versus $\gamma\gamma\to\rho^0\rho^0$ it is too early to conclude that a certain model, like the four-quark model, is successful. There are several models anticipating this effect, some which you have not mentioned. For instance, η exchange in $\gamma\gamma\to\rho^0\rho^0$ was calculated by using the experimental $\rho\to\eta\gamma$ width by C.Schmidt from ETH (unpublished) and I understand this could be a significant contribution.

2) You mentioned trouble with sum rules in relation with $T\to\gamma\gamma$ widths. It seems to me that this is specific to superconvergent $\gamma\gamma$ scattering sum rules (due to Greco and Grassberger and Kögerler). However, there is no similar problem if one considers the pseudoscalar meson - photon scattering amplitudes, from which the correct A_2, f^0, $f' \to \gamma\gamma$ widths were derived. I discussed this topic in the theoretical session.

Q: S.Brodsky (Stanford): Have you compared the $\gamma\gamma\to\rho^0\rho^0$ data with the QCD prediction at large mass ? The QCD predictions give significantly larger cross sections for $\rho\rho$ compared to $\pi\pi$ and have definite ρ-helicity and angular behaviours.

A: H.K.: As far as I know that has not yet been done.

Q: U.Maor (Tel-Aviv): Concerning the $\gamma\gamma\to\rho^0\rho^0$ data, I think that the first question to be asked is if the threshold enhancement observed is particular to $\gamma\gamma$ or just a phenomenon compatible with the rest of our knowledge on two-body cross sections near threshold coupled with the particular $\gamma\gamma$ kinematics. The conclusion of Alexander, Williams and myself is that apparently the $\rho^0\rho^0$ enhancement does not require any "exotic" explanation. In general, I suggest that data analysis advocating some "new physics" explanation should secure that the ratio of the "new physics" signal to the "old physics" background is good enough to make such an analysis sensible.

A: H.K.: I do not think that relating the $\rho^0\rho^0$ production to other processes via VMD-Regge helps one to understand the underlying physics of the effect.

Q: B.Stella (Rome): Could you comment somewhat more on the 2.1 GeV effect? I think that TASSO has now enough statistics to show some dynamical effect, like angular dependences, in order to exclude the interpretation of 2.1 being a kinematical (?) effect.

A: H.K.: The analysis of the narrow structure around 2.1 GeV includes all the data TASSO collected up to summer 82. The angular distributions do not show any special behaviour in the signal region. There is one property which seems to distinguish the signal from background: The events in the signal region look flatter than the events in the sidebands. Monte Carlo studies showed that this behaviour is expected if the signal is predominantly due to four pion phase space production rather than $\rho^\circ \rho^\circ$. It has also been checked that there are no obvious signals in other channels (but upper limits have not yet been evaluated).

TOTAL CROSS SECTIONS AND PHOTON STRUCTURE FUNCTIONS

Walter Wagner

1. Phys. Inst. RWTH Aachen,
Sommerfeldstr., D-51 Aachen

Introduction

Hadron production by the two photon mechanism in e^+e^- storage rings (Fig. 1)

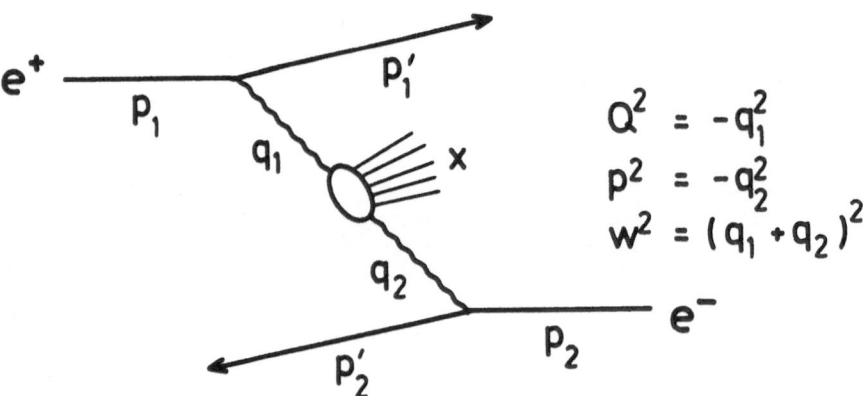

$$Q^2 = -q_1^2$$
$$p^2 = -q_2^2$$
$$w^2 = (q_1 + q_2)^2$$

Fig. 1 Kinematics of the two photon reaction

is certainly a complicated thing. The total differential cross section for unpolarized electron beams is given by the expression [1]

$$\frac{d\sigma}{d\Gamma} = L^{TT}\,\sigma_{TT} + (L^{TL}\,\sigma_{TL} + L^{LT}\,\sigma_{LT}) + L^{LL}\,\sigma_{LL}$$

$$+ \tilde{L}^{TT}\,\tau_{TT} \cdot \cos2\phi + \tilde{L}^{TL}\,\tau_{TL}\,\cos\phi \tag{1}$$

The luminosity functions L, $\tilde{L}$ for the transverse and longitudinal photon fluxes are well known and calculable in QED. The σ's are the cross sections of interest and the τ's are interference terms with ϕ being the angle between the electron and the positron scattering planes. Using the obvious relation σ_{TL} (W^2, q_1^2, q_2^2) = σ_{LT} (W^2, q_2^2, q_1^2) we are left with five measurable quantities as a function of three variables W^2, q_1^2, q_2^2.

This is quite an extensive experimental program and after almost five years analysis at DESY and SLAC we are just at the beginning. Almost all results have been obtained in the single tag mode (where only one of the outgoing electrons has been detected). In this set up the measurable quantity is the sum of $\sigma_{TT} + \varepsilon \sigma_{TL}$ which can be studied as a function of W and Q^2 ($= - q_1^2$, say). Up to now no attempt has been made to disentangle the two contributions σ_{TT} and σ_{TL} nor have the terms σ_{LL} and τ_{TL} ever been looked at.

The progress that has been made since the last $\gamma\gamma$ - conference in Paris is that the study has been extended to very large values of Q^2 (up to 300 GeV^2) and to the double tag mode. Here a first slight indication of the τ_{TT} term has been observed. Before I present the results let me start with a brief guided tour through the wide kinematic country.

Low Q^2 land. In the region where both q^2 are small, the process is naturally explained in the VDM picture:

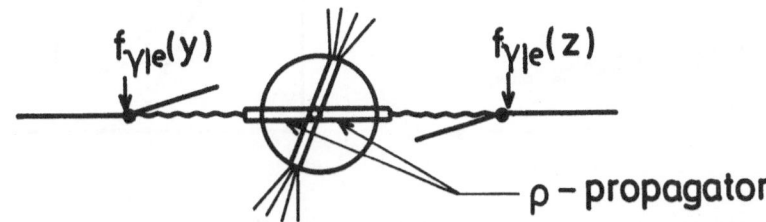

It factorizes into the probability to find (on each side) a photon inside an electron $f_{\gamma/e}(y)$, $f_{\gamma/e}(z)$ and the total cross section $\gamma\gamma \rightarrow X$, which has been estimated to be (240 + 270 GeV/W) nb [2].

$$d\sigma = f_{\gamma/e}(y)dy \ (240 + \frac{270}{W}) \ nb \cdot f_{\gamma/e}(z)dz \qquad (2)$$

If one of the photons gets moderate q^2 values (single tag), $f_{\gamma/e}(y)$ is replaced by $\Gamma_t(y, Q^2)$ and the ρ-propagator enters

$$\frac{d\sigma}{d\Omega \ dE'} = \Gamma_t \ (\frac{m_\rho^2}{Q^2+m_\rho^2})^2 \ (240 + \frac{270}{W}) \ nb \ f_{\gamma/e}(z)dz \qquad (3)$$

This picture can be generalized by including heavier vector mesons (ω, ϕ, ψ, ...).
In addition to this hadronic part there is a pointlike piece due to the direct coupling of the photons to a quark pair:

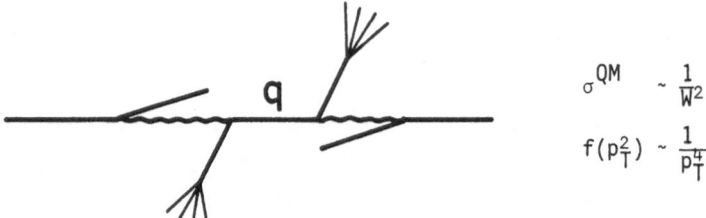

$$\sigma^{QM} \sim \frac{1}{W^2}$$

$$f(p_T^2) \sim \frac{1}{p_T^4}$$

This piece, though accounting only for a small fraction of the total cross section can dominate in two corners of phase space: at low invariant masses ($\sigma^{QM} \sim \frac{1}{W^2}$) and at large transverse momenta of the final state particles ($f(p_T^2) \sim \frac{1}{p_T^4}$ compared to $e^{-Ap_T^2}$ in VDM)

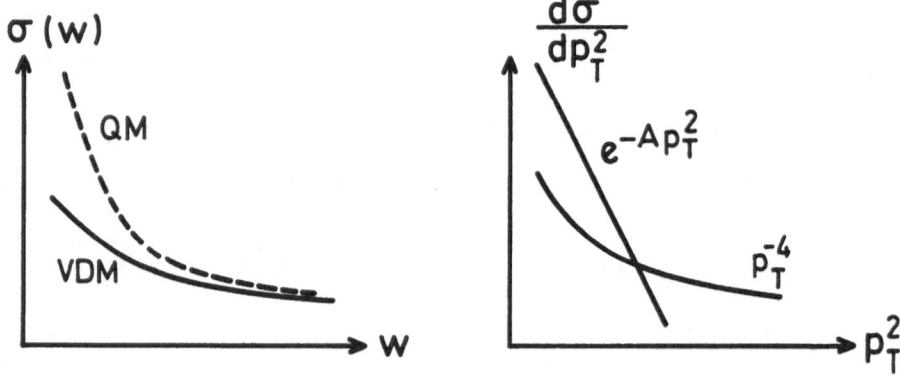

Fig. 2 Qualitative behaviour of the W dependence a) and the p_T^2 dependence b) in the low Q^2 region

This situation is sketched in Fig. 2.

High Q^2 land. If the tagged photon gets highly virtual, $Q^2 \gg m_\rho^2$, the process is more naturally interpreted as deep inelastic electron photon scattering:

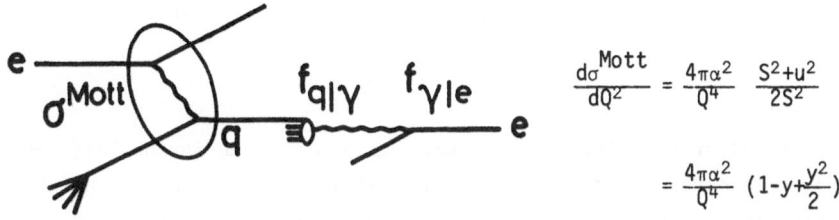

$$\frac{d\sigma^{Mott}}{dQ^2} = \frac{4\pi\alpha^2}{Q^4} \frac{S^2+u^2}{2S^2}$$

$$= \frac{4\pi\alpha^2}{Q^4} (1-y+\frac{y^2}{2})$$

In the quark picture the process is simply given by the elastic electron quark scattering multiplied by the probability to find a quark inside a photon $f_{q/\gamma}(x)$ times the probability to find a photon inside an electron $f_{\gamma/e}(z)$:

$$\frac{d\sigma}{dQ^2dx} = \frac{4\pi\alpha^2}{Q^4} (1-y+\frac{y^2}{2}) \underbrace{\sum_{q+\bar{q}} e_q^2 \cdot f_{q/\gamma}(x)}_{\frac{1}{x} F_2^\gamma (x)} \cdot f_{\gamma/e}(z)dz \qquad (4)$$

Here the photon structure F_2 has the very intuitive meaning of (x times) the momentum distribution of quarks inside a photon. If longitudinal terms are kept, eq (4) reads more precisely

$$\frac{d\sigma}{dQ^2dx} = \frac{4\pi\alpha^2}{Q^4} \frac{1}{x} \{ (1-y) F_2 + xy^2 F_1 \} f_{\gamma/e}(z)dz \qquad (5)$$

The formal connection of cross sections and structure functions is given by

$$2 x F_1 = \frac{Q^2}{4\pi^2\alpha} \sigma_{TT}$$

$$F_2 = \frac{Q^2}{4\pi^2\alpha} (\sigma_{TT} + \sigma_{TL}) \qquad (6)$$

and the scaling variables x, y, z are defined by

$$x = \frac{-q_1^2}{2q_1q_2} = \frac{Q^2}{Q^2+W^2} \qquad \simeq \frac{E_q}{E_\gamma}$$

$$y = \frac{q_1q_2}{p_1q_2} = 1 - \frac{E'}{E} \cos^2 \frac{\Theta}{2} \qquad \simeq \frac{E_\gamma^1}{E} \qquad (7)$$

$$z = \qquad\qquad\qquad\qquad\qquad \frac{E_\gamma^2}{E}$$

with the intuitive meaning of fractional momenta of one particle inside another.

Low Q^2 Region

The progress on total cross section measurements at low Q^2 made in the last two years one could call marginal. But there is an enormous amount of data available in different experiments which are analysed at present (PLUTO e.g. has about 40000 hadronic events in the no tag sample). The analysis is very complicated and the effect of fragmentation, especially in the low W range, has been extensively studied [3, 4]. Before I discuss the new results from PLUTO in the double tag mode let me briefly summarize the old measurements.

The total cross section has been analysed by the San Diego group at SPEAR in a double tag experiment [5] and by PLUTO [6] and TASSO [7] in the single tag mode.

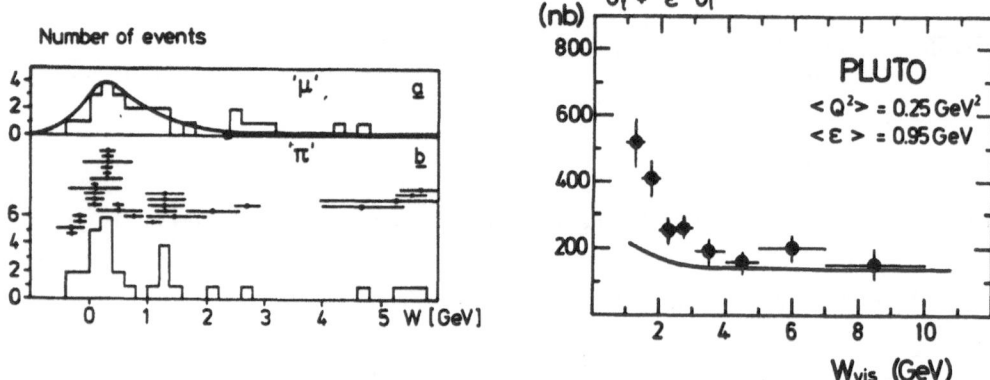

Fig. 3 Total $\gamma\gamma$ cross section results from SPEAR a) and PLUTO b)

The SPEAR results are presented in terms of $R_{\gamma\gamma}$, the ratio of the total hadronic and the muonic cross section (Fig. 3a)

$$R_{\gamma\gamma} = \frac{\sigma(\gamma\gamma \to \text{hadrons})}{\sigma(\gamma\gamma \to \mu^+\mu^-)} = 1.1 \pm 0.3 \tag{8}$$

The value of $R_{\gamma\gamma}$ in the kinematic range $0 < W < 2.5$ GeV, $0.07 < |q_1^2|$, $|q_2^2| < 0.3$ GeV^2 is very suggestive for the presence of a $1/W^2$ term. PLUTO presents the results as a cross section as function of the visible invariant mass W_{vis} (Fig. 3b). Above 2 GeV the shape is perfectly described by the VDM, whereas the normalization comes out about 30 % higher. If the enhencement below 2 GeV is interpreted in the spirit of the quark model, using a fit

$$\sigma_{\gamma\gamma}(W) = A \cdot \sigma^{VDM} + \frac{B}{W^2} \quad \text{(PLUTO)} \tag{9}$$

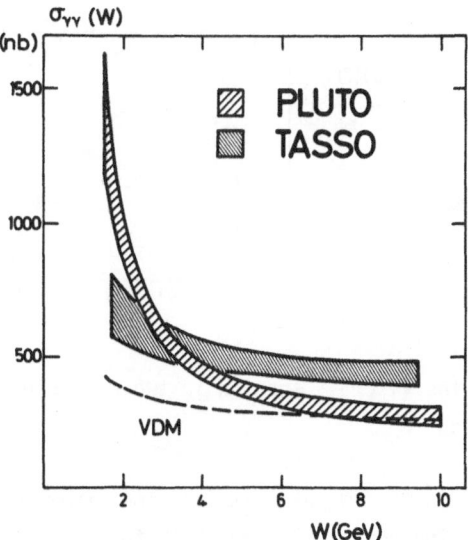

Fig. 4   Unfolded γγ cross section as a function of W. The shaded areas represent
the 1σ contours not including the systematic errors

the result is shown in Fig. 4. The figure also includes the TASSO data, which are
fitted by an ansatz

$$\sigma_{\gamma\gamma} (W) = A + \frac{B}{W} \qquad (TASSO) \qquad\qquad (10)$$

This figure, which is in the literature since 1980, gave rise to many speculations
and discussions and I want to add some remarks.

1. Fig. 4 only contains the statistical errors and the correlations introduced by
    the fit. The systematic errors, coming mainly from correlations between the
    W-dependence of $\sigma_{\gamma\gamma}$ and the fragmentation model, are not included in the
    figure.

2. The differences, which seem to be mainly introduced by the different fit
    approaches, are not as significant as they look in the 1σ contours. The two
    experiments can not exclude the parametrization of each other on a significant
    basis [8].

3. If the data are separated into two W_{vis} bins and plotted versus Q^2, the two
    experiments are in good agreement for W > 3.5 GeV (Fig. 5b)

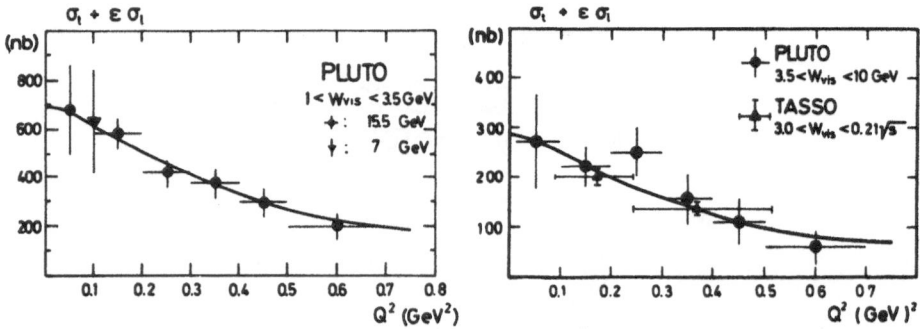

Fig. 5 Q^2 dependence of the $\gamma\gamma$ cross section in two W_{vis} bins. The curves show
the VDM expectation

Thus the whole difference in the two experiments comes from small W, where the PLUTO
data are slightly higher. The ansatz (9) for the unfolding pushes the cross section
up. This is due to the fact that only 15 % of the events with W_{vis} < 2 GeV come from
W < 2 GeV, whereas the rest of 85 % have larger W and enter that bin through particle losses. To conclude, the two experiments are in much better agreement than the
fits of Fig. 4 suggest. They differ in the significance of the $1/W^2$ term. The PLUTO
data are consistent with a large $1/W^2$ term but they cannot be regarded as an unambiguous proof of such a term.

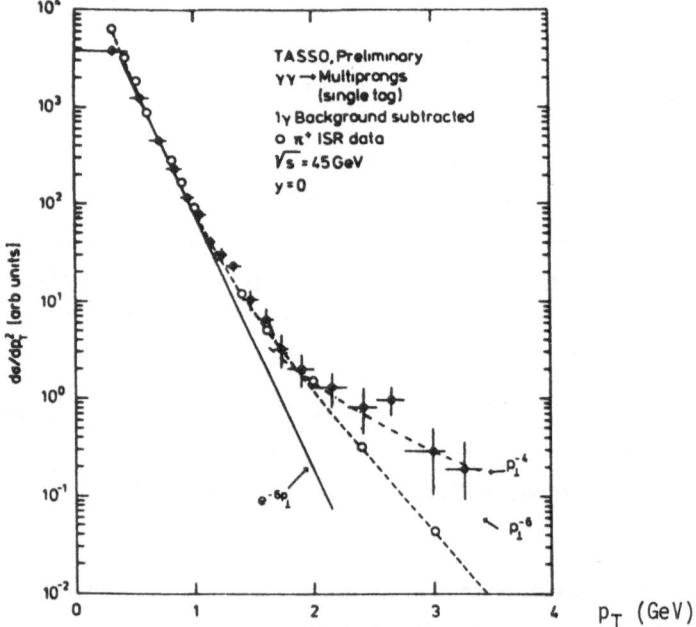

Fig. 6 Inclusive p_T^2 distribution of the charged particles (TASSO). Also included
are the ISR data

The Q^2 dependence agrees with a ρ - pole formfactor, but the data are also consistent with (and the TASSO data slightly favour) a weaker dependence as suggested by the generalized vector dominance model, GVDM [9].

The tail in the p_T distribution with an approximate p_T^{-4} behaviour has been observed very clearly. Above 1.5 GeV it sticks out significantly (Fig. 6) and can be regarded as a good indication for a hard component of the photon.

The new analysis of PLUTO uses 201 hadronic double tag events, where the tagging was done either in the small angle tagger (SAT, $<Q^2> \simeq 0.4$ GeV^2) or the large angle tagger (LAT, $<Q^2> \simeq 5$ GeV^2). The event structure as well as the absolute number of events perfectly agrees with a superposition of the generalized VDM and the QM prediction (213 events predicted, see Fig. 7a, b). The subsample of

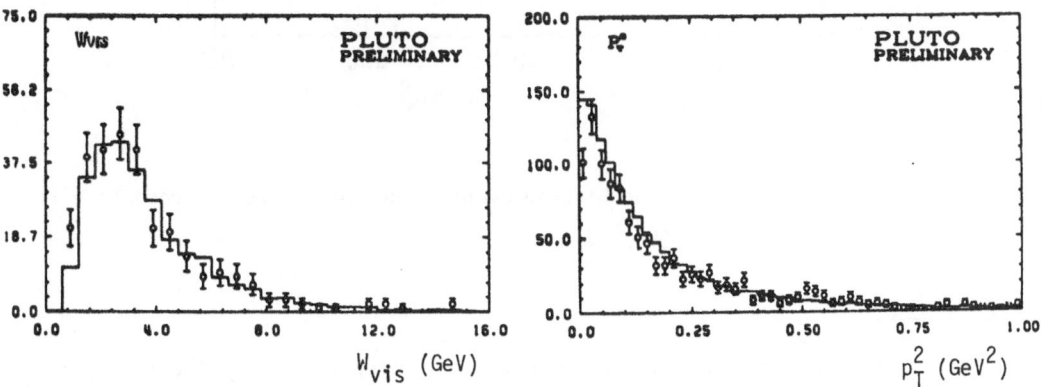

Fig. 7   Invariant mass distribution a) and transverse momentum distribution b) of the double tag events (PLUTO)

the data with one small and one large Q^2 (LAT - SAT coincidences) are interpreted in terms of the virtual photon structure function and are discussed in the next chapter.

Most of the data are coming from small angle tags (SAT - SAT coincidences), where the tagging system is completely ϕ symmetric. Thus the angle ϕ (= $\phi_{e^+} - \phi_{e^-}$) the angle between the e^+ and the e^- scattering planes in the $\gamma\gamma$ CMS can be measured with uniform efficiency. Fig. 8 shows the distribution of the angle ϕ after folding to the interval 0 - 90°. The data are fitted by the form A (1 + B cos 2ϕ). The result (solid curve)

$$B = - 0.31 \pm 0.12 \tag{11}$$

can be regarded as a first indication of the presence of a negative τ_{TT} term. The significance is marginal, about 2.5 standard derivations away from zero, and sign and magnitude agree with the QM prediction.

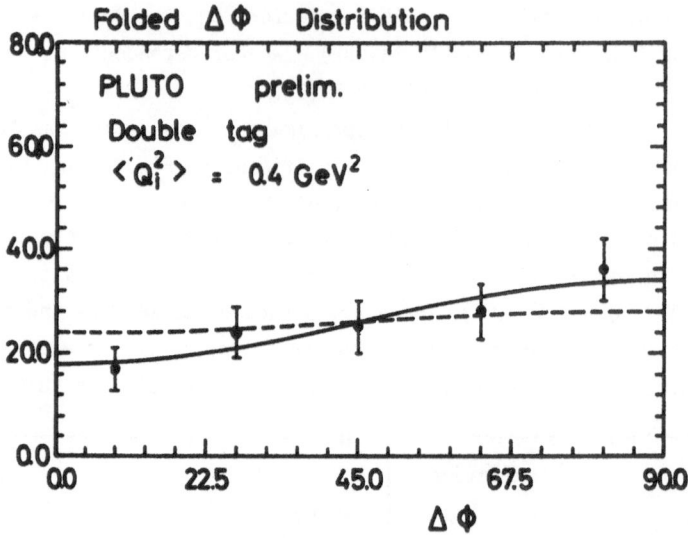

Fig. 8 Δφ distribution of the double tag events. The solid curve represents a fit
of the type A (1 + B cos 2φ)

High Q^2 Region

The investigation of the high Q^2 region made a big step forward in the last two
years. In 1980 we had roughly 10Q hadronic events in the Q^2 range from 1 - 15 GeV^2.

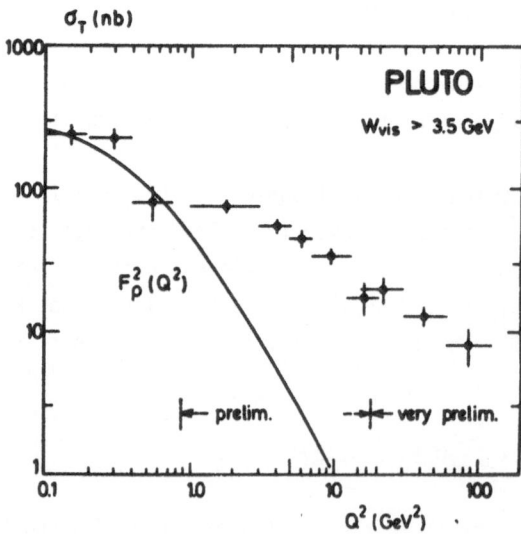

Fig. 9 Q^2 dependence of the total γγ cross section. The solid curve represents
the ρ pole formfactor

The first indication that these rather small values are already large enough came from the Q^2 dependence of the total cross section (Fig. 9), which was in definite disagreement with a simple ρ - pole formfactor above 2 GeV. The VDM, however, turned out to be flexible enough to describe the Q^2 dependence in a natural way by including higher vector mesons (GVDM). But the interpretation of this weak Q^2 fall off as the onset of the deep inelastic regime got support from the final state analysis. Fig. 10 shows the p_T^2 distribution of reconstructed jets for single tag events in different Q^2 regions. (The index T refers to the transverse direction in the $\gamma\gamma$ CMS) [10]. At small Q^2 ($Q^2 < 1$ GeV^2) the QM accounts only for a small fraction of the events and describes only the high p_T tail quantitatively. At large Q^2 a large frac-

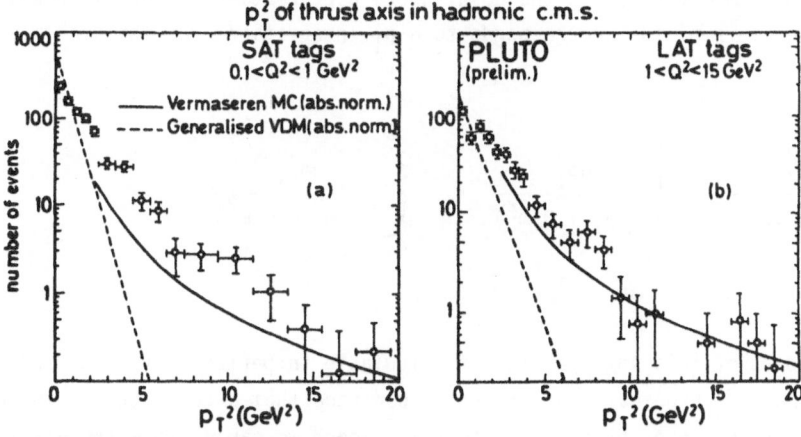

Fig. 10 p_T^2 distribution of the reconstructed jets compared to VDM and QM predictions for small a) and large b) Q^2

tion of the data follows the QM prediction. The natural explanation of this behaviour is that we start seeing the direct photon quark coupling so that the process is better understood as deep inelastic electron photon scattering. The quantity to measure in this case is the photon structure function F_2^γ (x, Q^2)

$$F_2^\gamma \ (x, Q^2) \ = \ 2 \ x \ \Sigma \ e_q^2 \ f_{q/\gamma} \ (x, Q^2) \tag{12}$$

Predictions. What are the models for F_2^γ, or in other words, what is the quark content of a photon?
In a very simplified picture the process proceeds as follows: At pointlike distances the photon splits into a $q\bar{q}$ pair and the two quarks fly apart. This vertex is straight forward calculable in the QM. At larger distances the gluon effects get

$$f_{q|\gamma}(x, Q^2):$$

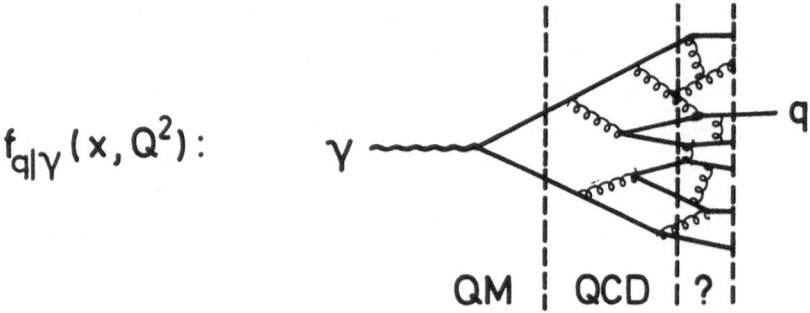

QM ┊ QCD ┊ ? ┊

more and more important up to the point where perturbative QCD breaks down. In momentum space this picture corresponds to a p_T integration over the available phase space:

$$\int_0^{Q^2} \frac{dp_T^2}{p_T^2 + 0(m^2)} = \int_0^{\Lambda^2} \frac{dp_T^2}{p_T^2 + 0(m^2)} + \int_{\Lambda^2}^{Q^2} \frac{dp_T^2}{p_T^2 + 0(m^2)} \tag{13}$$

$$\text{I} \qquad\qquad\qquad \text{II}$$

The first part of this integral is not calculable in perturbative QCD. This is bad but there may be three ways out. First we can hope that a full QCD, including lattice calculations, can solve this integral. Second we can go to extremely large Q^2 values so that the relative importance of part I vanishes. Third we can call this integral the hadronic piece of the photon and try to extract that piece from hadronic structure functions. The motivation for that is that a typical value of Λ is not very much different from the inverse hadronic radius.

The second part can be calculated in a straight forward way if Q^2 is large enough [11]. The result is a structure function that rises proportional to $\ln Q^2$:

$$F_2 \sim \ln \frac{Q^2}{\Lambda^2} \sim \frac{1}{\alpha_s} \quad \text{if} \quad \Lambda^2 \gg m^2$$

$$\tag{14}$$

$$F_2 \sim \ln \frac{Q^2}{0(m^2)} \qquad \text{if} \quad \Lambda^2 \ll m^2$$

The photon structure function is unique in many respects:

1. it is expected to rise with x unlike all other hadronic structure functions

2. it should have a strong positive scale breaking effect ~ ln Q^2

3. the absolute normalization of the dominant part is predicted by QCD

4. the absolute normalization is very sensitive to α_s and Λ ($1/\alpha_s$ as compared to $(1 + \alpha_s/4\pi)$ in e^+e^- annihilation).

Especially the last point made many people (experimentalists and theoreticians) enthusiastic about the photon structure function. A determination of Λ to a 30 % accuracy seemed to be possible from an experimental point of view (see discussion below).

However, this optimistic situation has changed recently. In addition to the problems at small x and in the region $x \simeq 1$ there are now singularities entering the game as severe as $x^{-5.33}$ poles [12]. These singularities wait for cancellations which are not calculable in the moment. Thus the whole beauty of the photon structure function would be destroyed. But it is not clear whether one has to be all that pessimistic [13]. We have got used to all sorts of singularities which we don't have to worry about. And we know from recent calculations by Uematzu and Walsh [14] that for the case of the virtual photon the singularities don't play a role if $- q_2^2 = p^2 \gg \Lambda^2$. In this case the denominator in eq. 13 gets an additional mass scale, p^2, and the integral gets the form

$$\ln \frac{Q^2}{p^2} \tag{15}$$

This is a safe QCD prediction but the parameter Λ hides behind the larger mass scale in the same way as it does for heavy quarks.

So let us hope that there will be a smooth transition from the case of the virtual structure function to the real structure function and that there finally will be a meaningfull QCD prediction to be compared with experimental measurements.

Fig. 11 summarizes the expectations for $F_2(x)$ (for a detailed discussion of the hadronic part see e.g. ref. [15]).

For small Q^2, say $Q^2 < 5$ GeV^2, the pointlike and the hadronic part add up to a more or less constant structure function, whereas for large Q^2, say $Q^2 > 20$ GeV^2, $F_2(x)$ rises with x.

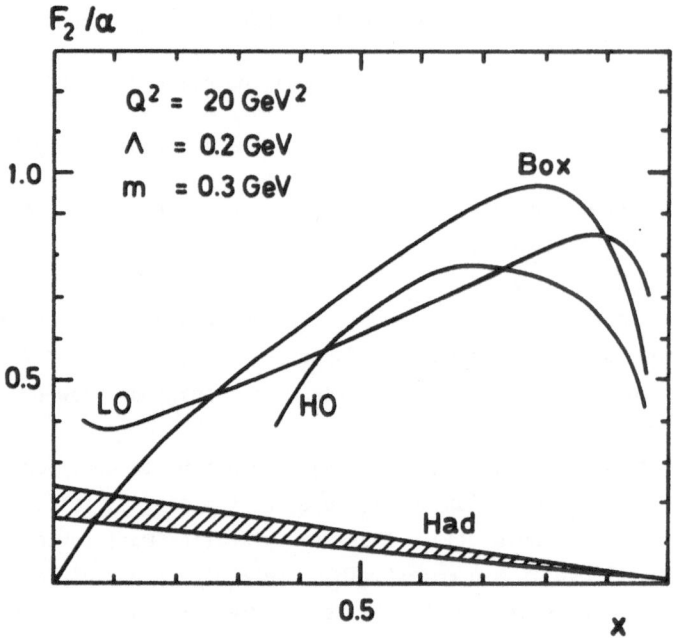

Fig. 11 Predictions for the photon structure function: Hadronic part (Had),
QCD calculations in leading order (LO) and higher order (HO) and the
quark model (Box)

First results. The first results on F_2 (x, Q^2), obtained by PLUTO three years ago
[16], show good agreement of the data with a simple superposition of QCD and the
hadronic part, if Λ is of the order of 200 MeV (see Fig. 12).

Though this measurement is based on roughly 100 events only, it has shown that the
photon has somewhat more than a hadronic component only.
Before I discuss the new experimental data I want to summarize the experimental
problems which we have to deal with.
Statistics. The experiments made a big step forward as far as statistics is concerned. The situation is summarized in Tab. 1. The 2000 events from PLUTO, e.g.
would allow a measurement of F_2 (x, Q^2) in 4 Q^2 and 5 x bins, with a statistical
error of about 10 % in each bin. It is also interesting to note that the Q^2 range
has extended up to 300 GeV^2!

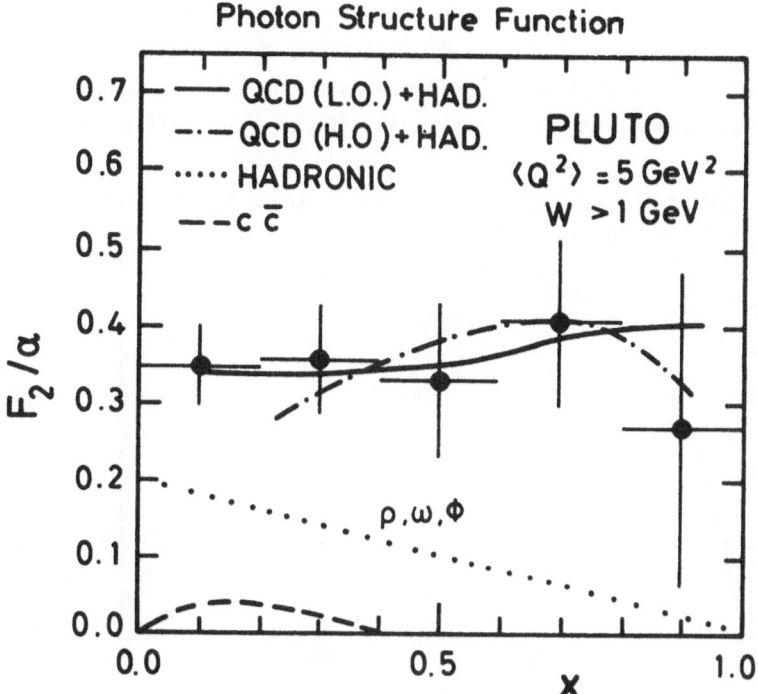

Fig. 12 First measurement of $F_2^\gamma(x)$ by PLUTO

experiment	events	Q^2 range	$\langle Q^2 \rangle$ [GeV^2]
CELLO	200	3 - 300	9
JADE	400	10 - 55	24
JADE	25	40 - 220	110
MAC	150		32
PLUTO	2000	1 - 18	6
PLUTO	120	15 - 150	45
TASSO	200	10 - 50	23

Tab. 1

Fragmentation. Only for an ideal detector with 4π coverage the process does not depend on the fragmentation model used. For any realistic detector, however, the detection efficiency depends on fragmentation and therefore the corrections are model dependent. Thus the event simulation for a bad detector could look indistinguishable using one model $F(x, Q^2)$ with a certain fragmentation and a completely different model $\tilde{F}(x, Q)$ with a different fragmentation.

A detailed Monte Carlo study for the PLUTO detector showed that the situation is not that bad. Due to a good coverage in forward directions the detection efficiency is quite satisfactory (larger than 50 % on average, see Fig. 13). It turned out that a variation of fragmentation parameters has small effects on the efficiency but can be

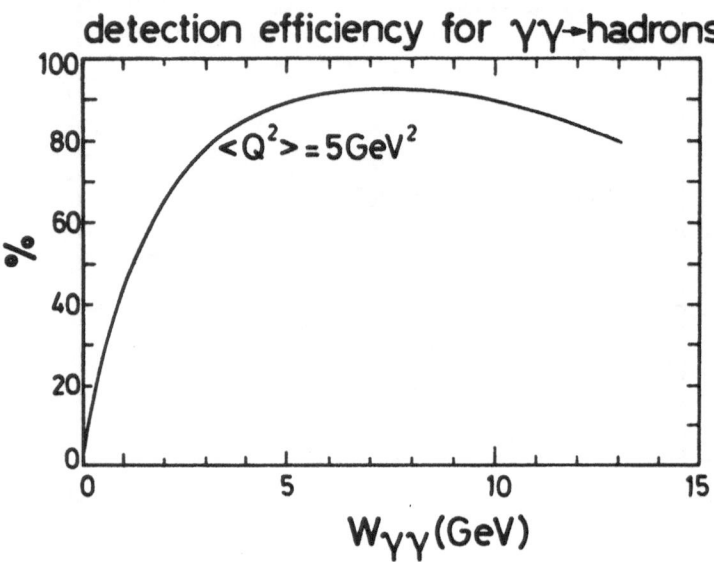

Fig. 13   Detection efficiency of the PLUTO detector

seen as a large effect in the fragmentation variables. This is demonstrated in Fig. 14 where the effect of the most sensitive variable, the transverse momenta of the par-

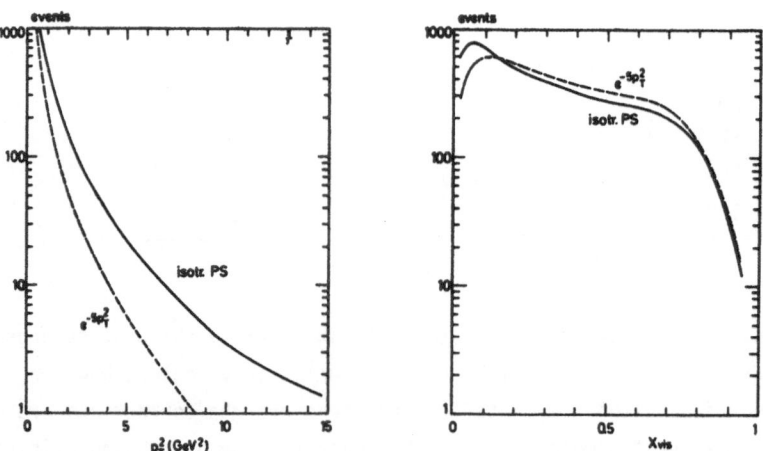

Fig. 14   Effect of the fragmentation variation on a) acceptance (x_{vis}), b) visible p_T^2 distribution

ticles, is studied. Two very extreme models have been compared, an isotropic phase space model (solid line) and a limited p_T phase space model, where the particles are collimated jet-like ($\sim e^{-5\ p_T^2}$) along the $\gamma\gamma$ axes (dashed line). The p_T^2 distributions differ by more than a factor of 2, whereas the effect on the x - distribution (due to differences in the efficiency) is only less than 20 %, except for very small x. Thus the remaining uncertainty after adjusting the p_T distribution to the data is estimated to be less than 10 %.

Unfolding. In addition to inefficiencies we have the problem of particle losses. Though the event may not be lost completely, the invariant mass of the visible final state, W_{vis}, is smaller than the true W, if one or more particles are not detected. Fig. 15 shows the distribution of the ratio W_{vis}/W as seen in the PLUTO detector.

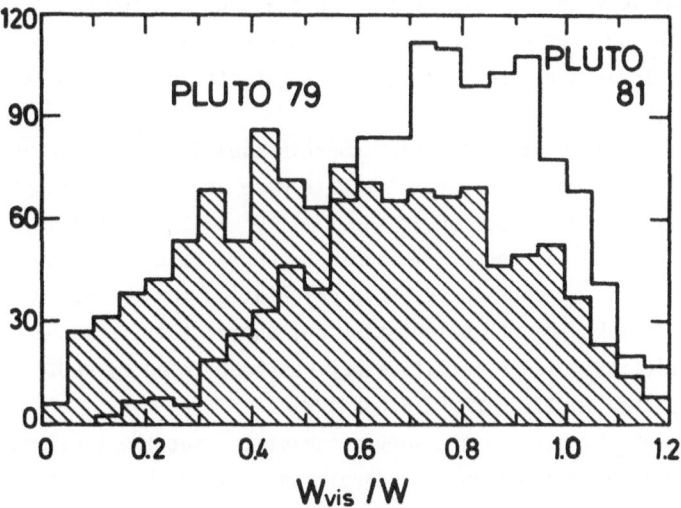

Fig. 15 Ratio of the visible invariant mass and generated invariant mass. The shaded area represents the detector without forward spectrometer

The shaded area represents the situation of PLUTO 1979 without a forward spectrometer. The inclusion of particle detection in the forward region obviously is of great help. The shift in W automatically leads to a shift in $x \rightarrow x_{vis} > x$ which is corrected for in an unfolding procedure. This procedure was presented by V. Blobel at this conference and is described in ref. [17]. A preliminary estimate of the systematic error, introduced by unfolding, is about 10 % or less.

Target mass effects. If the target photon is not restricted to the mass shell by using a good antitagging, the correct treatment of the target mass - $q_2^2 \equiv p^2$ is essential. For $p^2 \gg \Lambda^2$ the process has been calculated in ref. [14]. The main effect is a change in the log term to $\ln (Q^2/p^2)$. A smooth interpolation to moderate p^2 would thus lead to

$$\ln \frac{Q^2}{\Lambda^2 + p^2} \tag{16}$$

However it could be that eq (16) overestimates the effect, as a straightforward QED - calculation results in

$$\ln \frac{Q^2}{\underbrace{m^2 \frac{1-x}{x} + x^2 p^2}_{\Lambda^2}} \qquad \rightarrow \qquad \ln \frac{Q^2}{\Lambda^2 + x^2 p^2} \tag{17}$$

If one neglects p^2 in eq. 16 the effect is only minor for an antitag $\Theta < 30$ mrad (5 % in α_s and 10 % in Λ). But for a loose antitag ($\Theta < 100$ mrad) it results in a 15 % change in α_s , or 50 % change in Λ. Thus one either needs a good antitag or a reliable calculation for the target mass effects in the transition region $p^2 \simeq \Lambda^2$.

Heavy quark. For small Q^2 the c - quark contribution is negligible and restricted to small x. As Q^2 increases, however, it gets more and more important and accounts for roughly half the cross section in the limit $Q^2 \rightarrow \infty$. Though this part is not a good tool to determine α_s or Λ, it should be known to an accuracy of about 10 % or better. As long as we don't have a QCD calculation for the heavy quarks the best thing to do is to take the QED calculation and subtract it from the data (or add it to the QCD predictions). An alternative way has been tried by PLUTO, where the W - range was restricted to $1.5 < W < 3.5$ GeV [22].

Hadronic component. Even if we are very optimistic about the treatment of the hadronic component of the photon as an additive part to the structure function, we have to worry about uncertainties. The best guess one can try for F_2^{Had}, based on experimental measurements is given by

$$F_2^{Had} (x, Q^2) = \alpha \cdot (0.2 \pm 0.05) (1-x) \tag{18}$$

where scale breaking effects are neglected. This Q^2 independent term can be compensated by another Q^2 independent term coming from QCD by changing Λ to $\Lambda \pm \Delta\Lambda$

$$\Delta F_2^{QCD} = h (x) \cdot \ln (\frac{\Lambda \pm \Delta\Lambda}{\Lambda})^2 = \Delta F_2^{Had} \tag{19}$$

$$\frac{\Delta\Lambda}{\Lambda} = \frac{\alpha \cdot 0.05 (1-x)}{2 \cdot h(x)} \qquad \neq f (Q^2)$$

Thus the ignorance about the hadronic component generates an uncertainty in Λ which is independent on Q^2 and which is $\simeq$ 40 % at x = 0.4 and 10 % at x = 0.8. To conclude, a determination of Λ to ± 30 % accuracy seems to be feasible from the experimental point of view. Whether or not this value of Λ is meaningful in the view of the theoretical problems with the singularities has to be answered by theorists.

New results. There are of the order of 3000 hadronic events collected in 5 different experiments, and at least 10 different analyses have been reported at this conference. I am not going into any details of the trigger conditions, data selection and analysis criteria, but refer to the summary of the parallel sessions instead. However the analyses have many common features:

Definition of the tag electron by requiring a minimum energy cluster of about E/2 in one of the shower counters.
Definition of the hadronic final state by requiring at least 3 particles. Sometimes the conditions are more restrictive and 3 or 4 charged particles are required, sometimes the class with two charged particles plus neutrals are added.
Antitag is normally applied as far as the detector allows for it.
Q^2 is always determined from the electron and W from the hadronic final state (except for a special analysis of double tag events from PLUTO).

The experiments differ in the way of presenting data. All experiments except PLUTO compare the x_{vis} distribution with model predictions including all detector effects. PLUTO presents the data as an unfolded structure function F(x). The comparison with models can then be done analytically. There is no consistent treatment of the hadronic part, which is neglected in the analyses of JADE, TASSO and MAC, added as F_2^{Had} = 0.11 · α (1-x) by CELLO and added as F_2^{Had} = 0.2 · α (1-x) in the PLUTO analysis.

In Fig. 16 the PLUTO result is shown. The unfolded structure function F_2^{γ} (x) is compared to leading order QCD predictions for different values of Λ. The data are in nice agreement with the superposition of the hadronic part plus the QCD part using $\Lambda \simeq$ 200 MeV. One also gets a feeling for the sensitivity to a variation of Λ. It is interesting to note that the data are also described by the QM plus hadronic piece but that a rather large quark mass of about 300 to 500 MeV is required (current quark masses of the order of 10 MeV can be ruled out).
In Figs. 17 - 21 the x_{vis} distributions are shown for CELLO [18], TASSO [19], JADE [20], MAC [21] and PLUTO. In each case the electron has been measured in the endcap shower counter. The data are in good agreement with the QM but also with leading order QCD calculations, which are slightly favoured. The TASSO data demonstrate nicely that at larger Q^2 values the charm contribution is definitely needed.

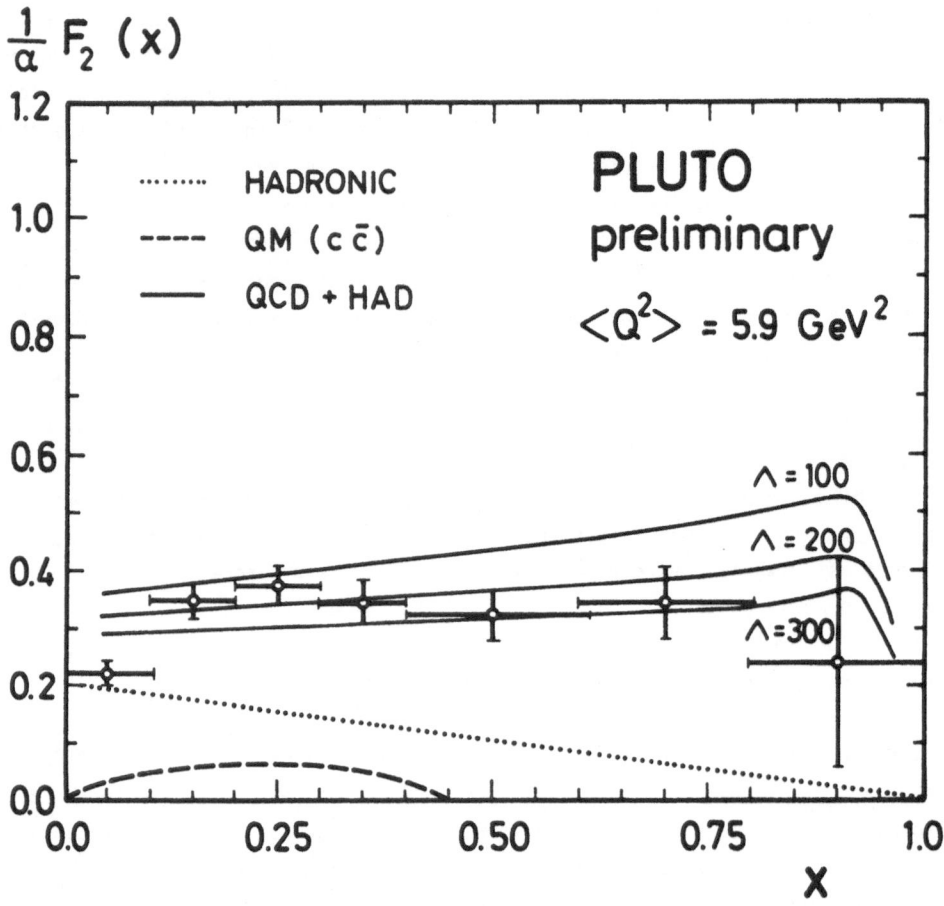

Fig. 16 Unfolded structure function $F_2^\gamma(x)$ at $\langle Q^2 \rangle$ = 5.9 GeV^2. The curves are
QCD predictions using different values of Λ (PLUTO)

In the first three figures the data have the tendency to be above the curves at
low x, indicating some room for hadronic contributions even at large Q^2. (In the
case of CELLO there is room for more than the used contribution α · 0.11 (1-x)).
The MAC data are compared with the shape of the QM, where the normalization is
adjusted to the data. From the good agreement one can conclude that at these large
Q^2 values (32 GeV^2) F_2^γ definitely rises with x. The same effect can be seen by
the comparison of the high Q^2 PLUTO data with a constant structure function
(Fig. 21).
The JADE group presented a result at a breathtaking average Q^2 value of 110 GeV^2,
based on roughly 25 events. The electron has been measured at very large angles,
around 90°, in the barrel shower counter. The event distribution is plotted in
Fig. 22 and perfectly agrees with the leading order QCD calculation, with Λ=200 MeV.

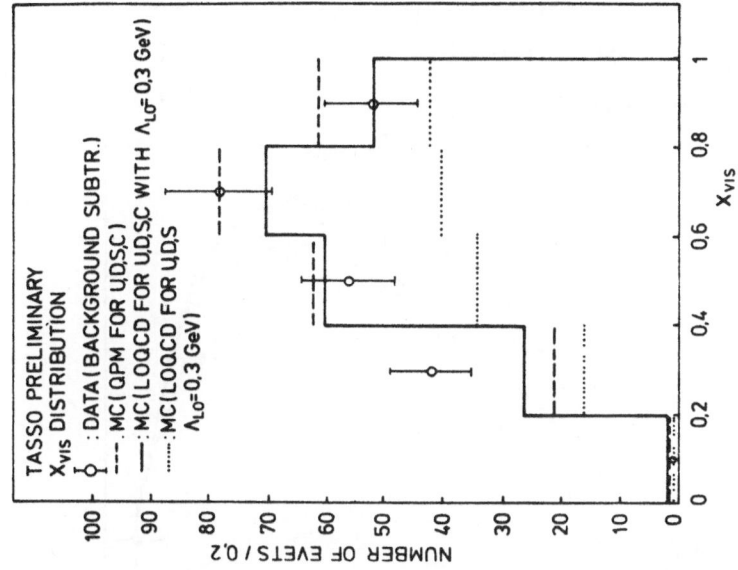

Fig. 18 Visible x distribution compared with
model predictions (TASSO)

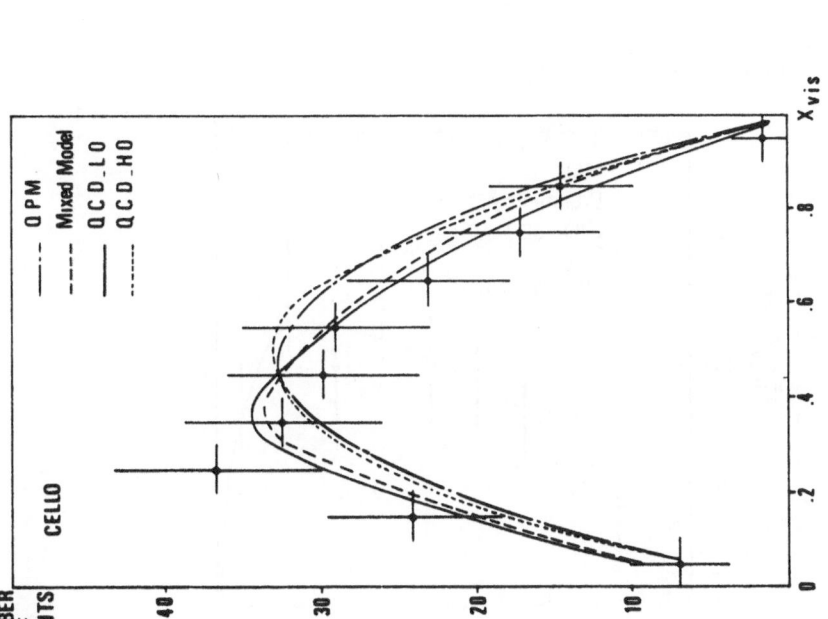

Fig. 17 Visible x distribution compared with
model predictions (CELLO)

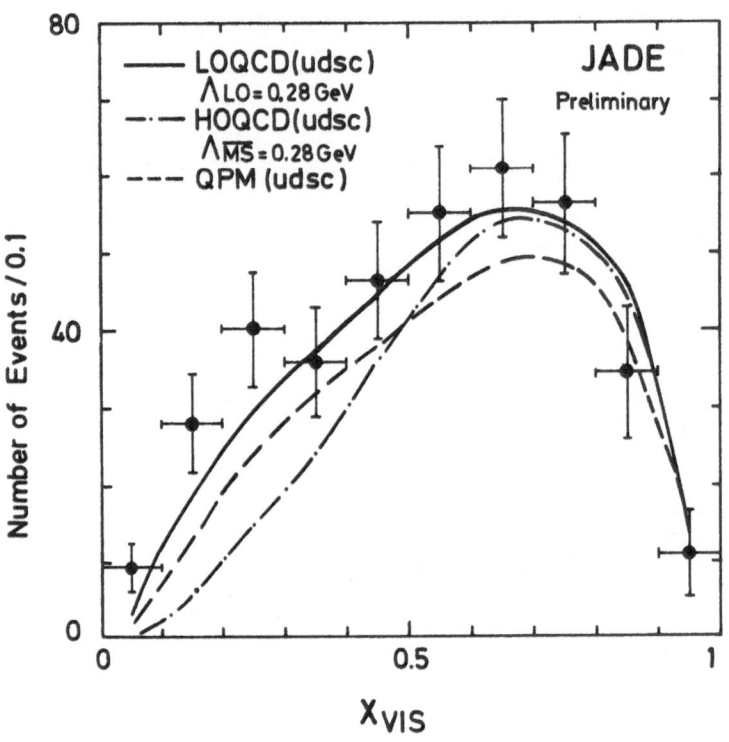

Fig. 19 Visible x distribution compared with model predictions (JADE)

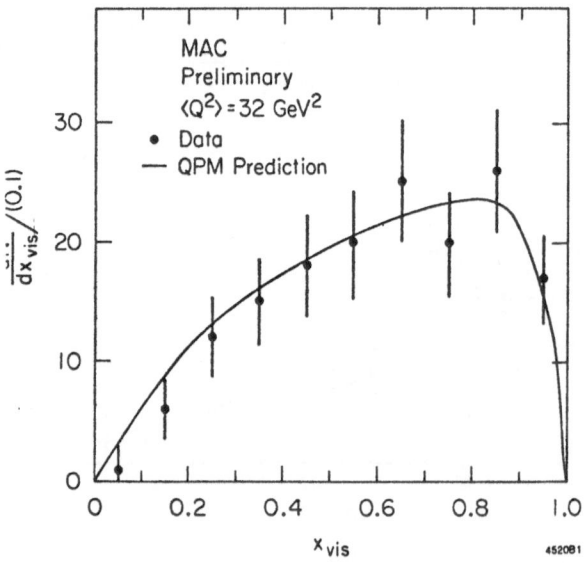

Fig. 20 Visible x distribution compared to the quark model expectation. The curve
is normalized to the data (MAC)

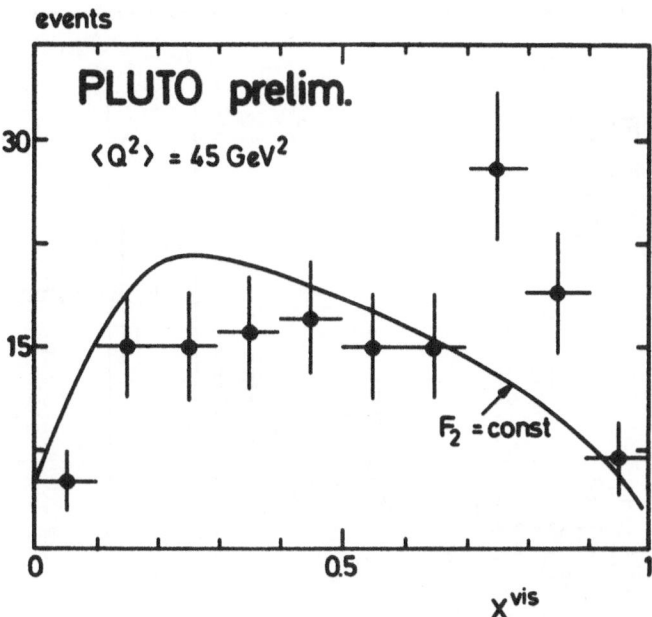

Fig. 21 Visible x distribution compared to the model F = const. (PLUTO)

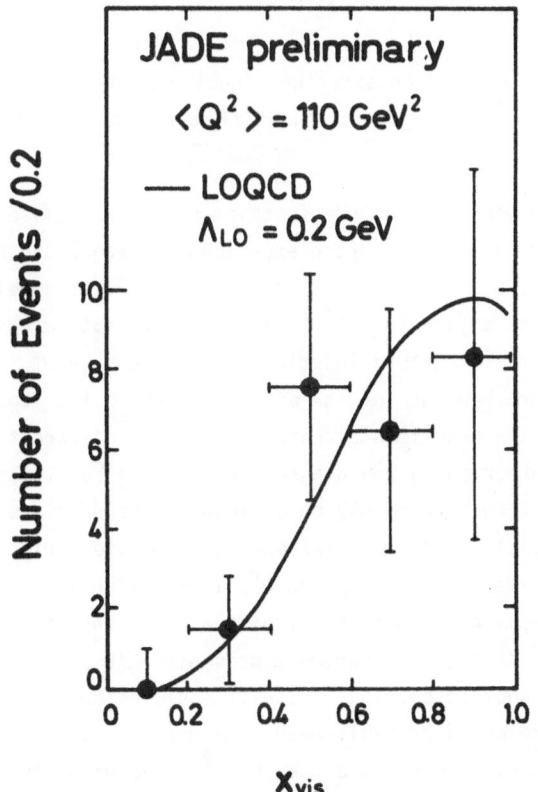

Fig. 22 Visible x distribution compared to leading order QCD, Λ = 0.2 GeV (JADE)

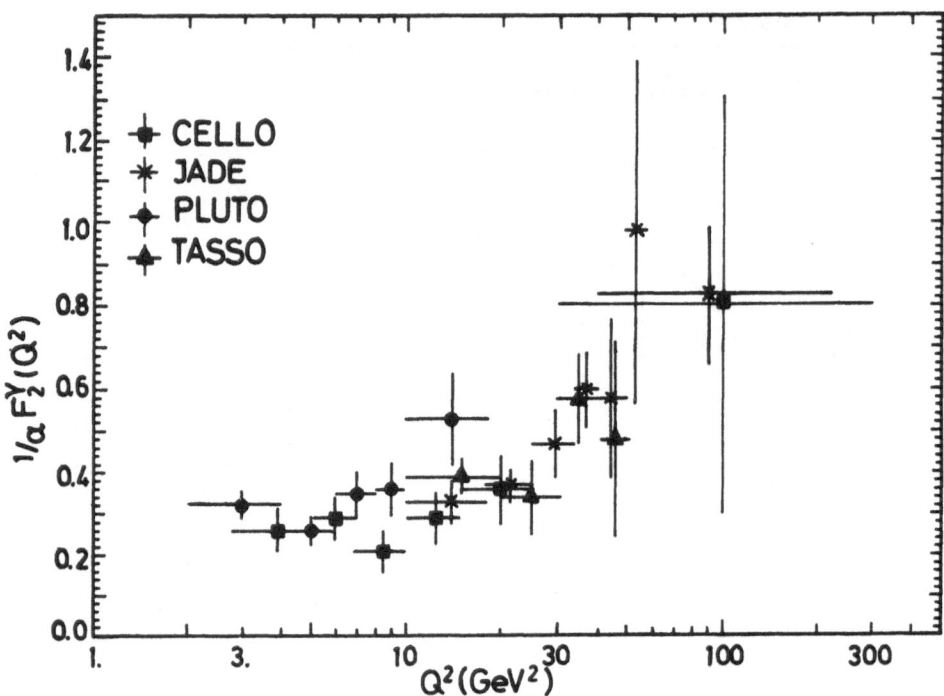

Fig. 23 Q^2 evolution of the photon structure function. The data are averaged over slightly different x ranges for the different experiments

The Q^2 evolution of the photon structure function has been analysed over a phantastic range from 2 < Q^2 < 300 GeV^2. Each experiment averaged the data over a certain range in x_{vis}, e.g. 0.2 < x_{vis} < 0.8, or x_{vis} > 0.3 or something similar, and plotted the result as a function of Q^2. The compiled data are shown in Fig. 23. There is no doubt that the structure function rises with Q^2, but it is very hard to judge from this figure how much of the effect is due to the logarithmic scale breaking. First of all the experimental cuts in x_{vis} are different in each experiment. But even if the experiments would agree on the same cuts, there are still x - Q^2 correlations left which can only be avoided if the data are compared at fixed x (or a small x-bin). These correlations can simulate a rise with Q^2 as the mean value of x in a large x-bin depends on Q^2. In addition to that a large increase of $F_2(Q^2)$ is simply due to the onset of the charm threshold. Thus one possible scenario would be that each experiment presents $F_2(Q^2)$ at a fixed x, with the charm contribution subtracted. These results could be compared to a universal curve. As this procedure would probably need some more statistics, the experiments use a different technique. They compare the $F_2(Q^2)$ measurements with model calculations where the detector effects, Q^2-x correlations and charm contributions were taken into account. They all find consistency with an increase as predicted by QCD using $\Lambda \simeq 200$ MeV. It would be nice to see the effects disentangled in the near future to be able to judge the sensitivity of QCD test in the scale breaking of the photon structure function.

Finally, the first measurement of the structure function of a virtual photon target has been presented by PLUTO in the double tag mode. The target photon was tagged at small but finite angles resulting in a mean value of $<p^2> \equiv -<q_2^2> = 0.4$ GeV^2. The Q^2 range of the probe is similar to that in the single tag analysis, resulting in $<Q^2> = 6$ GeV^2. If both electrons are detected the kinematics of the process is completely known and the variables Q^2, p^2 and W can be determined from the scattered electrons. This procedure is almost independent on the fragmentation and the final state hadrons are only used for event definition (separation against the QED processes $\gamma\gamma \to$ ee, eeγ). Only in the rare cases when W^2 comes out negative (due to finite energy resolution in the tagging systems) the final state hadrons are used to determine W.

The result is shown in Fig. 24 as a function of x and p^2. The data are in nice agreement with a superposition of the hadronic and pointlike component, where it makes almost no difference whether the pointlike component is calculated in the QM or in QCD. It is nice to see that the structure function drops with p^2 as expected (see eq. 15). For the p^2 dependence of the hadronic term a simple ρ - pole formfactor was assumed.

The comparison of the measurements and the models is only straightforward for the QM calculation as this calculation exists for the complete process ee $\to$ eex. The QCD calculations, however, are done for electron scattering off a (real or virtual) photon target [14], where the spin average is done over the incoming electron and photon. This leads to a formula with only two structure functions, F_2 and F_1, whereas the cross section for unpolarized electrons is given (after ϕ integration) by a four term formula (eq. 1). In the QM one can show that for Q^2, $p^2 \gg m_q^2$ the measurable cross section corresponds to [23]

$$\sigma \simeq \frac{4\pi^2\alpha}{Q^2} \ (F_2 + \frac{3}{2} F_1) \qquad\qquad (20)$$

Therefore in Fig. 24 the longitudinal structure function (times 3/2) was added to F_2 in the QCD calculation. For the future it would be helpful to have a QCD calculation for the whole process which could be directly compared to the measurements.

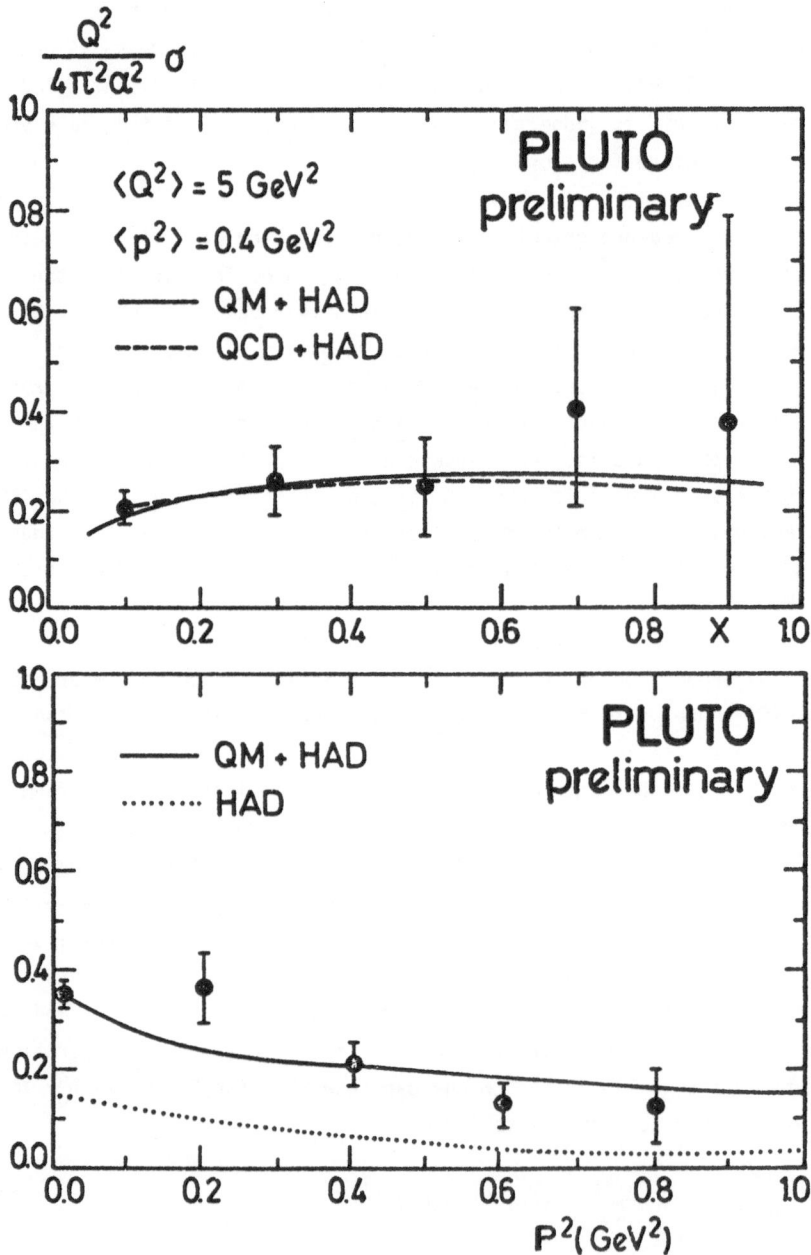

Fig. 24 Virtual photon structure function measured in the double tag mode (PLUTO)
a) as a function of x averaged over Q^2 and p^2 and b) as a function of p^2
averaged over Q^2 and x

Conclusions

An enormous progress has been made during the last two years in the measurement of the photon structure function. The pointlike component of the photon has definitely shown up. Despite all the problems with QCD calculations the data are described by the very simple model 'hadronic part + leading order QCD, $\Lambda \simeq 200$ MeV' in many respects: the shape in x as well as the absolute normalization of $F_2^\gamma(x)$ agrees very well at moderate $<Q^2>$ values of 5.9. The Q^2 evolution has been studied up to several hundred GeV and is also consistent with this model. A detailed QCD test from the Q^2 evolution of F_2^γ, however, would require about ten times the statistic. I think, its worth the effort!

Acknowledgement

I am grateful to all my colleagues at PETRA and PEP for providing me with all the information and material needed for this review. I wish to thank Chris Berger for organizing such a nice and fruitful meeting.

References

1   V.M. Budnev et al., Phys. Reports 15C, (1975) 181

2   J.L. Rosner, ISABELLE Physics Prospects, BNL Report 17522 (1972) 316,
    unpublished

3   Ch. Berger, Proc. Int. Colloquium on Photon Photon Interactions, Paris (1981)
    ed. by G.W. London

4   N. Wermes, Ph.D. Thesis, Bonn - IR - 82 - 27 (1982)

5   C.J. Biddick et al., Phys. Lett. 97B, (1980) 320

6   PLUTO collaboration, Ch. Berger et al. Phys. Lett. 89B, (1979) 120
    PLUTO collaboration, Ch. Berger et al. Phys. Lett. 99B, (1981) 287

7   E. Hilger, DESY 80/75

8   F. Raupach, Ph.D. Thesis, PITHA 81/05

9   I.F. Ginzburg and V.A. Serbo, Phys. Lett. 109B (1982) 231

10 N. Wermes, these proceedings

11 E. Witten, Nucl. Phys. B120 (1977) 189

12 D. Duke, these proceedings

13 W. Bardeen, discussion remark in ref. 12

14 T. Uematzu and T.F. Walsh, Nucl. Phys. B199 (1982) 93

15 W. Wagner, PITHA 83/03

16 PLUTO collaboration Ch. Berger et al., Phys. Lett. 107B (1981) 168

17 H. Spitzer, these proceedings

18 CELLO collaboration, H.J. Behrend et al., DESY 83/018

19 F.J. Kirschfink, talk given in the parallel session, summarized in ref. 17

20 JADE collaboration W. Bartel et al., Phys. Lett. 121B (1983) 203
    T. Nozaki, talk given in the parallel session, summarized in ref. 17

21 MAC collaboration, private communication

22 F. Raupach, talk given in the parallel session, summarized in ref. 17

23 This was pointed out by J. Dainton and S. Maxfield

Discussion

T. Nozaki (DESY):   What value of Q^2 did you use for estimation of Λ determination? I want to comment that the VDM contribution decrease as a function of Q^2. And if we use the prediction of Frazer and Gunion we get a factor 2 smaller VDM contributions at $Q^2 > 25$ GeV^2. This might reduce the ambiguity of Λ determination in the high Q^2 region.

W. Wagner: I used an uncertainty of the hadronic part of $0.05 \cdot \alpha \ (1-x)$ and assumed that this uncertainty is independent on Q^2. In this case the uncertainty of Λ is independent on Q^2. Going to large Q^2 only helps for α_s, not for Λ! Only if we once have a reliable calculation of the VDM part, more accurate than the above mentioned uncertainty, we can determine Λ more precisely. So far I would regard the differences in the theoretical predictions to be a good estimate for the uncertainties.

T. Nozaki (DESY):   Did MAC collaboration applied any special cut to reduce the VDM component in their analysis?

W. Wagner: I don't know about any special cut against VDM background. You probably reflect to the excellent agreement of their data with the QM prediction. But you should note that the QM has been normalized to the data. So its not clear that they don't have room for some VDM component.

QED PROCESSES IN TWO PHOTON REACTIONS [*]

M. Pohl
III. Phys. Inst. A
R W T H Aachen, Germany

ABSTRACT

I review experimental results on the reactions $e^+e^- \rightarrow e^+e^-e^+e^-$ and $e^+e^- \rightarrow e^+e^-\mu^+\mu^-$ from PETRA and PEP. Recent high statistics measurements are compared to QED predictions in the form of diagrammatic leading order (α^4) calculations and to the equivalent photon approximation. The leading order calculation describes the data well over the full kinematic range covered by experiments ($0.1 \lesssim Q^2 \lesssim 100$ GeV^2/c^2). The "two photon interaction"graph is found to saturate the observed cross section. Only for small masses of the produced leptonic system, first indications of a bremsstrahlung type background are observed at the percent level. The leptonic structure function $F_2^\gamma(x, Q^2)$ is measured and also agrees with the behavior expected from QED.

[*] Invited talk given at the 5th International Workshop on Photon Photon Collisions, Aachen, April 1983.

1.   INTRODUCTION

Lepton pair production by the two photon process

$$e^+e^- \rightarrow e^+e^-e^+e^- \tag{1}$$

$$e^+e^- \rightarrow e^+e^-\mu^+\mu^- \tag{2}$$

is one of the few examples of a higher order (α^4) QED reaction that can be studied at large momentum transfers (Q^2). A measurement of these processes at high energies therefore provides a test of QED complementary to the more conventional studies of higher orders at low Q^2 [1] or first order at maximum Q^2 [2].

Because of their logarithmic growth with the cms energy of the initial state ($\sqrt{s}$), the total cross sections are large even at high energies. In present collider experiments, reaction (2) even dominates the first order process $e^+e^- \rightarrow \mu^+\mu^-$ by several orders of magnitude. Most of this cross section, however, goes unobserved, since the bremsstrahlung nature of the process strongly aligns the reaction products with the incoming beams. Cross section measurements of two photon reactions therefore generally involve large extrapolation of the accessible kinematic range. A detailed analysis of data on these calculable "gauge reactions" allows one to study the reliability of such extrapolations.

It is mainly because of these two reasons that almost all experiments at PETRA and PEP are presenting high statistics measurements of reactions (1) and/or (2). The kinematic variables used in characterizing the data are displayed in Fig. 1. When the momentum Q_1^2 transferred to both electrons is small, only the produced lepton pair can be observed. In this "no tag" case, the mass W of the system and the transverse momentum $P_\perp$ of the particles are measured. In the comparatively rare case where one of the photons is sufficiently off shell such that the radiating electron is "tagged" in the detector, also its momentum transfer can be measured and the kinematics are completely determined to the extent that the second photon is real.

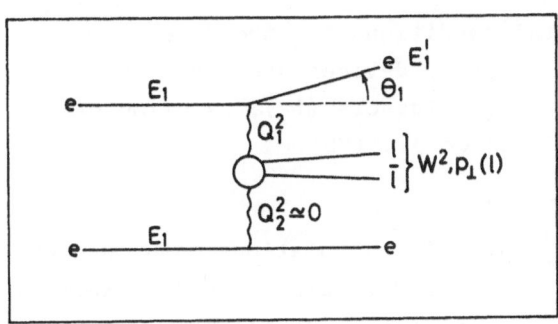

Fig. 1

The experimental conditions and statistics for the data included in this report are summarized in Table I. The Q^2 range covered by the experiments is shown in Fig. 2. It can be seen that detectors are rather complementary and that the accessible momentum transfers range all the way from $Q^2 \simeq 0.1$ GeV^2/c^2 to 100 GeV^2/c^2.

Experi-ment	$\int L dt$ pb^{-1}	$< \sqrt{} >$ GeV	Final State	Number of Events no tag	tag	Tag Device
CELLO	7.5	34	eeee		130	EC
			eeμμ		111	EC
MARK-J	85.8	34	eeμμ	2861	(132)	CD
PLUTO	40	35	eeμμ	987	537	SA
					345	LA
					100	EC
TASSO	19.3	33	eeμμ		127	EC
MAC	25	29	eeμμ	2763		-
PEP-9/ TPC	3	29	eeμμ		421	SA

Table I

Statistics and experimental conditions for the data included in this report. Tagging devices used are small angle taggers (SA), large angle taggers (LA), calorimeter end caps (EC), and central detectors (CD).

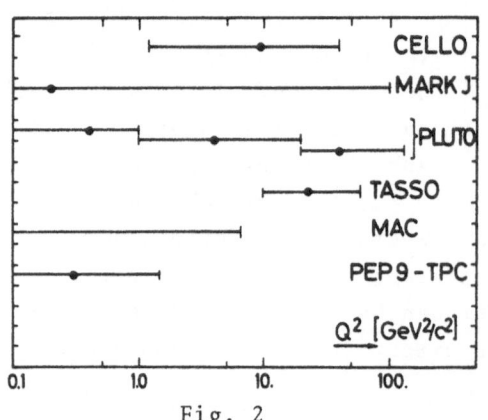

Fig. 2

The QED calculations used to compare to the data are twofold. Exact dia-grammatic calculations to lowest order[3] exist in form of a Monte Carlo integration program and event generator and allow for detailed studies taking into account the detector response.

The types of diagrams considered are schematically shown in Fig.3. According to QED, the lepton pair production is dominated by the "two photon" graph with charge parity C = +1. A contribution at the percent level comes from the t-channel bremsstrahlung diagram with C = -1. Bremsstrahlung from the s-channel is negligible. Bremsstrahlung contributions may be found experimentally by looking at low mass lepton pairs or by searching for a small charge asymmetry in the

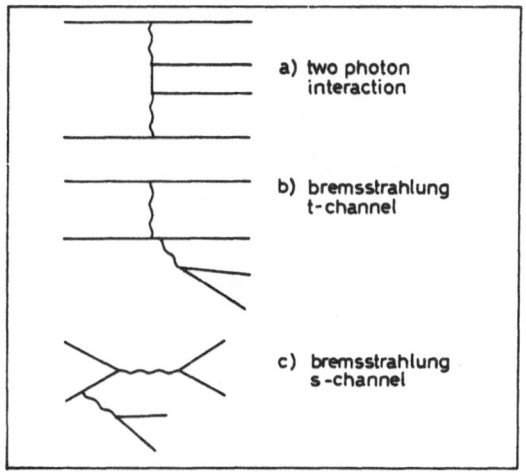

a) two photon interaction

b) bremsstrahlung t-channel

c) bremsstrahlung s-channel

Fig. 3

produced leptons due to interference between the two graphs with opposite C parity. Calculations of higher order corrections (α^5) are in progress[4]. First indications are that they are very small and can safely be neglected at the present level of experimental accuracy.

Given the dominance of the two photon interaction diagram, the cross section can also be calculated in the equivalent photon approximation (EPA)[5]. In that case the photon flux[6] coming from an incident electron is convoluted with the cross section for eγ scattering. If both photons are quasi-real, the photon flux from both electrons may be convoluted with the cross section for $\gamma\gamma$ scattering (double equivalent photon approximation, DEPA)[7].

2.　GENERAL PROPERTIES OF $e^+e^- \rightarrow e^+e^-\ell^+\ell^-$

Since two photon interactions are a bremsstrahlung phenomenon, both initial state photons are of low energy and strongly aligned with the incoming $e^\pm$ beams. Moreover, the lepton pair production cross section in the $\gamma\gamma$ center of mass system

$$\frac{d\sigma(\gamma\gamma \rightarrow \ell\bar{\ell})}{d\cos\theta^*} \sim \frac{1 + \cos^2\theta^*}{1 - \cos^2\theta^*} \qquad (3)$$

is strongly peaked along the photon direction. The observable fraction of cross section in a storage ring experiment is therefore at the percent level. Fig. 4 shows the observed cross section as a function of $\sqrt{s}$ for $e^+e^- \rightarrow e^+e^-\mu^+\mu^-$ in the MARK-J experiment. It becomes

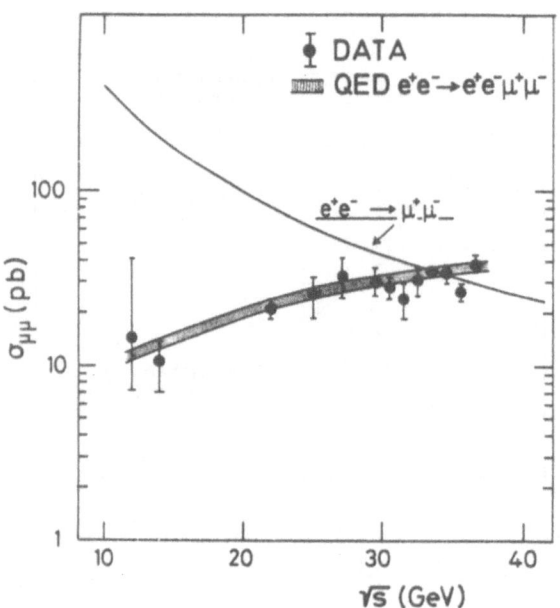

Fig. 4

comparable to the one photon exchange cross section (of which ∿ 60% are observable) only at the highest PETRA energies. The two processes can, however, easily be separated even if no tagged electron is observed. Fig. 5 shows the momentum spectrum of muon pairs measured in MARK-J. The peak at $P_\mu/P_{beam} \simeq 1$, due to the one-photon process, is clearly separated from the low energy muons produced by two photon interactions.

Fig. 6 compares the acolinearity and acoplanarity distributions observed in two-photon muon pairs to the QED prediction including detector resolution. While the boost of the $\gamma\gamma$ cms tends to make the muons acolinear, they stay largely coplanar since under "no tag" conditions the transverse momentum carried away by the electrons is limited.

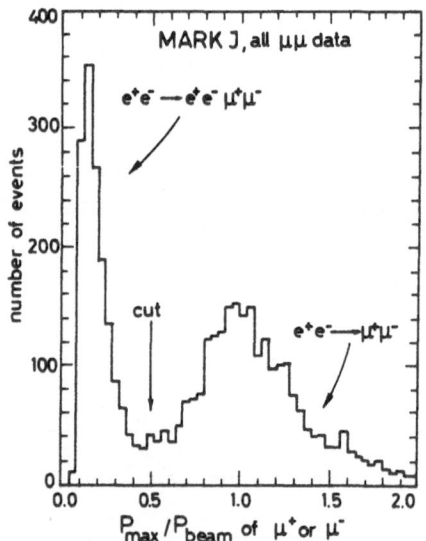

Fig. 5

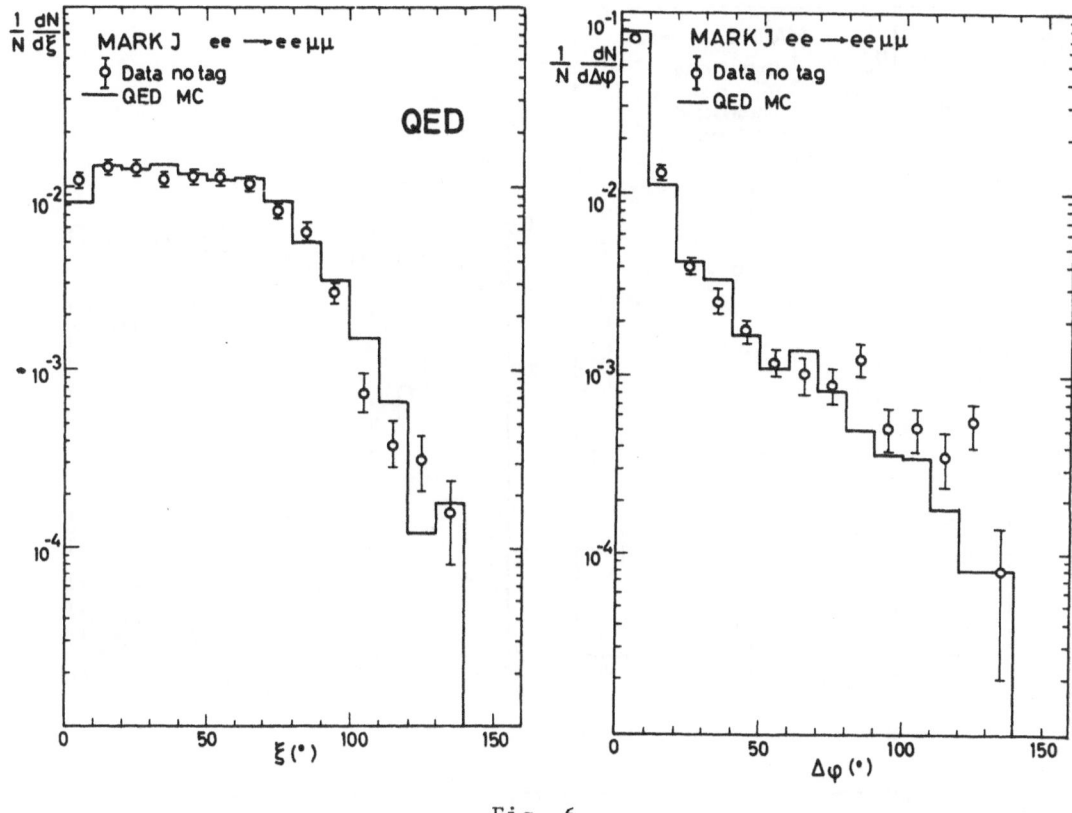

Fig. 6

3) UNTAGGED LEPTON PAIR PRODUCTION

In the case where none of the initial electrons is scattered into
the detector, it is difficult to disentangle the two photon process
$e^+e^- \rightarrow e^+e^-e^+e^-$ from radiative Bhabha scattering $e^+e^- \rightarrow e^+e^-\gamma$. Data
under "no tag" conditions are therefore only available for muon pair
production.

Fig. 7 shows the uncorrected mass distribution of two photon muon
pairs observed by MARK-J and PLUTO at PETRA ($\sqrt{s} \simeq 34$ GeV) and by MAC
at PEP ($\sqrt{s} = 29$ GeV). The data are compared to a QED (α^4) Monte Carlo
calculation[3] which takes into account the detector acceptance and
resolution. The agreement is very satisfactory up to the highest masses
accessible. No indication for a new state with $C = + 1$ decaying into
$\mu^+\mu^-$ is seen in the mass range covered.

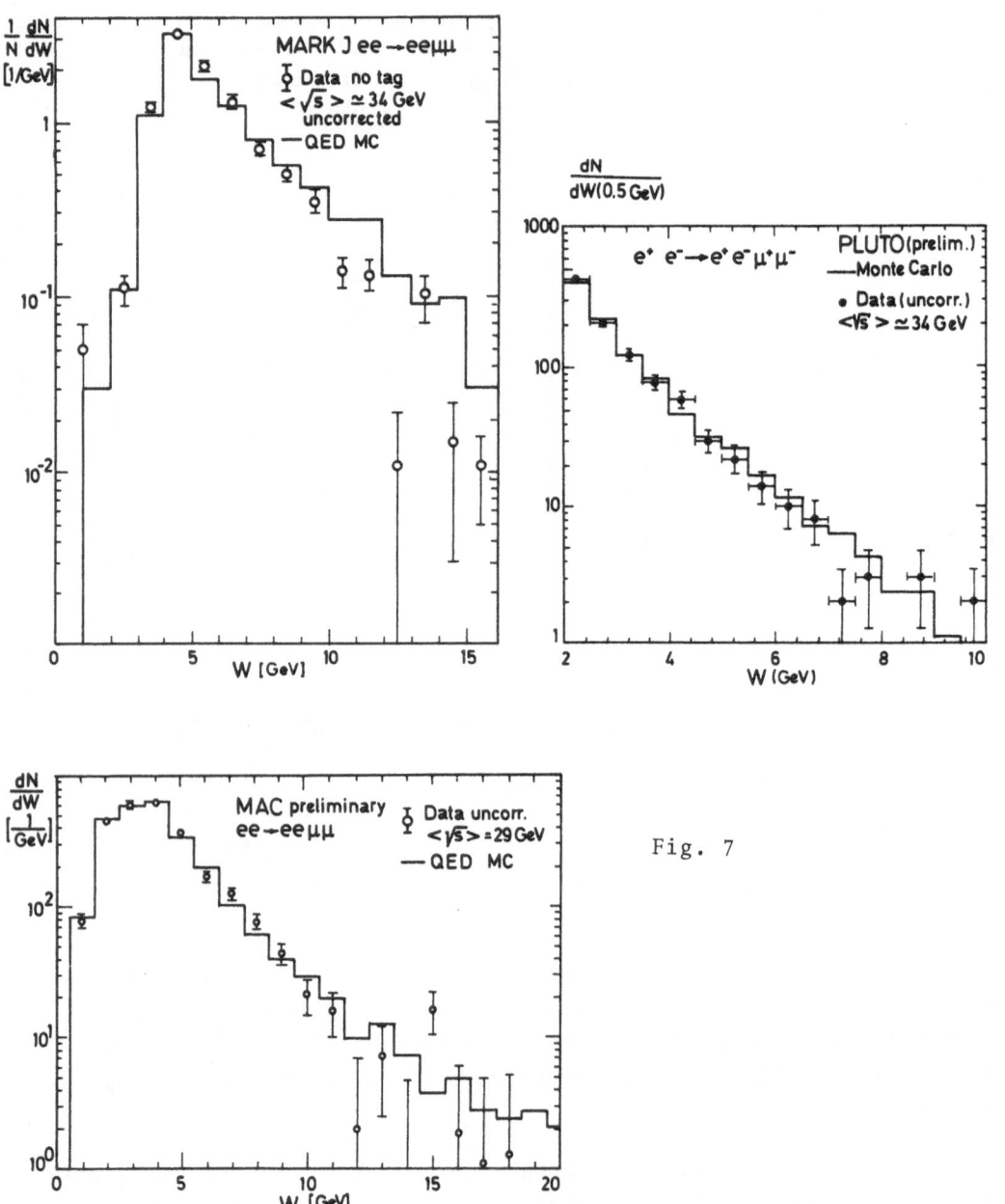

Fig. 7

The agreement between data and leading order QED Monte Carlo after detector response simulation allows one to apply a bin-wise acceptance correction to the data. This has been done for the data in Fig. 8 which displays the differential cross section as a function of the muon

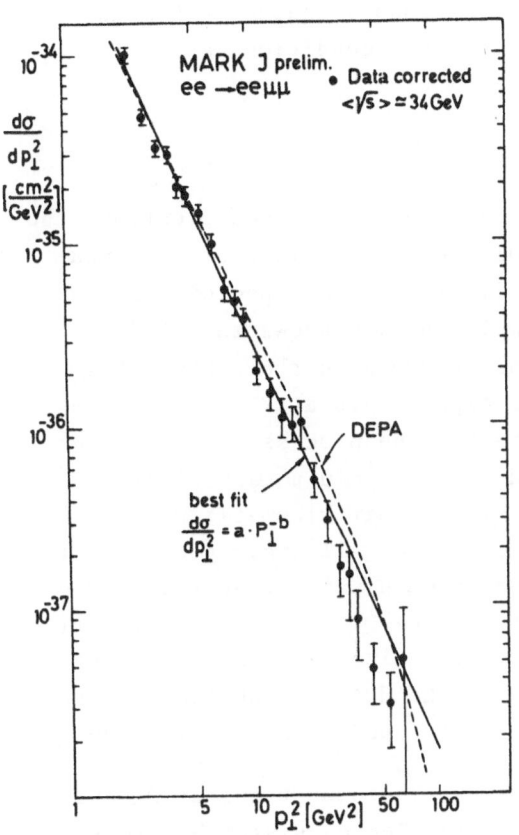

Fig. 8

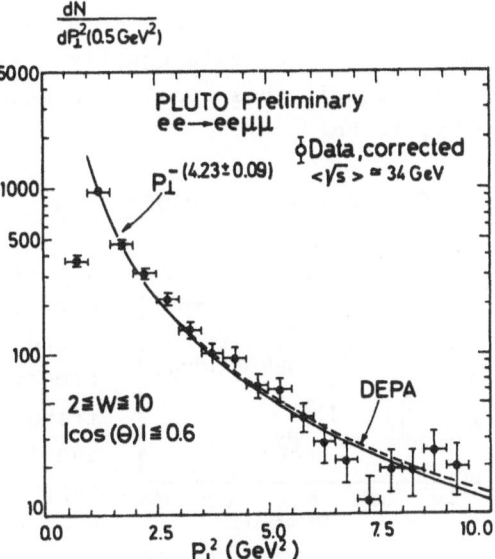

transverse momentum $P_\perp$ with respect to the beam line. The data fit well to a power law

$$\frac{d\sigma}{dP_\perp^2} = a \cdot P_\perp^{-b} \qquad (4)$$

In the case of the MARK-J data, taken at an average Q^2 of 0.2 GeV^2/c^2 and for $|\cos\theta_\mu| < 0.86$, the best fit parameters are

$$a = 5.43 \pm 0.90 \text{ (stat)} \pm 0.50 \text{ (syst)} \quad nb/(GeV^2/c^2)$$
$$b = 4.47 \pm 0.21 \text{ (stat)} \pm 0.25 \text{ (syst)} \qquad (5)$$

QED predicts an exponent $b = 4.54$ in good agreement with the observation. The preliminary PLUTO data have an average Q^2 of 0.01 GeV^2/c^2 and cover $|\cos\theta_\mu| < 0.60$. The exponent is fitted to be

$$b = 4.23 \pm 0.09 \text{ (stat)} \qquad (6)$$

also in good agreement with the QED prediction of 4.19 calculated for their conditions.

Also shown in Fig. 8 are the predictions of a DEPA calculation[7] which describes the data equally well. The double equivalent photon approximation is thus applicable under "no tag" conditions as it is expected to be.

4) TAGGED LEPTON PAIR PRODUCTION

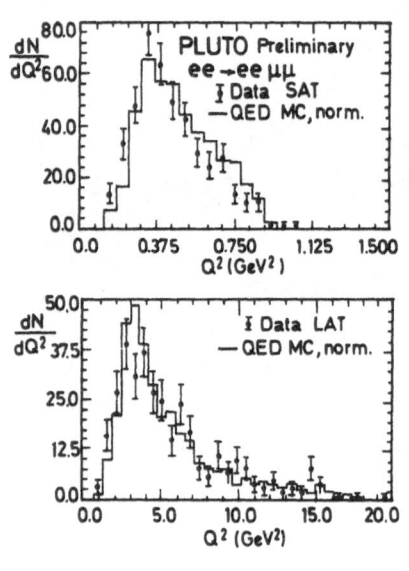

In the case of a tagged electron, the mass Q^2 of the projectile photon is measured, while the target photon remains almost real. Fig. 9 shows the Q^2 distributions observed in the CELLO, PLUTO and TASSO experiments at PETRA. In the CELLO data[8], the processes $e^+e^- \rightarrow e^+e^- e^+e^-$ and $e^+e^- \rightarrow e^+e^- \mu^+\mu^-$ are plotted together since their kinematic properties are very similar. All other groups restrict themselves to muon pair production. Fig.9 substantiates the Q^2 range of the experiments already indicated in Fig. 2. The agreement of the data to the leading order QED Monte Carlo predictions is very good.

Since the momentum transfer Q^2 and the mass of the lepton system W^2 are simultaneously measured, the "scaling" variable

$$x = \frac{Q^2}{Q^2 + W^2} \qquad (7)$$

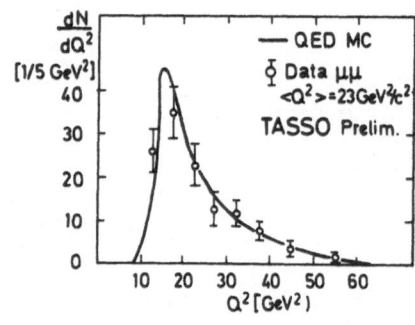

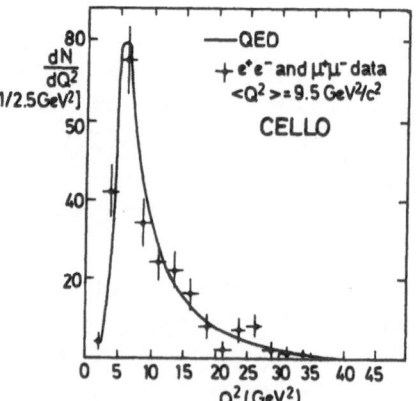

Fig. 9

can be infered. Experimental results are shown in Fig. 10. The agreement with QED is again generally good. An interesting aspect is seen in the PLUTO data where the high Q^2 data are compared to an absolute QED prediction considering <u>only</u> the "two-photon" graph of Fig. 3a. An excess of events at $x \simeq 1$ ($W^2 \simeq 0$) is observed. It can be attributed to events from virtual bremsstrahlung, Fig. 3b, which are of course expected to have a small muon pair mass. The candidates observed represent roughly 1% of the observed cross section which is also the correct order of magnitude for bremsstrahlung contributions.

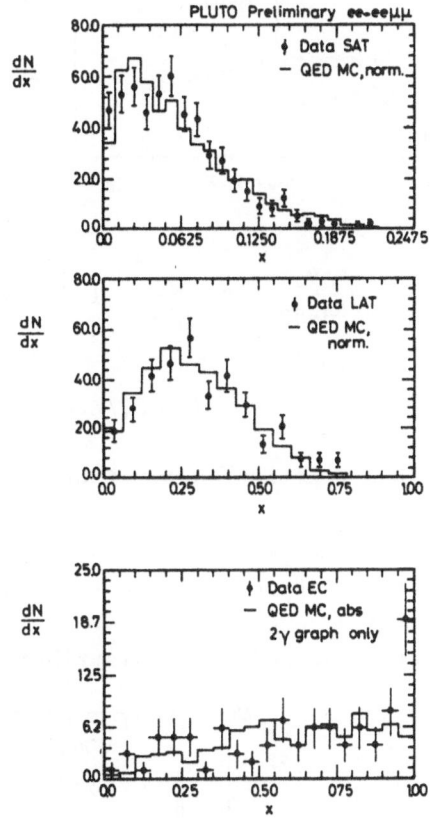

Fig. 10

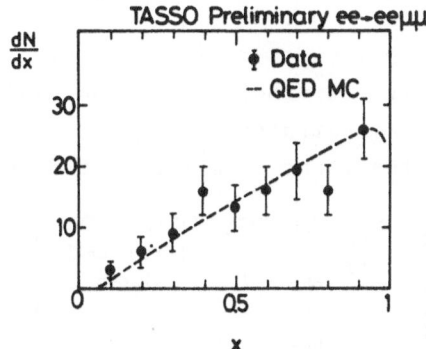

5) "LEPTONIC STRUCTURE" OF THE PHOTON

As mentioned briefly in the introduction, the lepton pair production by two photons is an important gauge reaction used to study and gain confidence in methods subsequently applied to other two photon reactions, like $e^+e^- \rightarrow e^+e^- q\bar{q}$, which are less well understood theoretically as well as experimentally. This especially applies to the structure function formalism. Unlike the hadronic structure function of the photon, its leptonic structure function can be reliably calculated in QED. Moreover, all relevant final state particles are observed and their momenta measured such that no experimental problems like unfolding the "true" kinematics from the visible ones occur.

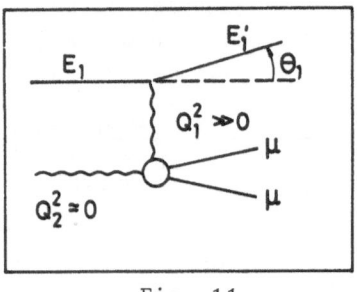

Fig. 11

For the sake of this test we pretend that we know nothing about the point-like coupling of the two photons to the lepton pair. We treat the process as deep inelastic scattering of an electron off a quasi-real photon, as shown in Fig. 11. We parametrize[5] the cross section in terms of structure functions $F_1(x, Q^2)$ and $F_2(x, Q^2)$. It then takes the general form

$$\frac{d^2\sigma}{dxdQ^2} = \frac{4\pi\alpha^2}{Q^4 x} \{ (1-y)F_2^\gamma + xy^2 F_1^\gamma \}$$ (8)

where

$$y = 1 - E_1'/E_1 \cos^2 \frac{\theta_1}{2}$$ (9)

is the relative energy transfer.

It is clear from equation (8) that experiments will have no sensitivity to F_1, unless low tagging energy thresholds E_{min}^1 and minimum angles θ_{min} are accepted. This is demonstrated in Fig. 12 with a Monte Carlo study of the y distributions resulting from the F_2 and xF_1 term

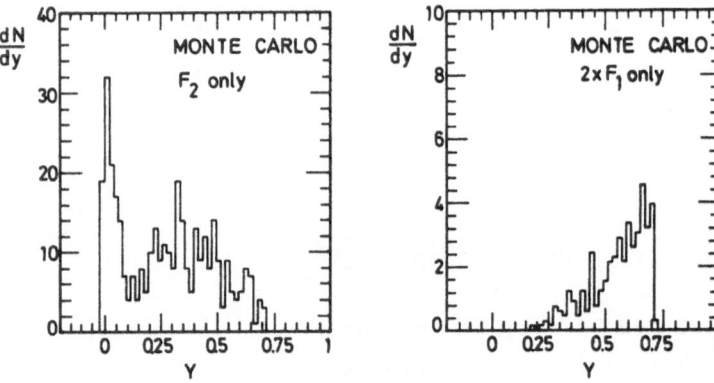

Fig. 12

in equation (8). An experimental y distribution from PEP-9 is shown in Fig. 13. It is evident that under present conditions, only F_2 can be efficiently extracted from the data. Fig. 14 shows the results from CELLO and PEP-9 at $<Q^2>$ of 9.5 GeV^2/c^2 and 0.3 GeV^2/c^2, respectively.

Since the measured F_2 is
averaged over the Q^2 range
accepted, the observed
difference is explained by $\frac{dN}{dy}$
the Q^2 dependence of the
structure function.
Neglecting the mass of the
target photon, QED predicts[5]

$$F_2^\gamma(x,Q_1^2) = \frac{\alpha}{\pi} \, x \, \{ [x^2 + (1-x)^2]$$

$$\ln \, \frac{Q_1^2 (\frac{1}{x} - 1)}{m_\mu^2} - 1 + 8x(1-x) \}$$

(10)

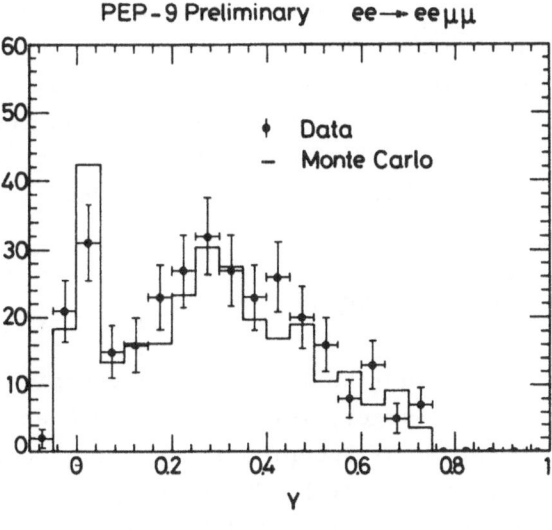

Fig. 13

The corresponding curves are also
shown in Fig. 14. The differences
between the two measurements are thus
quantitatively understood by the $\ell n Q^2$
dependence of the structure function.
The agreement with QED is good.

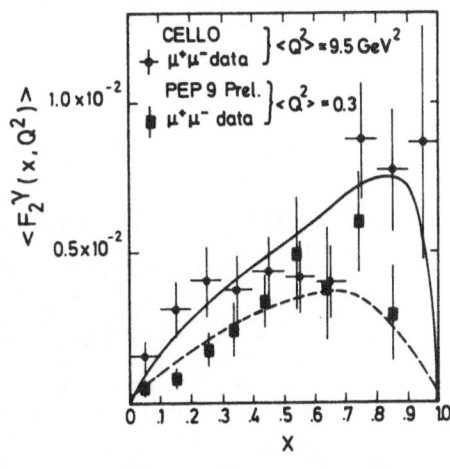

Fig. 14

CONCLUSIONS

The recently accumulated high statistics data on

$$e^+e^- \rightarrow e^+e^- \, e^+e^-$$

and

$$e^+e^- \rightarrow e^+e^- \, \mu^+\mu^-$$

from experiments at PETRA and PEP agree with the expectations from
leading order (α^4) calculations in QED. The data cover a Q^2 range
$0.1 \lesssim Q^2 \lesssim 100 \text{ GeV}^2/c^2$, and are taken under "no tag" as well as
"single tag" conditions. The no tag cross sections are also well
described by an analytical calculation based on the double equivalent

photon approximation. Under most conditions, the "two-photon interaction" graph alone adequately accounts for the data. Only at $x \simeq 1$,
a "bremsstrahlung" background at the percent level is observed. The
structure function approach gives an appropriate description of the data
on $e\gamma \rightarrow e\mu\mu$. The measured "muonic" structure function $<F_2^{\gamma}(x,Q^2)>$ agrees
with the behavior expected from QED.

ACKNOWLEDGEMENTS

I would like to thank the organizers of this conference, especially
Prof. Ch. Berger, for creating such an informal athmosphere at this
meeting. I am indepted to my colleagues from the PETRA and PEP experiments, especially Profs. E. Hilger, F.J. Kirschfink, W. Ko, K. Lau,
G. Swider, A. Tylka and C. Williams for providing me with preliminary
data prior to publication. I also thank the directors of DESY, Prof.
V. Soergel and Prof. P. Söding for their hospitality.

REFERENCES

(1) J. Builey et al., Nucl. Phys. B150, 1 (1979).
     R.S. Van Dyck, P.B. Schwingberg and H.G. Dehmelt, Phys. Rev. Lett.
     38, 310 (1979).
     K. von Klitzing, G. Dorda and M. Pepper, Phys. Rev. Lett. 45, 494
     (1980).
(2) H.J. Behrend et al., Z. Phys. C16, 301 (1983).
     W. Bartel et al., Phys. Lett. 108B, 160 (1982) and DESY 83-035.
     B. Adeva et al., Phys. Rev. Lett. 48, 1701 (1982).
     Ch. Berger et al., Phys. Lett. 99B, 292 (1981).
     R. Brandelik et al., Phys. Lett. 117B, 365 (1982).
     E. Fernandez et al., Phys. Rev. Lett. 50, 1238 (1983).
(3) R. Bhattacharya,J.Smith and G. Grammer, Phys.Rev.D15, 3267 (1977).
     J.A.M. Vermaseren, Proc. of the Int. Workshop on $\gamma\gamma$ Collisions,
     Amiens 1980, G. Cochard and P. Kessler Edts, Lecture Notes in
     Physics 134, 35 (1980).
     S. Kawabata, see J. Field, Proc.of the 4th Int.Workshop on Photon-
     Photon Interactions, G.W. London Edt., Paris 1981, p. 447.
(4) J.A.M. Vermaseren, Contribution to this conference.
     Ph. Daverveldt, Contribution to this conference.
(5) see V.M. Buchner,I.F. Ginzburg,G.V. Meledin and V.G.Serbo, Phys.
     Rep. 15, 181 (1975).
(6) J.H. Field, Nucl. Phys. B168, 477 (1980); B176, 345 (1980).
     Ch. Berger and J.H. Field, Nucl. Phys. B187, 585 (1981).
(7) A. Coureau, CAL 82/19 (1982).
(8) CELLO Collaboration, H.J. Behrend et al., DESY 83/017 (1983).

FIGURE CAPTIONS

Fig. 1   Kinematic variables used to study lepton pair production by
          two photon interactions.

Fig. 2 Q^2 range covered by the data included in this report.

Fig. 3   Types of Feynman graphs taken into account in the diagrammatic
          calculations of Ref. 3.

Fig. 4   Observed cross section for $e^+e^- \to e^+e^- \mu^+\mu^-$ (no tag) from
          MARK-J as a function of center of mass energy.

Fig. 5   Maximum muon momentum measured by MARK-J in all events with
          two muons. The one photon and two photon production processes
          are clearly separated.

Fig. 6   Acolinearity (ξ) and acoplanarity ($\Delta\phi$) distributions for two
          photon muon pairs from MARK-J.

Fig. 7   Observed mass distribution for two photon muon pairs from
          MARK-J, PLUTO, and MAC, compared to the QED prediction
          including detector acceptance and resolution.

Fig. 8   Corrected cross section $d\sigma/dP_\perp^2$ for two photon muon pairs
          (no tag) from MARK-J and PLUTO. The data are compared to
          a DEPA calculation and fitted to a power law.

Fig. 9 Q^2 distribution of single tag data from CELLO, PLUTO (small
          angle, large angle and end cap calorimeter tags) and TASSO.
          Data from CELLO include electron pair production.

Fig. 10   x distribution of single tag data from PLUTO and TASSO.

Fig. 11   Kinematic variables used to study eγ scattering.

Fig. 12   A Monte Carlo study from the PEP 9 group showing the con-
          tributions of the F_2 term and the x F_1 term to be observed
          y distribution. Note the difference in vertical scale for
          the two graphs.

Fig. 13   Observed y distribution from PEP 9 compared to the QED prediction.

Fig. 14   The muonic structure function F_2^γ (x, Q^2) as measured by
CELLO ($<Q^2> \simeq 9.5$ GeV GeV^2/c^2) and PEP 9 ($<Q^2> \simeq 03.$ GeV^2/c^2),
averaged over the Q^2 range of the experiments. The data are
compared to the QED prediction, equ. 10, neglecting the target
photon mass.

DISCUSSION

Question   M. Gorn (Munich):   If you had to measure the muon mass
                                 from the leptonic structure function F_2,
                                 what would be the uncertainty?

Answer   M. Pohl: I don't know the answer since nobody
                                 has tried that up to now. You can see,
                                 however, from the error bars in Fig.14
                                 and the logarithmic dependence of F_2
                                 on m_μ that the accuracy would not be
                                 great.

Question   S. Brodsky (SLAC):   One of the best ways to verify that the
                                 τ is a normal lepton is $\gamma\gamma \rightarrow \tau^+\tau^-$. Have
                                 such events been identified?

Answer   M. Pohl: No. There is also little hope to do
                                 that in the lepton pair final states
                                 because of the suppression by the
                                 leptonic branching ratio of the τ and
                                 the indistinct event signatures.

Comment   J.H. Field: The speaker is correct in stating that
                                 there is no possibility in extracting
                                 the $\tau\tau$ signal from the ee and $\mu\mu$ pair
                                 final states. This can, however, be done
                                 by using the hadronic decay modes of the
                                 τ leading to a 3 + 1 final state topo-
                                 logy.

Comment   J. Haissinski (Orsay): The radiative corrections are much
                                 larger in the tagged event case than in
                                 the no tag case. In the former case,
                                 real photon emission by the incoming
                                 electron (the one that is scattered at
                                 wide angle and tagged) has to be taken
                                 into account, not so much because σ_{tot}
                                 is changed, but because it modifies the
                                 various kinematical distributions. Such
                                 corrections have been applied by the
                                 CELLO collaboration.

Question P. Kessler   :   In the CELLO experiment on $e^+e^- \rightarrow e^+e^-$
          (College de France) $\ell^+\ell^-$, a rather high asymmetry has been
found between ℓ^+ and ℓ^-. Can you
comment on that? Have the other groups
found similar asymmetries?

Answer   M. Pohl: QED indeed predicts a small (a few %)
forward-backward charge asymmetry
coming from an interference between the
graphs in Fig. 3a and 3b. CELLO seems
to find an asymmetry in their tag data
that is higher, but compatible with
this prediction. They are, however,
not ready to quote a number. In the
MARK-J no tag data, i.e. at low Q^2, the
observed asymmetry is very small and
compatible with zero.

A THEORETICAL REVIEW OF THE
PHOTON STRUCTURE FUNCTION

Dennis W. Duke
Physics Department, Florida State University
Tallahassee, Florida 32306/USA

I. Introduction

We are seeing at this workshop the results of almost incredible advances on the experimental side of two photon physics. We are also lucky, perhaps, that there have been exciting and important new advances on the theoretical side as well. The main improvements are that the reasons for the somewhat bothersome higher order predictions of negative values of the structure function $F_2^\gamma(x,Q^2)$ are now more thoroughly understood, and that the range of applicability of the QCD predictions for F_2^γ is more carefully delineated and appreciated. The implications of this new understanding for the old experimental goal of using F_2^γ to measure Λ are not, as we shall discuss, altogether positive.

II. A Brief History

The prediction and experimental discovery of approximate Bjorken scaling in deep inelastic electron nucleon scattering has taught us to think in terms of pointlike constituents of hadrons dominating the physics, at least in certain kinematic regions. In 1971 the extension of these ideas to two photon physics was suggested by Walsh[1] and by Brodsky, Kinoshita and Terazawa[2]. These authors suggested that the reaction

$$e^+ e^- \rightarrow e^+ e^- + hadrons$$

would offer an opportunity to measure the structure functions of (nearly) real target photons when probed by another highly virtual photon by tagging one lepton at a large angle and the other lepton at a small angle. When the hadronic properties of the target photon are interpreted in terms of the vector meson dominance (VMD) model, then one expects to observe approximate Bjorken scaling for the photon structure function, exactly in analogy with the nucleon target case.

In 1973 Walsh and Zerwas[3] pointed out that the target photon could actually couple directly to its pointlike quark-parton constituents and provide anomalous contributions, e.g. $F_2^\gamma(x,Q^2)$ has a nonscaling contribution proportional to $\ln Q^2/m^2$, and $F_L^\gamma(x,Q^2)$ is nonzero in distinction to the usual expectations for spin 1/2 partons. This subject was later studied by Kingsley[4] and by Worden.[5]

In 1974 it was discovered by Politzer[6] and by Gross and Wilczek[7] that nonabelian gauge theories (e.g. QCD) are asymptotically free and offer an explanation of the approximate Bjorken scaling observed in deep inelastic electron nucleon scattering, as well as quantitative predictions for the pattern of scaling violations to be expected. The predictions of QCD for the hadronic or VMD part of the photon structure functions were then studied by Ahmed and Ross[8] in 1975.

The treatment of the photon structure functions in QCD was the subject of a remarkable paper by Witten[9] in 1977. Witten showed, within the operator product expansion (OPE), how the hadronic and pointlike parts of the photon structure functions are unified by operator mixing between the usual quark and gluon operators and the photon operator. The main new result was that the photon structure function is absolutely calculable in QCD for asymptotically large values of Q^2.

In 1979 Witten's OPE analysis was translated into diagrammatic and/or Altarelli-Parisi-Lipatov[10] language by Llewellyn Smith[11], Frazer and Gunion[12], and DeWitt, Jones, Sullivan, Willen and Wyld.[13]

The completion of the calculations of the higher order QCD corrections to deep inelastic scattering in 1978 led Bardeen and Buras[14] in 1979 to extend Witten's OPE analysis beyond the leading order. In 1980 Duke and Owens[15] extended slightly the Bardeen-Buras calculation and pointed out that the higher order corrections actually drove the perturbatively calculable part of F_2^γ to negative values for x<0.1 and x~1.

For experimentally accessible values of Q^2, these unphysical predictions of negative cross sections are not alleviated by simply adding the hadronic VMD pieces in the usual naive way.

In 1981 Uematsu and Walsh[16)] showed that when the target photon mass $-p^2$ is not zero and bounded by $\Lambda^2 << p^2 << Q^2$, then the hadronic part of F_2^γ is perturbatively calculable and the predictions, including higher order corrections, are no longer negative for small x (although there is still a problem for large x~1).

Based on the Uematsu-Walsh result, Bardeen[17)] explained the origin and resolution of the small x problem in his 1981 Bonn Conference talk. In addition, Frazer[18)] has discussed the behavior of $F_2^\gamma(x,Q^2)$ for x~1 based on Frazer and Rossi[19)], and Chase.[20)]

III. Witten's Breakthrough

Prior to the work of Witten, the photon structure function was generally understood in terms of two separate pieces – a hadronic piece presumably described by VMD, and a pointlike piece described by the box diagram. Neither of these pieces is perturbatively calculable in QCD, since we cannot calculate either the hadronic matrix elements of the photon for VMD or the dynamical mass scale(s) necessary to regulate the infrared divergences of the box diagram. In terms of moments of $F_2^\gamma(x,Q^2)$, for example, we have then

$$M_n^\gamma(Q^2) = \int_0^1 dx \; x^{n-2} \; F_2^\gamma(x,Q^2)$$
$$= \sum_{i=+,-,NS} C_n^i(Q^2/^2,g^2,) <\gamma|O_n^i|\gamma>$$
$$+ M_n^{box}(Q^2). \tag{1}$$

In Eq.(1) the C_n^i are each of $O(\alpha^0)$ while the $<\gamma|O_n^i|\gamma>$ are each of $O(\alpha^1)$. Also, to $O(\alpha)$ we have

$$M_n^{box}(Q^2) = \frac{3f\langle e^4\rangle\,\alpha}{\pi}\; x\left[(1-2x+2x^2)\;\ln\frac{Q^2(1-x)}{m^2 x}\right.$$
$$\left. -1+8x+8x^2\right].$$

Witten realized that in the OPE framework, we should have instead

$$M_n^{box}(Q^2) = C_n^\gamma(Q^2/\mu^2,g^2,\alpha)\langle\gamma|O_n^\gamma|\gamma\rangle, \qquad (2)$$

where now $C_n^\gamma\sim O(\alpha)$ and $\langle\gamma|O_n^\gamma|\gamma\rangle\sim O(1)$. In Eq.(2) we note the appearance of a twist-two operator O_n^γ constructed from the photon tensor $F_{\mu\nu}$ in the same way that O_n^G is consructed from the gluon tensor $G_{\mu\nu}$.

The essential new idea - mixing between the operators $O_n^{+,-,NS}$ and O_n^γ - is expressed by the anomalous dimension matrix[9]

$$\begin{bmatrix} \gamma_n^{++} & \gamma_n^{--} & 0 & K_n^+ \\ \gamma_n^{-+} & \gamma_n^{--} & 0 & K_n^- \\ 0 & 0 & \gamma_n^{NS} & K_n^{NS} \\ 0 & 0 & 0 & 0 \end{bmatrix}.$$

Applying the renormalization group treatment to $M_n^\gamma(Q^2)$ in the usual way, we get[9,14]

$$M_n^\gamma(Q^2) = \sum_{i,j} C_n^i(1,\bar{g}^2,\alpha)\; M_n^{i,j}\;\langle\gamma|O_n^j|\gamma\rangle$$
$$+ \sum_i C_n^i(1,\bar{g}^2,\alpha)\; X_n^i\;\langle\gamma|O_n^\gamma|\gamma\rangle$$
$$+ C_n^\gamma(1,\bar{g}^2,\alpha)\;\langle\gamma|O_n^\gamma|\gamma\rangle, \quad i,j=+,-,NS \qquad (3)$$

Now using $\langle\gamma|O_n^\gamma|\gamma\rangle=1$ to $O(\alpha^0)$, we finally get

$$M_n^\gamma(Q^2) = \sum_i A_n^i(\mu^2)\left(\frac{\alpha_s(Q^2)}{\alpha_s(\mu^2)}\right)^{d_n^i}$$
$$+\sum_i \frac{a_n^i}{\alpha_s(Q^2)}\;\frac{1}{d_n^i+1}\left[1-\left(\frac{\alpha_s(Q^2)}{\alpha_s(\mu^2)}\right)^{d_n^i+1}\right]$$
$$+\sum_i \frac{b_n^i}{d_n^i}\left[1-\left(\frac{\alpha_s(Q^2)}{\alpha_s(\mu^2)}\right)^{d_n^i}\right] \; + \; c_n^\gamma \qquad (4)$$

In Eq.(4) μ^2 is an arbitrary renormalization point, the A_n^i are unknown hadronic matrix elements of the photon, the d_n^i are proportional to the one-loop anomalous dimensions (e.g. $d_n^{NS} = \gamma_n^{NS}/2\beta_0$) , and the a_n^i, b_n^i are exactly calculable (indeed, calculated) numbers.

Now for Q^2 asymptotically large, we have $\alpha_s(Q^2) \sim 1/\ln Q^2$ so for fixed n we see that $1/\alpha_s$ dominates $(\alpha_s(Q^2)/\alpha_s(\mu^2))$ if $Q^2 >> \Lambda^2$ and we have

$$\lim_{\substack{Q^2 \to \infty \\ n \text{ fixed}}} M_n^\gamma(Q^2) = \frac{1}{\alpha_s(Q^2)} \sum_i \frac{a_n^i}{d_n^i+1} + \sum_i \frac{b_n^i}{d_n^i} + c_n^\gamma \qquad (5)$$

Eq.(5) is the source of the often heard statement: " QCD makes an absolute, one parameter(Λ) prediction for $F_2^\gamma(x,Q^2)$". The shape of $F_2^\gamma(x,Q^2)$ implied by Eq.(5) is noticeably different from that due to the box graph alone (see fig. 1). This change is due to the renormalization or mixing effects of the hadronic operators. Note that the leading logarithm (LL) result of Witten is much smaller than the LL part of the box at large x, and that the LL result of Witten has developed a small peak at x~0. This overall behavior is easily understood by separating in the calculation the parts proportional to $<e^4>$ and $<e^2>^2$, respectively.[15] The former are usually termed "valence" contributions because the struck quark originates directly from the parent target photon (see figs. 2,3). The latter are called "sea" contributions because the struck quark was generated by evolution, or radiation from the valence quarks. The parallel with the nucleon target case is then rather obvious.

IV. Higher Order Breakdown and Its Resolution

As we have just discussed, the asymptotic predictions for $M_n^\gamma(Q^2)$ are calculable with only the QCD scale Λ needed in order to make an absolute prediction. It came as somewhat of a discouraging surprise, however, when it was found[15] that, taken as it stands, the result

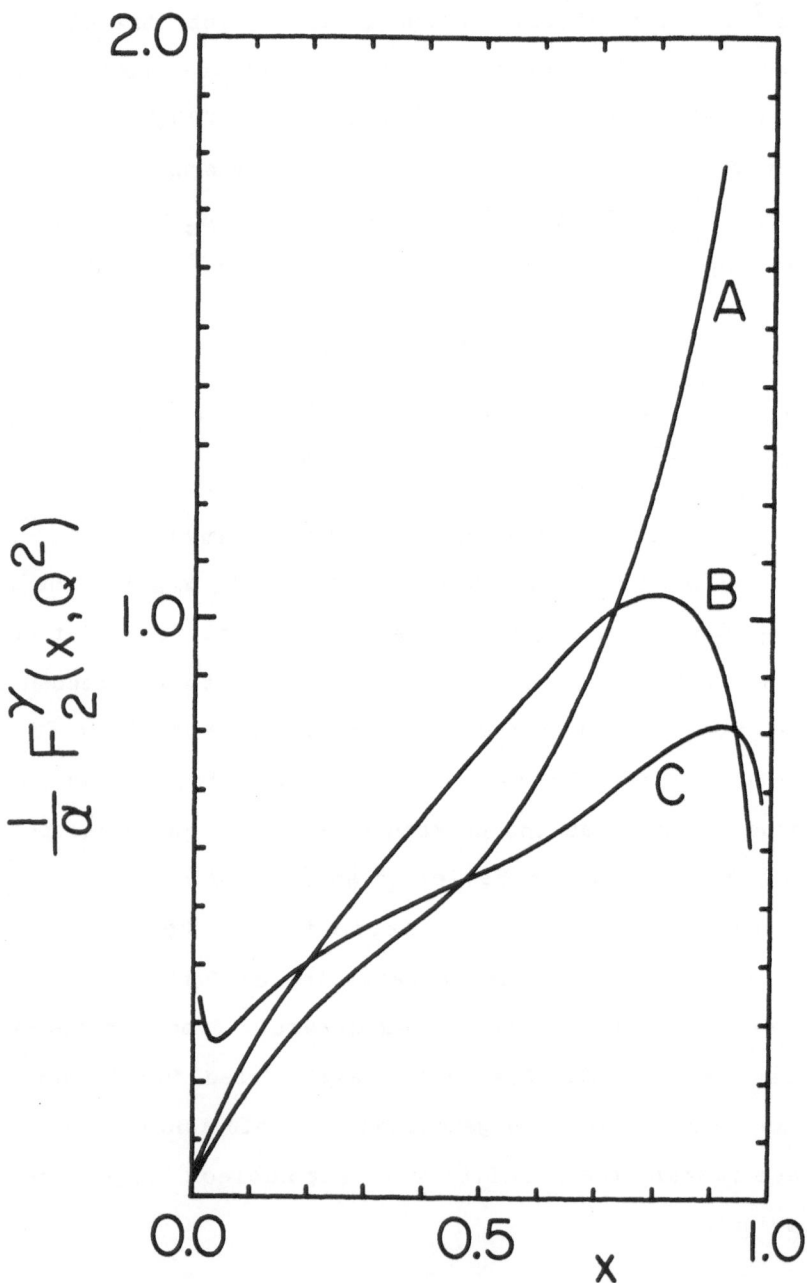

Fig. 1. A) The leading log part of the box graph.
B) The full box graph contribution.
C) Witten's leading log result.

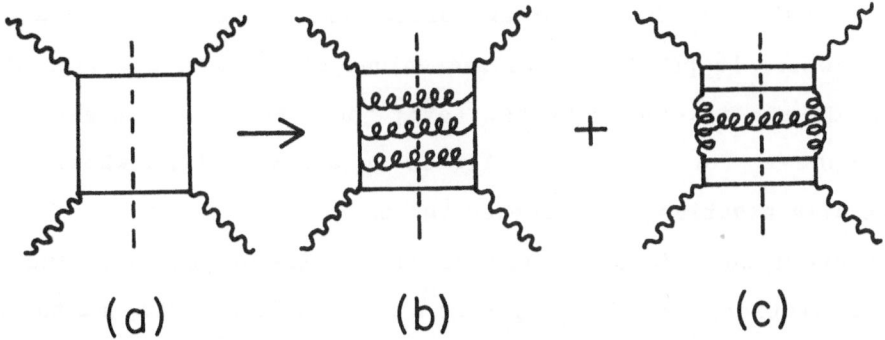

Fig. 2.   A schematic demonstration of how QCD evolution
          turns the box graph(a) into a valence part(b)
          and a sea part(c).

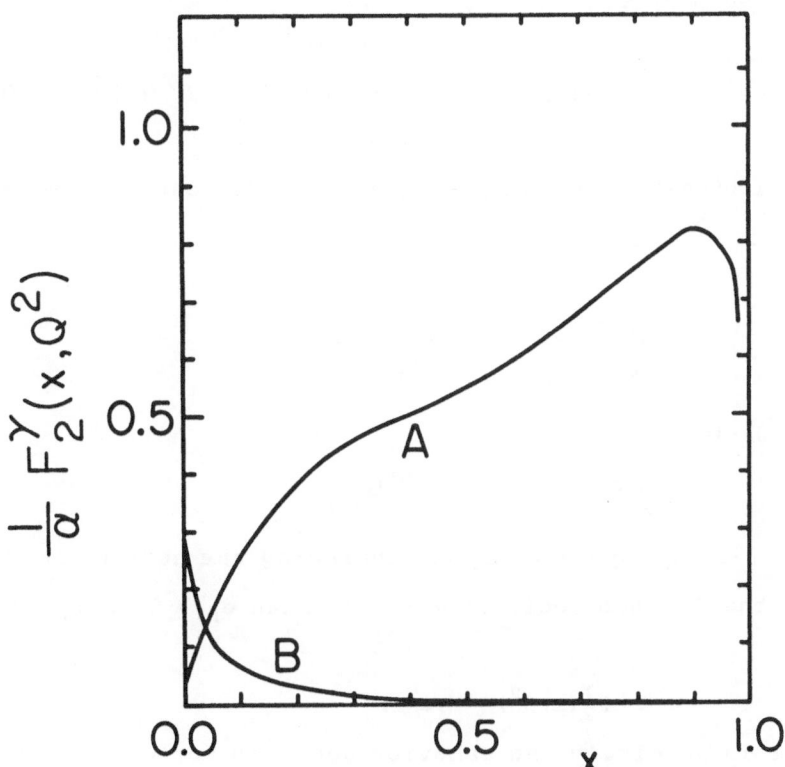

Fig. 3.   A) The valence part of Witten's leading log result.
          B) The corresponding sea part.

Eq.(5) including the higher order corrections is really quite sick for x~0, x~1 where it predicts negative values of $F_2^\gamma(x,Q^2)$ (see fig. 4). For x~0 the damage is caused by the $\langle e^2\rangle^2$ "sea" term, which has a large negative spike. For x~1 the problem is in the $\langle e^4\rangle$ term whose moments are actually negative for large enough n.

In order to understand the origin of these problems and hence their resolution, we need only look a bit more carefully at the steps leading from Eq.(4) to Eq.(5) above.

A. x~0

In order to reconstruct $F_2^\gamma(x,Q^2)$ we must perform an inverse Mellin transform on $M^\gamma(n,q^2)$:

$$F_2^\gamma(x,Q^2) = \frac{1}{2\pi i} \int_{c-i\infty}^{c+i\infty} dn \; x^{-n+1} \; M^\gamma(n,Q^2) \qquad (6)$$

Now the rightmost singularity in n of $M^\gamma(n,Q^2)$ controls the leading x~0 behavior of $F_2^\gamma(x,Q^2)$, i.e. if

$$\lim_{n\to n_0} M^\gamma(n,Q^2) \sim \frac{c}{n-n_0} + \text{regular},$$

then Eq.(6) implies

$$\lim_{x\to 0} F_2^\gamma(x,Q^2) \sim cx^{1-n_0}.$$

Suppose we apply Eq.(6) to Eq.(5) including the next to leading terms b_n. Then the dominant pole in n occurs when $d_n^- = 0$ or $n_0 = 2$, leading to

$$F_2^\gamma(x,Q^2) \sim -c/x.$$

This is precisely the behavior observed for the sea term in fig. 4. In addition, the LL result of Witten has a pole when $d_n^- + 1 = 0$ or $n_0 = 1.5962$ leading to the positive spike $\sim x^{-0.5962}$ observed in the LL sea (see fig. 3).

It is clear, however, that these predictions of spikes are entirely spurious. To see this, look at Eq.(4) and note that the poles

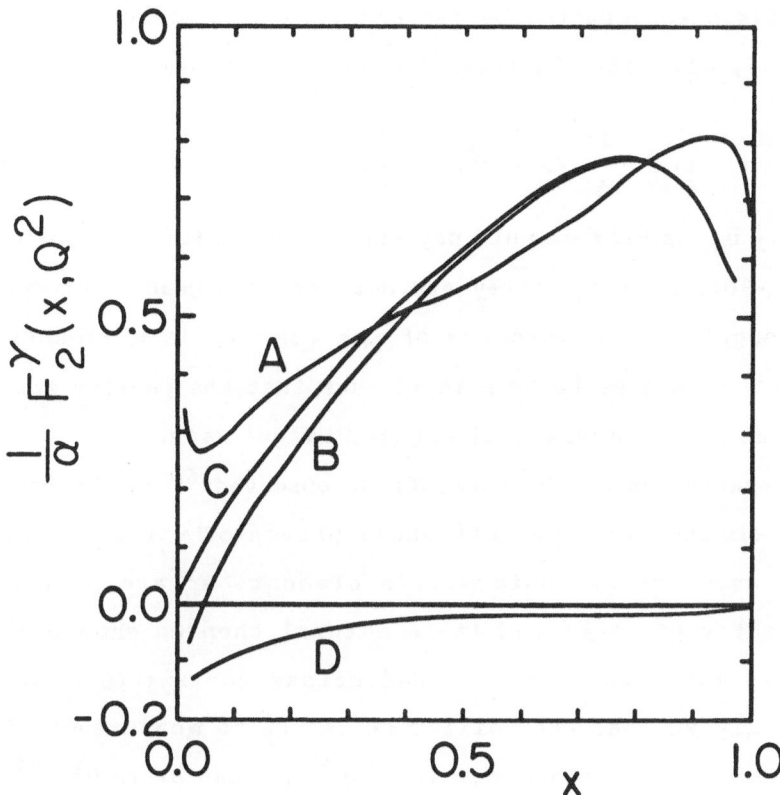

Fig. 4.   The LL result of Witten A) compared to the higher order
           calculation of Bardeen and Buras B). Curve C) is the
           valence part and curve D) is the sea part of the higher
           order calculation.

1/d are always accompanied by factors $(1 - x^d)$, where $x = \alpha_s(Q^2)/\alpha_s(\mu^2)$ and $d = d_n^i + 1$, $d_n^i, \ldots$ etc. Instead of a pole we actually get for $d \to 0$,

$$\lim_{d \to 0} \frac{1}{d} (1 - x^d) = -\ln x.$$

This simply means that we must pay strict attention to the order of the limits $Q^2 \to \infty$, $n \to n_0$ - they are not interchangeable. Of course this kind of nonuniform convergence of the moments is well-known in the nucleon or hadronic sector[21]; it is just that the penalty for ignoring it seems much more severe in the photon target case.

The behavior noted above was first observed[16] in the case of deep inelastic scattering on an off-shell photon of virtual mass $-p^2$. In that case even the hadronic matrix elements A_n^i are perturbatively calculable for $p^2 \gg \Lambda^2$ and it is natural then to choose $\mu^2 = p^2$ in Eq.(4). This leads to positive predictions for $F_2^\gamma(x, Q^2)$ for $x \sim 0$ but unfortunately we lose the ability to use F_2^γ to measure Λ! In effect the dominant logarithm becomes $\ln Q^2/p^2$ instead of $\ln Q^2/\Lambda^2$. This is illustrated in fig. 5.

Now for real photons[17] $(p^2 < \Lambda^2)$ there is still some freedom in the method in which the $1/d_n$ poles are regulated. To see this note that Eq.(4) can be rewritten trivially as

$$M_n^\gamma(Q^2) = \sum_i \tilde{A}_n^i(\mu^2) \left(\frac{\alpha_s(Q^2)}{\alpha_s(\mu^2)} \right)^{d_n^i}$$

$$+ \frac{1}{\alpha_s(Q^2)} \sum_i \frac{a_n^i}{d_n^i + 1} + \sum_i \frac{b_n^i}{d_n^i} + c_n^\gamma,$$

where the $\tilde{A}_n^i$ differ from the already unknown A_n^i only by a finite, and hence harmless, renormalization. Thus there are poles in the $\tilde{A}_n^i$ but for $p^2 < \Lambda^2$ we have no particular reason to assume the exact form of Eq.(7). For example, a set of simple poles $f(n)/n - n_0$ with an arbitrary $f(n)$ that gives the correct residue would be a possibility. A simple scheme along these lines has been worked out by Antoniadis and Grunberg[22]. In their scheme there is one free parameter t which

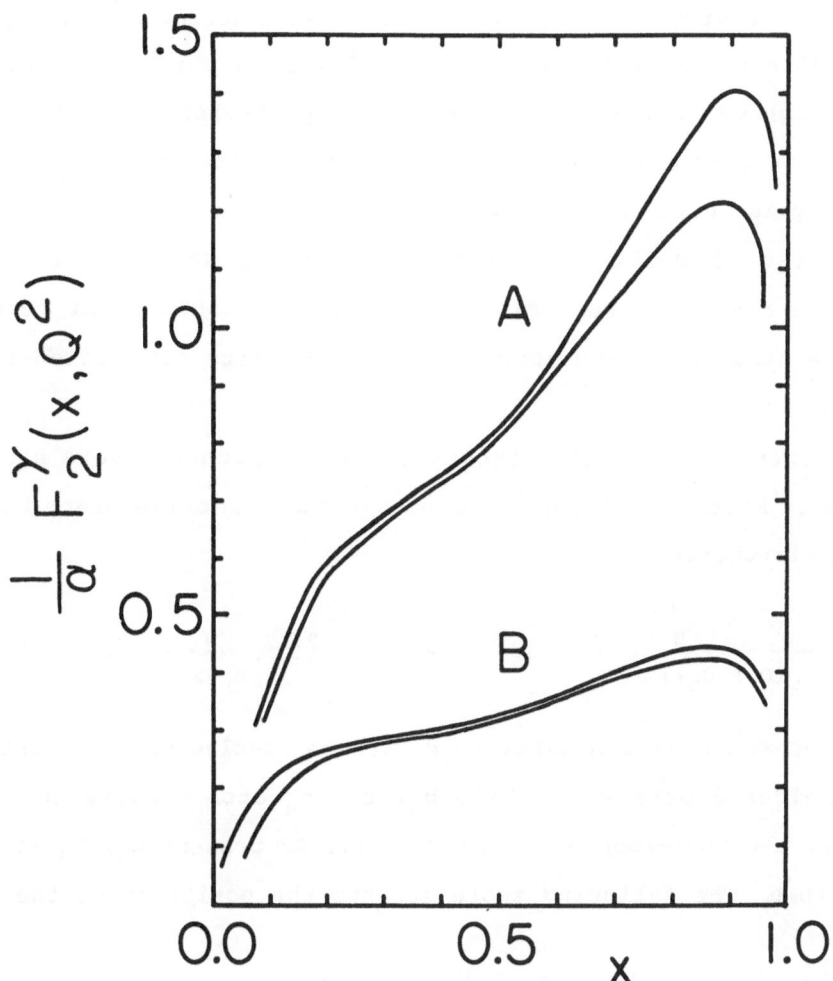

Fig. 5. The result of Uematsu and Walsh for A) Q^2=2000 and
B) Q^2=20 GeV^2. The upper(lower) curve in each
case corresponds to Λ=100(500) MeV.

summarizes in a simple way our nonperturbative ignorance. An example of their results is shown in fig. 6, where the main point to observe is that although we have solved the negativity problem, we unavoidably induce a significant uncertainty in the prediction. In this case the uncertainty is still concentrated in $x < 0.1$ where we had the problem in the first place. Optimistically then we may hope that our predictions are still reliable for $x > 0.1$, but this point requires still more discussion and study before a definite conclusion can be reached.

The extension of this discussion to yet higher orders has been initiated by Rossi[23], who has shown that the pointlike piece has the following structure:

$$\frac{1}{\alpha_s(Q^2)} \frac{a_n}{d_n+1} + \frac{b_n}{d_n} + \alpha_s(Q^2)\frac{c_n}{d_n-1} + \alpha_s^2(Q^2)\frac{e_n}{d_n-2} + \ldots \; .$$

In this expression the denominators d_n always involve only the the one-loop anomalous dimensions, while b_n, c_n, e_n etc. involve two-loop, three-loop, and four-loop etc. calculations. At present c_n, e_n etc. are not available. The following table locates the positions of the poles in n.

d_n	$n_0^{(-)}$	$n_0^{(NS)}$	$n_0^{(+)}$
-1	1.5926	.3099	-
0	2.0000	1.000	-
1	5.326	5.250	2.386
2	26.52	26.58	4.402

We see that at higher orders the singularities quickly move to large n_0, implying stronger and stronger singularities x^{1-n_0}. Whether or not these singularities contaminate the large-x region is a question of the size of the calculable, but as yet uncalculated, residues.

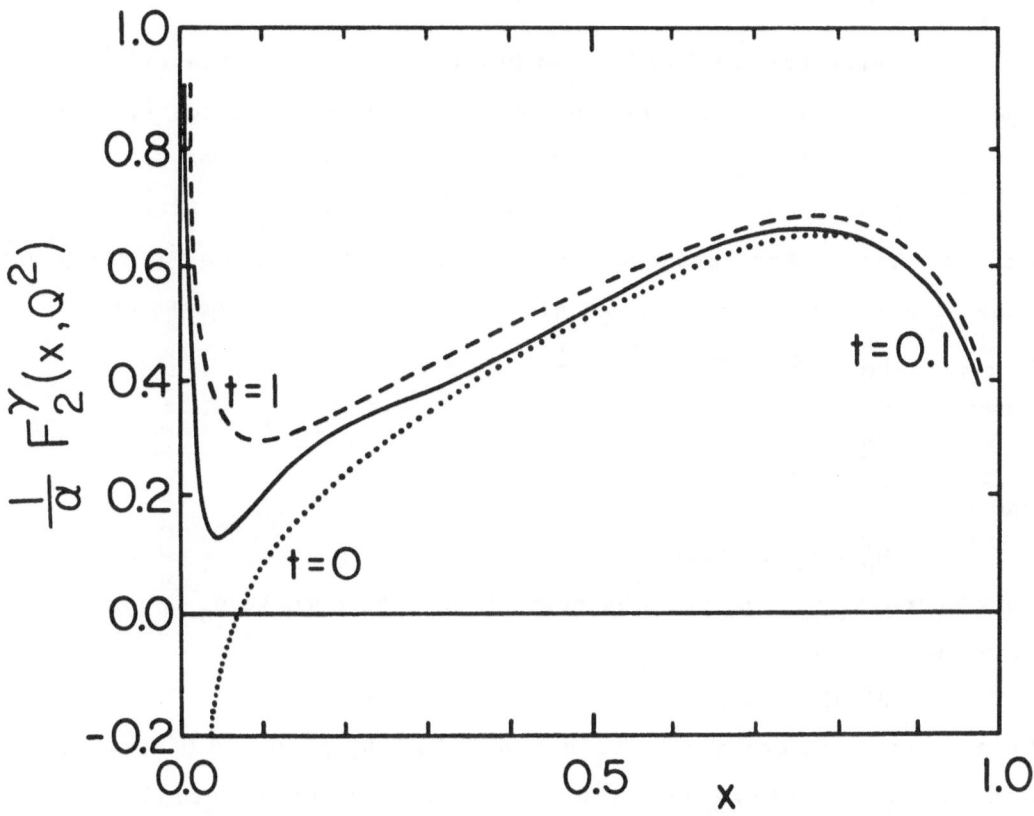

Fig. 6. The reult of Antoniadis and Grunberg for the unregular-
ized case(t=0, dotted line) and two values(t=0.1,solid
line and t=1.0,dashed line) of the regularization
parameter.

B. x~1

For large enough n and fixed Q^2, the explicit calculation shows that b_n^{NS} and hence $M_n^\gamma(Q^2)$ becomes negative. Hence at large x, $F_2^\gamma(x,Q^2)$ is negative, which is a nonsensical prediction for a cross section.

The reason for this bad behavior is illustrated in fig. 7 for the Uematsu-Walsh case, although the general trend follows for the $p^2=\Lambda^2$ case as well. The valence piece has the general form from Eq.(3)

$$M_n^V \sim \langle e^4 \rangle \; \frac{a_n}{\alpha_s} + b_n' + B_n^\gamma$$

where

$$\frac{a_n}{\alpha_s} + b_n' \sim C_n^{NS} \; x_n^{NS}$$

and

$$B_n^\gamma \sim C_n(1,\bar{g}^2,\alpha).$$

In x space $a_n/\alpha_s + b_n'$ becomes curve A of fig. 7 while B_n^γ becomes (see curve B)

$$B^\gamma(x) \sim x \left[(1-2x+2x^2) \; \ln(1/x^2) \; -2+6x-6x^2 \right].$$

This is just the box graph evaluated for $m^2=0$, $p^2 \neq 0$ and $Q^2=p^2$, and it clearly is negative for x>0.6 even though it is still a cross section[24]. The negativity in this case results simply from dropping $O(p^2/Q^2)$ and $O(m^2/Q^2)$ factors which would insure the correct threshold behavior. In an OPE analysis these factors would arise from higher twist contributions. We see that the theoretical predictions at large x are probably very complicated and potentially unreliable, at least at the leading twist two level.

C. Heavy Quarks

The situation regarding heavy quark contributions to F_2^γ is really no different in principle than the treatment in the usual nucleon target deep inelastic scattering[25]. The _quantitative_ effect, however, is much larger for F_2^γ, especially for c and eventually t quarks. Within the OPE heavy quarks have been treated by Hill and Ross[26], but their results have not been used phenomenologically so far, and apparently the subject is plagued by subtleties[27]. The more mundane approach of simply using the box graph is being improved by including first order

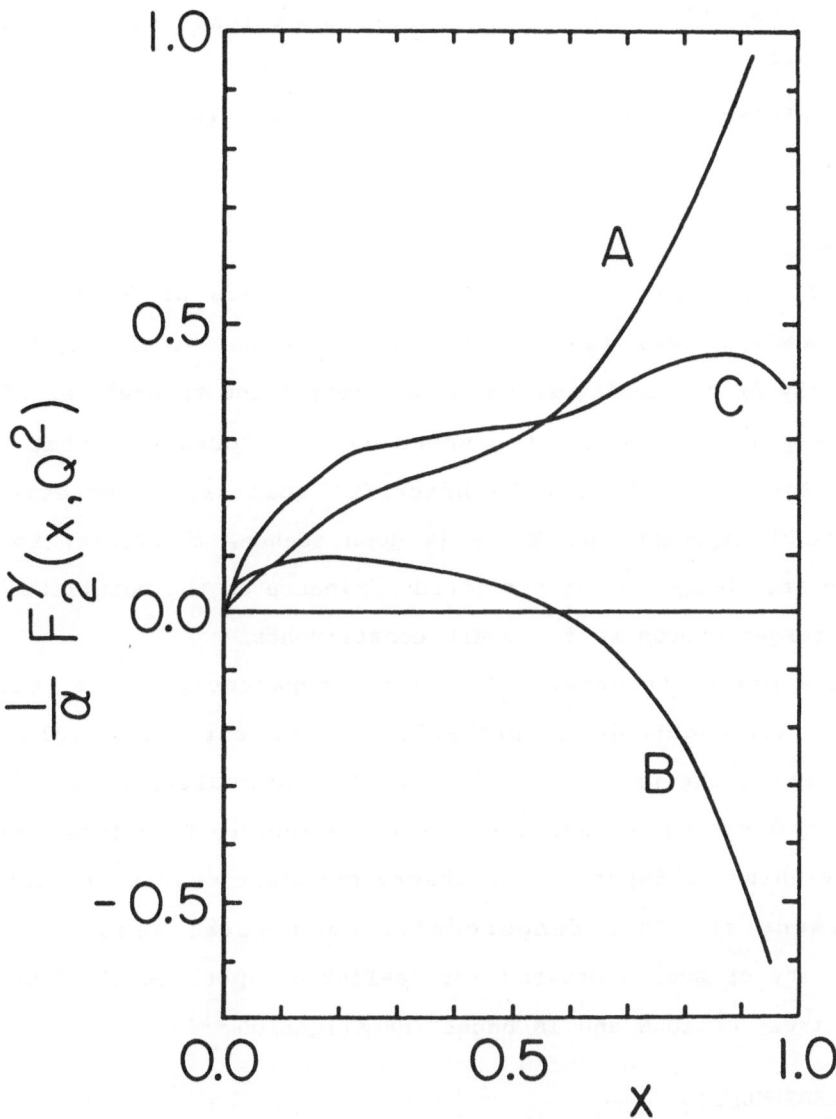

Fig. 7.   The breakdown of the valence piece into its Q^2 dependent
(curve A) and Q^2 independent (curve B) parts. Curve C is
the total.

QCD effects and should be of some use at least not too far above threshold[28].

Hopefully this area will soon receive more theoretical attention.

V. Summary

The hope that experimental measurements of F_2^γ will provide exceptionally "clean" tests of QCD is, of course, still very much alive and worthy of continuing strong experimental and theoretical effort. It is surely no accident and indeed very encouraging that all the experimental results are in beautiful qualitative agreement with theoretical expectations. There is good reason, therefore, to believe that we are observing the predicted dominance of the pointlike coupling of the target photon to its quark constituents.

As a method to measure Λ , we are apparently not so well off as previously thought. We are probably reduced, for finite values of Q^2, to emulating the program of deep inelastic nucleon scattering, i.e. measuring Λ via the Q^2 evolution of the structure functions. This would require using as input to the theory boundary conditions taken from experiment or, more dangerously, some model such as VMD. The feasibility of such a program for realistic experimental situations is not entirely obvious and is under investigation.[29]

ACKNOWLEDGEMENTS

I would like to thank S. J. Brodsky, W. R. Frazer, C. K. Lee, J. F. Owens, P. Zerwas and, most of all, W. A. Bardeen for many helpful discussions and explanations. Also, I would like to thank Prof. Ch. Berger and his colleagues for hosting such an impressive and enjoyable conference.

REFERENCES

1) T. F. Walsh, Phys. Letters 36B,121(1971).
2) S. J. Brodsky, T. Kinoshita, and H. Terazawa, Phys. Rev. Letters 27,280(1971).
3) T. F. Walsh and P. Zerwas, Phys. Letters 44B,195(1973).
4) R. L. Kingsley, Nucl. Phys. B60,45(1973).
5) R. P. Worden, Phys. Letters 51B,57(1974).
6) H. D. Politzer, Phys. Rev. Letters 30,1343(1973).
7) D. J. Gross and F. Wilczek, Phys. Rev. Letters 30,1343(1973).
8) M. A. Ahmed and G. G. Ross, Phys. Letters 59B,369(1975).
9) E. Witten, Nucl. Phys. B120,189(1977).
10) G. Altarelli and G. Parisi, Nucl. Phys. B126,298(1977); L. N. Lipatov, Sov. J. Nucl. Phys. 20,94(1975).
11) C. H. Llewellyn Smith, Phys. Letters 79B,83(1978).
12) W. R. Frazer and J. F. Gunion, Phys. Rev. D20,147(1979).
13) R. J. DeWitt et al., Phys Rev. D19,2046(1979); D20,1751(E)(1979).
14) W. A. Bardeen and A. J. Buras, Phys. Rev. D20,166(1979); D21,2041(E)(1980).
15) D. W. Duke and J. F. Owens, Phys. Rev. D22,2280(1980).
16) T. Uematsu and T. F. Walsh, Nucl.Phys. B199,93(1982).
17) W. A. Bardeen, Proceedings of the 1981 International Symposium on Lepton and Photon Interactions at High Energies,Bonn, ed. W. Pfiel.
18) W. R. Frazer, Fourth International Colloquiom on Photon-Photon Interactions, Paris, 1981, ed. G. London(World Scientific, Singapore, 1981).
19) W. R. Frazer and G. Rossi, Phys. Rev. D25,843(1982).
20) M. K. Chase, Nucl. Phys. B189,461(1981).
21) A. DeRujula et al., Phys. Rev. D10,1649(1974).
22) I. Antoniadis and G. Grunberg, Nucl. Phys. B213,445(1983).
23) G. Rossi, Lawrence Berkeley Laboratory report LBL-15912, 1983.
24) Exact results are found in V. Budnev et al., Phys. Rep. 15C,181(1975).
25) E. Witten, Nucl. Phys. B104,445(1976).
26) C. T. Hill and G. G. Ross, Nucl. Phys. B148,373(1979).
27) See, e.g. W. Kummer, Vienna preprint, January, 1983 and references therein.
28) I. Schmitt et al., these proceedings.
29) D. W. Duke and J. F. Owens (in preparation). Also, M. Gluck and E. Reya, Dortmund preprint DO-TH 83/05(1983).

DISCUSSION

Q. J. H. Field - Paris

Is not your final conclusion that F_2^{γ} is not so useful for measuring Λ unduly pessimistic ?

After all, for the case of the nucleon structure function perturbative QCD is able to calculate only the Q^2 dependence of relatively small corrections to a structure function which is a priori uncalculable. On the other hand, for the photon structure function there are uncalculable (but probably small) corrections to a structure function which is calculable both in absolute shape and in terms of Q^2 evolution by perturbative QCD.

A. D.W. Duke

For truly asymptotic values of Q^2, your reasoning is correct. For $Q^2 < 100$ GeV^2 or so, where the transition from Eq.(4) to Eq.(5) of the text is not so harmless, the shape of F_2^γ looks about right but the absolute normalization of the theoretical prediction has been lost, not to be regained until Q^2 is very large.

Q. S. Brodsky - SLAC

Consider fixed quark mass m with $x, Q^2 \neq 0$ and let $\Lambda \to 0$. Does this give the box answer ? Assuming this is the case, perhaps the charm component of the photon structure function can be used to check QCD, although not to measure Λ.

A. D. W. Duke

I think that the answer to the question is yes, and I think that developments along this line will be very important to further test QCD.

Q. Ch. Berger - Aachen

Do you encourage us to go through the pain of measuring the Q^2 dependence of F_2^γ ?

A. D. W. Duke

Even under the not so optimistic circumstances that I have discussed, the photon structure function still enjoys a very favored position in all of perturbative QCD. It is, as far as we know, one of the quantities least sensitive to hadronization and other such nonperturbative and/or uncalculable effects. So every effort should be made to measure F_2^γ as well as possible.

C. W. A. Bardeen - Fermilab

I would like to comment on the theoretical status of the photon structure function. Duke has emphasized the existence of singularities in the pointlike component which must be cancelled by similar singularities in the hadronic component. These singularities were responsible for the large higher order corrections found at small x by Bardeen and Buras. The cancellation of these large effects by the hadronic component should actually improve the predictive power of perturbative QCD as the higher order terms in α_s are now properly suppressed relative to the leading orders. Similar singularities encountered in the pointlike component in yet higher orders should also be suppressed by analogous cancellations. During the next few years the remaining ambiguities in the hadronic component may be resolved through real solutions to QCD (lattice calculations, etc.) which go beyond perturbation theory.

Finally, I think the computation of the charm quark contribution including perturbative corrections should be reliable at low Q^2 and can be matched to the asymptotic forms using reasonable physical analysis at high Q^2.

The measurement of the photon structure function has already shown remarkable agreement in shape and in magnitude with the fundamental predictions of perturbative QCD. In the future we can expect great improvement in both our experimental and our theoretical understanding of these processes.

γγ AND eγ COLLISIONS AT

FUTURE HIGH ENERGY COLLIDERS

J.H. FIELD *
L.P.N.H.E. Université Pierre et Marie Curie
4, Place Jussieu – F75230 Paris

ABSTRACT

After briefly reviewing previous work on two photon collisions at future high energy colliders (e^+e^-, e^-p, pp, $p\bar{p}$) a comparative survey is made of PETRA, LEP, SLC and HERA from the view points of luminosity, acceptance and energy range. A more detailed study is then presented of a $0°$ tagging system in the proton beam line of HERA. Bremsstrahlung background and the separation of two photon production and diffractive electroproduction events are briefly discussed. It is concluded that HERA gives a unique possibility to study single and double tag two photon physics for large centre of mass energies > 10 GeV.

The interest of observing $W^{\pm}$ and $Z°$ production in virtual photon e, p collisions at HERA is also mentioned.

In the second part of the talk, devoted to the more distant future, the possibility of producing real γγ or γe collisions in very high energy linear colliders such as SLC or VLEPP is discussed. Both the technical realisation, by Compton back scattering of high intensity laser beams, and the new domain of physics which would be opened to experiment are considered. Large rates of single $Z°$ and $W^{\pm}$ production are expected in γe collisions well as measurable rates for W^+W^- pair production in γγ collisions. The W production cross sections are sensitive to the fundamental gauge boson vertices γWW and γγWW.

---

* On leave of absence from DESY Hamburg.

1. INTRODUCTION

A list of some future (and existing) collider projects is given in TABLE 1. This
list, though certainly not exhaustive, is at least representative as it contains
examples of all types of colliders forseen until now :

(i) e^+e^- storage ring

(ii) $e^\pm e^-$ one pass linear collider

(iii) $p\bar{p}$ single ring collider

(iv)   pp   double ring collider

(v)   ep   double ring collider

A priori,two photon physics is possible at any collider, though all experimental results published till now (with one exception) have come from e^+e^- storage rings. By
far the largest number of studies have been devoted to the LEP machine [1],[2],[3] ,
[4], [5] . For a comprehensive survey of the various physics topics accessible at
LEP the interested reader is referred to Davier's contribution to the 1981 Paris two
photon workshop [5] . No further discussion will be given here except to compare in
a global way LEP with PETRA, SLC and HERA with respect to luminosity and energy range.

The linear colliders (ii) such as SLC [6] and VLEPP [7] are of interest not so
much for "classical" 2γ physics using virtual photons but because of the unique
possibilities they offer for generating high intensity real photon beams (in arbitary
polarisation states) which may be used to make very high energy (W > 100 GeV) real
$\gamma\gamma$ and $e\gamma$ collisions. The technique used to generate the real photon beams, as well
as the new physics topics which can be investigated by their use, are discussed below in sections 6 and 7. In these sections I have drawn largely from previous work
carried out both at SLAC [8] and Novosibirsk [9] [10] [11] on the technical problem
of generating the real photon beams by Compton back scattering of Laser beams. This
subject has also been reviewed by Kessler [12] . For the physics of very high energy
γe and $\gamma\gamma$ collisions the work of Renard [13] and Ginsberg et al [14] has largely
been used.

Two photon physics in pp and $p\bar{p}$ collisions has been reviewed from the theoretical
view-point by Donnachie [15] and from the experimental one by Tao [16]. Most theoretical work has been devoted to calculations [17] , [18] , [19] of the background due to
the 2γ process $pp\to\mu^+\mu^-pp$ in the Drell Yan [20] production of μ pairs. One experiment at the ISR [21] has detected 2γ production of $\mu^+\mu^-$ pairs in pp collisions
(this is the "exception"referred to above). More recent calculations of $\mu^+\mu^-$ production
in pp, $p\bar{p}$ and ep collisions have been performed by Vermaseren [22] .

As for the production of hadronic final states in 2γ collisions with pp or p$\bar{\text{p}}$ colliders it has been generally concluded [15], [16], [23] that the two photon collision processes would be completely overwhelmed by pomeron-pomeron fusion (Fig. 1a) or by double gluon exchange graphs (Fig. 1b). It is interesting to note that, as

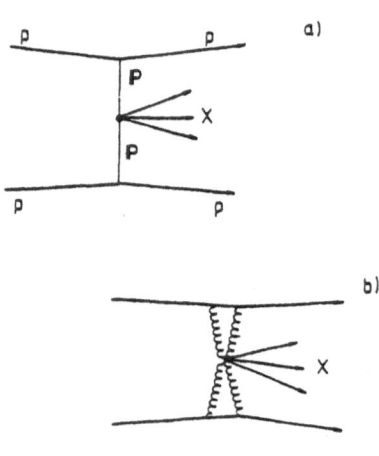

Fig. 1 - a) *Pomeron pomeron fusion graph.*
b) *Double gluon exchange graph.*

already pointed out by Sens [24], the experimental data from the I.S.R. [25] on the reaction pp → ppX, where the scattered protons are detected at small scattering angles ($t \simeq 0.1$ (GeV/c)^2) do not indicate dominance of pomeron pomeron scattering as suggested, for example, in Ref. [15]. The observed mass spectra for the system X at total centre of mass energies of 23 and 30 GeV/c^2 (Fig. 2) show no dominant production of states of spin parity 0^+, 2^+, ... (for example f_o) as required for pomeron pomeron fusion. Rather, copious production of ω_o, ρ_o, A_2 is seen. There is a small f_o signal. Since however the f_o cross section shows the same energy dependance as

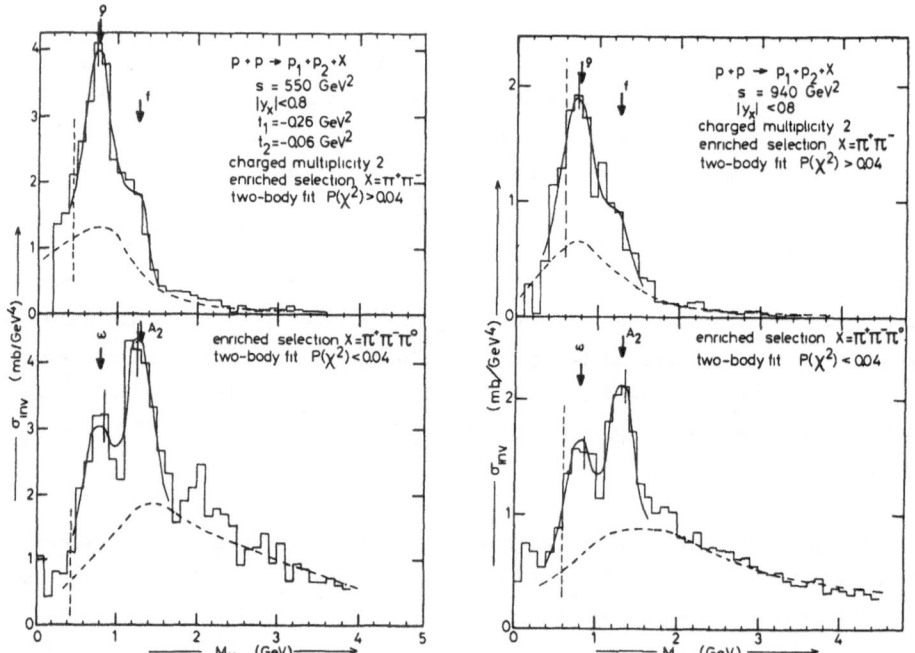

Fig. 2 - *Missing mass spectra of the reaction* p + p → p + p + X. *ISR experiment Ref.* [25]. *The dash-dot line indicates the estimated non-resonant background.*

the ρ_0, ω_0 production it can not be produced predominantly by pomeron pomeron fusion. It may turn out that the pomeron, which is a useful concept for describing diffractive scattering in the language of Regge-pole exchange, has little relevance in processes such as that show in Fig. 1a. Because of the short range $\simeq$ 1f (or equivalently large momentum transfer $\simeq$ 1 m_π) associated with confinement effects in QCD the importance of a two double gluon exchange graph such as that shown in Fig. 1b in the very low t region where one or both protons are elastically scattered is unclear. To the writer's knowledge no QCD calculations of such processes have so far been attempted.

Before leaving the subject of pp and $p\bar{p}$ collisions the essential conclusions of Ref. [16] may be restated. A very high luminosity pp collider such as the original ISABELLE project would have $\simeq$ 2X the LEP luminosity for two photon physics in the region of energy overlap and a much larger energy range (W_{MAX} = 800 GeV/c^2 compared to 260 GeV/c^2). Because of the unknown level of purely hadronic background processes such as these indicated in Fig. 1a, 1b it was concluded that 2γ collision events may be difficult to isolate, but that the hadronic background processes are themselves of considerable physical interest. In contrast the luminosity of the CERN $p\bar{p}$ collider and the planned fermilab $p\bar{p}$ collider are too low to be of any interest for two photon physics, except perhaps for the detection of the 2μ final state as a luminosity monitor.

The writer's opinion is that 2 photon physics remains a very interesting option for a high luminosity pp collider. The separate lattices of the two colliding proton beams contain horizontal bends which, as in the case of HERA (see below) can be used to facilitate 0° tagging. My guess is that processes such these shown in Fig. 1a,b will not constitute an overwhelming background, in any case, for the 2 photon events of the greatest physical interest e.g. those with high transverse momentum jets, or exclusively produced meson pairs in the final state.

The possibility of using an ep collider to study 2γ collisions was first suggested by Coignet [26], and was briefly considered also by Kessler [23]. Now that the HERA proposal [27] is available a more quantitative re-appraisal of its potential for two photon physics is possible. In Section 2 below the luminosity and available kinematic range of HERA are compared with PETRA, LEP and the SLC while in Section 3 a first study is made of 0° tagging in the proton beam line. In both cases the conclusions reached are encouraging and indicate the interest of further, more detailed studies. Section 4 contains estimates of the background due to diffractive electroproduction processes and beam-beam bremsstrahlung in the proposed 0° tagging system. In Section 5 some estimates of the production rate of Z° and W$^\pm$ bosons in γ e and γ p collisions at HERA are briefly reviewed.

2. A COMPARISON OF PETRA, LEP, SLC AND HERA

To make a realistic comparison of an existing machine (PETRA) with the proposed LEP, SLC and HERA colliders an effective luminosity (arbitrarily taken to be $\frac{1}{3}$ of the design luminosity) is assumed for LEP, SLC and HERA while the luminosity measured over a 65 day run is taken for PETRA.

The collision energies and effective luminosities $\bar{\mathcal{L}}$ are then as indicated in Table 2.

Using the equivalent photon approximation, the two photon luminosity differential in the effective mass W of the colliding 2γ system can be written as :

$$\frac{d\mathcal{L}_{\gamma\gamma}^{tot}}{dW} = \frac{dL}{dW} \, \bar{\mathcal{L}} \tag{1}$$

Where $\frac{dL}{dW}$ is a differential two photon luminosity function given by :

$$\frac{dL}{dW} = 4 \left(\frac{\alpha}{\pi}\right)^2 \frac{1}{W} \, \ln\left(\frac{2E}{m_1}\right) \ln\left(\frac{2E}{m_2}\right) f\left(\frac{W}{2E}\right) \quad \ldots \tag{2}$$

where $f(z) = (2+z^2)^2 \ln\frac{1}{z} - (1-z^2)(3+z^2) \simeq 4 \ln\frac{1}{z} - 3$ if $z \ll 1$.

E is the beam energy in the centre of mass system of the colliding beams and m_1, m_2 are the masses of the beam particles. Eqn(2) does not take into account, for the proton case, the effect of form factors. In Ref. (16) it is shown that for $\mu^+\mu^-$ production in $\bar{p}p$ collisions at 2E = 540 GeV the total virtual photon flux is reduced by a factor ~ 4 masses of 10 GeV and a factor of ~ 7 for masses of 20 GeV. For HERA these factors will be reduced to roughly $(4)^{\frac{1}{2}} = 2$ or $(7)^{\frac{1}{2}} = 2.6$ respectively. Most of this loss of flux occurs at large q^2 values. In the small angle tagging system at HERA considered below the maximum q^2 value is $\simeq 0.2 (\text{GeV/c})^2$. Parameterising the proton form factor as [19] :

$$\frac{1}{(1 + \frac{q^2}{0.7})^2}$$

The maximum suppression factor expected is ~ 1.7. In order to give a realistic estimate of the tagging efficiency in the small angle region (Eqn 15 below) where form factor effects are small, the tagging efficiency is normalised to the total flux of transverse photons without form factor corrections. Multiplying this tagging efficiency by the total flux given by Eqn(2) then gives a good estimated of the tagged luminosity. I should be borne in mind however that due to the form factor effects the untagged luminosity may be, depending on the mass of the produced system, up to a factor of two smaller than given by Eqn 2.

A more useful parameter for comparing different machines is the acceptance corrected luminosity given by :

$$\frac{d\mathcal{L}^{Acc}_{\gamma\gamma}}{dW} = \frac{d\mathcal{L}^{TOT}_{\gamma\gamma}}{dW} \ A(W,E,\theta_c) \ ... \tag{3}$$

where $A(W, E, \theta_c)$ is an acceptance function which depends upon W,E and the lower polar angle limit θ_c for the acceptance of produced particles.

The precise definition of A will depend on the fraction of the solid angle in the CM system that is required to be observed. Consider the production of ultra-relativistic particles in the two photon collision at an angle θ in the lab system (Fig.3)

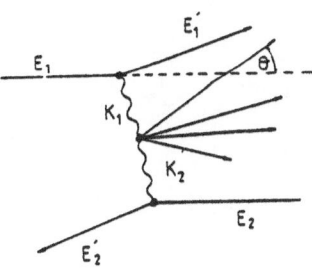

Fig. 3 - *Kinematical definitions in the lab system.*

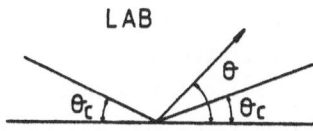

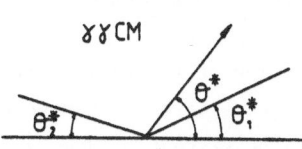

Fig. 4 - *Angular acceptance intervals in the lab and* γγ *CM systems.*

For a given angular cut in the lab system : $\theta_c < \theta < \pi - \theta_c$ (a symmetrical detector is assumed) and a given range of production angles in the two photon centre of mass system : $\theta_1^* < \theta^* < \pi - \theta_2^*$ (Fig. 4) there is a maximum boost given by the relativistic velocity β_c such that all particles produced in the indicated range of θ^* lie within the angular cut in the lab.

For fixed values of θ_c and θ_1^* it is found that :

$$\beta_c = \frac{\cos\theta_1^* \left[\sqrt{R(\ 1 + (R-1)\cos^2\theta_1^*)} - 1 \right]}{1 + R\cos^2\theta_1^*} \tag{4}$$

$$\cos\theta_2^* = \frac{(\gamma_c^2 - 1)\ \sin^2\theta_c + \cos\theta_c}{\gamma_c^2\ \sin^2\theta_c + \cos^2\theta_c} \ , \ R = \frac{\tan^2\theta_1^*}{\tan^2\theta_c}$$

Taking for example $\theta_c = 10°$ to correspond to the minimum angle in a solenoidal detector at which detection and momentum analysis by magnetic deflection is feasible, and $\theta_1^* = 45°$ to have a reasonably large acceptance in the γγ centre of mass sytem it is found that :

$$y_c = \frac{1}{2}\ \ln\ \frac{(1+\beta_c)}{(1-\beta_c)} \ = \ 1.6 \ , \ \theta_2^* = 6°$$

Acceptance = 72 % 4π   in γγ C.M.

Here y is the laboratory rapidity of the γγ system. This can also be written in terms of the scaled energies of the two virtual photons $x_1 = K_1/E_1$, $x_2 = K_2/E_2$ (see Fig. 3) as

$$y = \frac{1}{2} \ln \frac{x_1}{x_2} \qquad (5)$$

In Eqn (5) and in the following discussion the transverse motion of the γγ system is neglected.

The luminosity function L in Eqn (1) can be written, in either of the following differential forms :

$$\frac{d^2L}{dx_1 dx_2} = (\frac{\alpha}{\pi})^2 (\ln\frac{2E}{m_1}) (\ln\frac{2E}{m_2}) \left[1+(1-x_1)^2 \right] \left[1+(1-x_2)^2\right] \frac{1}{x_1} \frac{1}{x_2} \qquad (6)$$

$$\frac{d^2L}{dydz} = 2(\frac{\alpha}{\pi})^2 (\ln\frac{2E}{m_1}) (\ln\frac{2E}{m_2}) \left[1+(1-ze^y)^2\right] \left[1+(1-ze^{-y})^2\right] \frac{1}{z} \qquad (7)$$

where z = W/2E

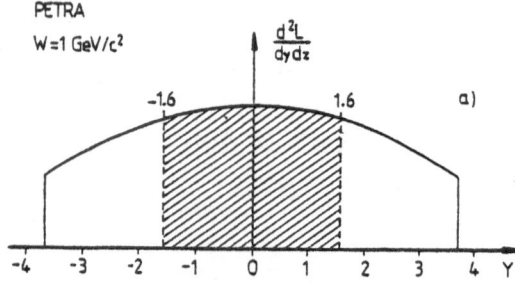

PETRA
W = 1 GeV/c²

a)

Note that x_1 and x_2 are invariant with respect to Lorentz transformations along the beam direction if the masses of the beam particles and the virtual photons are neglected. They thus have the same values in the lab, the colliding beam centre of mass system and the γγ centre of mass system.

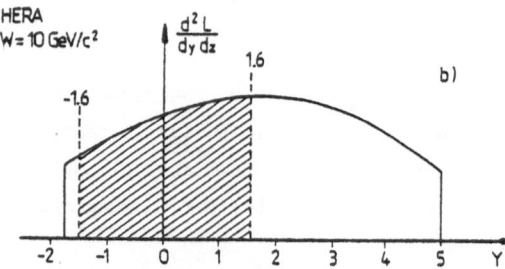

HERA
W = 10 GeV/c²

b)

The distribution of the luminosity function in terms of the laboratory rapidity of the γγ system is shown in Fig. 5a for PETRA with W = 1.0 GeV/c² and in Fig. 5b for HERA with W = 10 GeV/c². The shape of the rapidity distribution is similar in the

Fig. 5 - Differential luminosity function versus lab rapidity.
a) PETRA, E = 20 GeV W = 1 GeV/c²
b) HERA, E_e = 30 GeV, Ep = 820 GeV, W = 10 GeV/c². The cross hatched area indicates region accepted by the cuts 10° <θ< 170°, 6°<θ*<45° for ultra-relativistic produced particles.

two cases, but for HERA the centre is displaced by $\simeq 1.6$ units of rapidity because
of the boost between the lab and the ep centre of mass system. The region of accepted luminosity corresponding to the cuts defined above is cross hatched. It is clear
that the boost between the lab and the overall centre of mass systems at HERA causes
a small loss of acceptance in this case. Simply a different part of the rapidity distribution (corresponding to 2γ systems moving in the same direction as the incoming
electron beam) is sampled. However, this is no longer true for larger masses
$\geqslant 13 \, \text{GeV/c}^2$ when the acceptance defined by the rapidity cut $|y| < 1.6$, is limited to
50 %. This will be further discussed below. By integrating Eqn (7) over the accepted
range of rapidity it is easy to obtain an analytical expression [28] for the acceptance function $A(W, E, \theta_c)$. Here, for simplicity, it is noted that the rapidity distributions are roughly flat so that :

$$A(W, E, \theta_c) = A(W, E, y_c) \simeq \frac{y_c}{y_{MAX}} = \frac{y_c}{\ln(\frac{1}{z})} \tag{8}$$

where Eqn. (5) and the relation :

$$z^2 = x_1 x_2 \tag{9}$$

have been used.

To compare the energy (W) range of the different machines and to see in a clear way
the effect of acceptance cuts it is particularly useful to use the Courau plot $|29|$.
If x_1, $x_2 \ll 1$ Eqns (6) and (7) simplify to :

$$d^2 L = K \times d \left[\ln(\frac{1}{x_1}) \right] d \left[\ln(\frac{1}{x_1}) \right] \tag{10}$$

or

$$= K' \times d \left[\ln(\frac{1}{z}) \right] dy \tag{11}$$

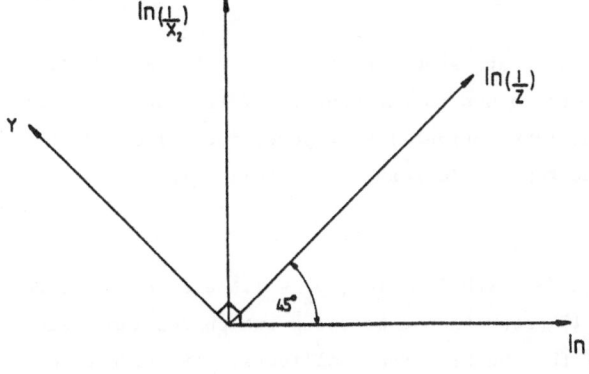

So the luminosity function is
proportional, in this approximation, to the area in a plot of
either $\ln(\frac{1}{x_1})$ versus $\ln(\frac{1}{x_2})$ or
$\ln(\frac{1}{z})$ versus y. Using Eqns (5) and
(9) it can be seen that these two
plots are related to each other by a
45° rotation (Fig. 6).

Fig. 6 - The Courau Plot

As the effect of an angular acceptance cut translates naturally into a rapidity cut (Eqn (8) above) different angular cuts correspond to lines parallel to the diagonal of the $\ln(\frac{1}{x_1})$ versus $\ln(\frac{1}{x_2})$ plot. The kinematic range defined by the possible values of the mass of the 2 photon system, W, is delimited by straight lines perpendicular to the diagonal. The neglect of x_1, x_2 as compared to unity in Eqns (6) and (7) is only important near the extremes of the rapidity plateau (see Figs 5a,b) where the differential luminosity drops by a factor $\leqslant 2$ as compared to the centre. Figs 7,8 and 9 are Courau plots in which HERA is compared with, respectively, PETRA, LEP $(\frac{1}{6})$ or SLC and LEP (2). Because of the boost between the overall centre of mass system and the lab the accepted region for HERA is displaced below the diagonal.

For masses less than $E_{CM} e^{-y_c} \simeq 13$ GeV/c^2 (y_c = 1.6) HERA loses little acceptance as compared to the e^+e^- colliders. However it is clear that for larger masses the acceptance is limited to 50 % as here the y_{MAX} allowed by kinematics is $< y_c$ and exactly the area below the diagonal is accepted. This results from the arbitary choice of y_c which is just equal to the rapidity difference between the lab and ep CM systems.

It is clear from Fig. 7 where PETRA (E_{cm} = 40 GeV) is compared with HERA E_{CM} = 314 GeV) that the effective energy range for two photon physics increases only very slowly with E_{CM}. The new kinematic region appears with only a relatively low luminosity. Fig. 10 indicates the great importance of forward acceptance for two photon physics. The Courau plot for HERA is shown with the acceptance regions corresponding to angular cuts θ_c of 2.6, 10°, 30° indicated separately. The 30° cut is typical of the first generation PETRA solenoïdal detectors. 2.6° corresponds to the minimum acceptance angle specified in the HERA proposal [27] . Also shown in Fig. 10 is the region accessible to a $\simeq 0°$ tagging system in the proton beam line, to be discussed below. It can be seen that tagging system covers the interesting high mass region from 10 to 40 GeV/c^2 with a good efficiency. It's acceptance is much worse in the low mass region from 1 to a few GeV/c^2 where the effects of diffractive electroproduction background to the 2γ process are expected to be most severe (see below).

The luminosity functions defined by Eqn 2 are shown in Fig. 11 for PETRA, LEP (1/6) or SLC, LEP(2) and HERA. LEP (1/6) corresponds to the probable first running configuration of LEP, with only 1/6 of the (conventional) R.F. power installed [30]. LEP (2) is the maximum energy machine that would result from full utilisation of superconducting R.F.

The acceptance functions given by Eqn (8) with θ_c = 10°, y_c = 1.6 are shown for the different machines in Fig. 12. Fig. 13 presents the accepted two photon luminosities, given by Eqn (3), as a function of W for the different colliders. Effective ee or ep luminosities are taken from TABLE 2.

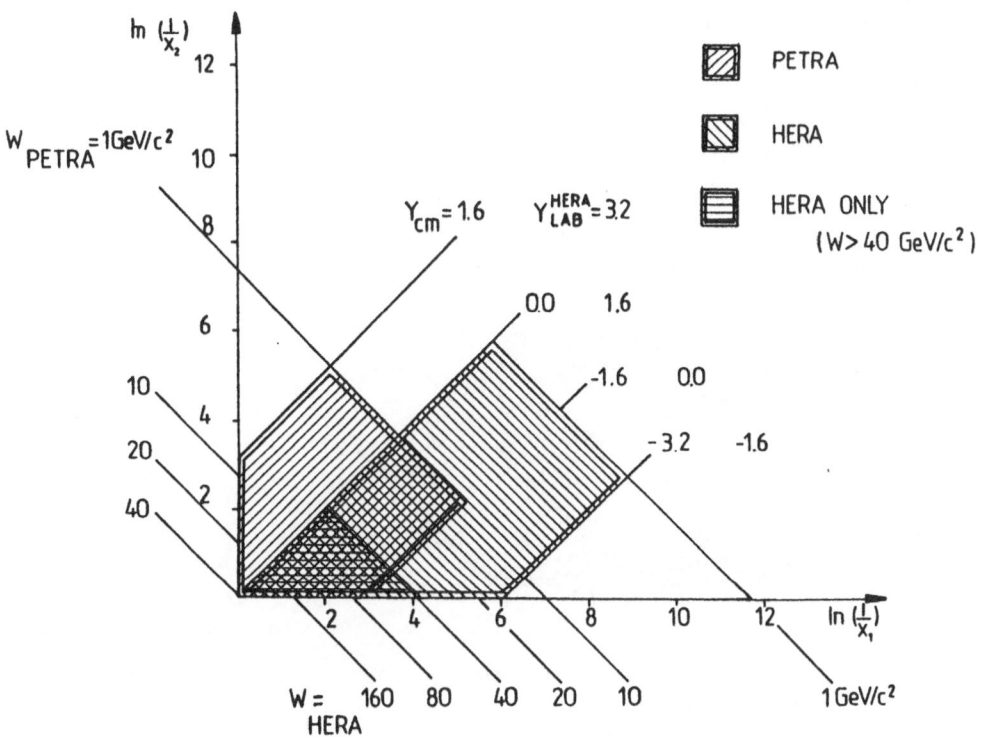

Fig. 7 - PETRA/HERA Comparison.

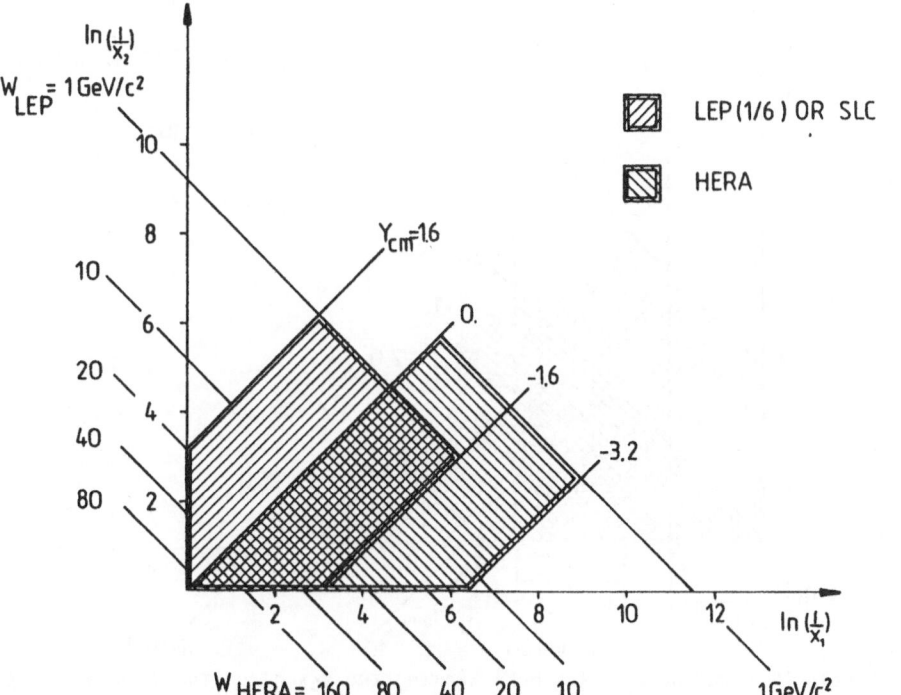

Fig. 8 - LEP(1/6) or SLC/HERA Comparison.

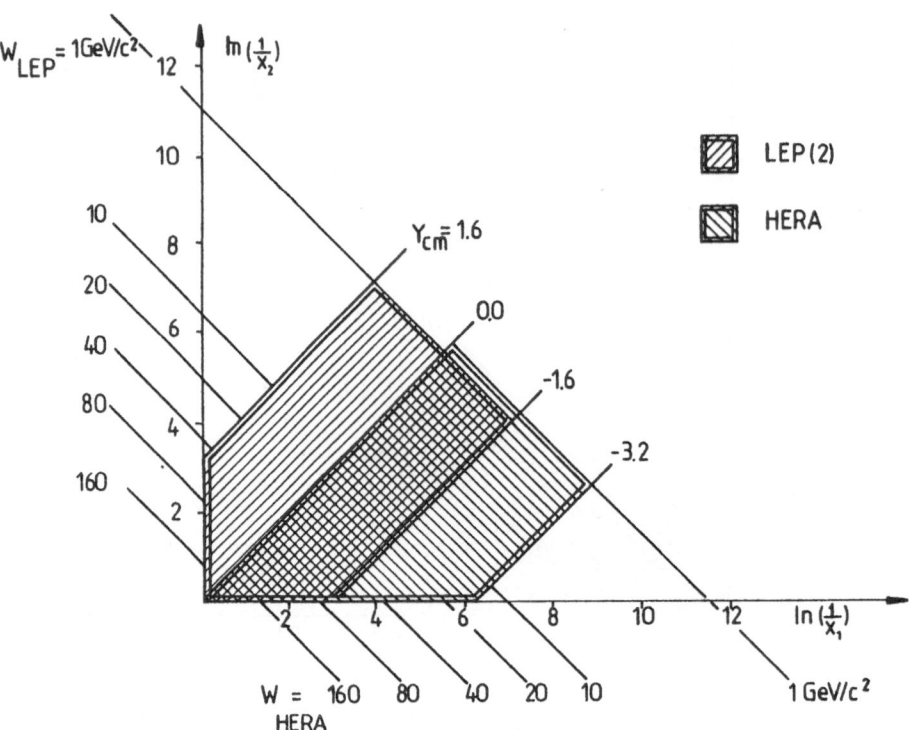

Fig. 9 - LEP(2)/HERA comparison.

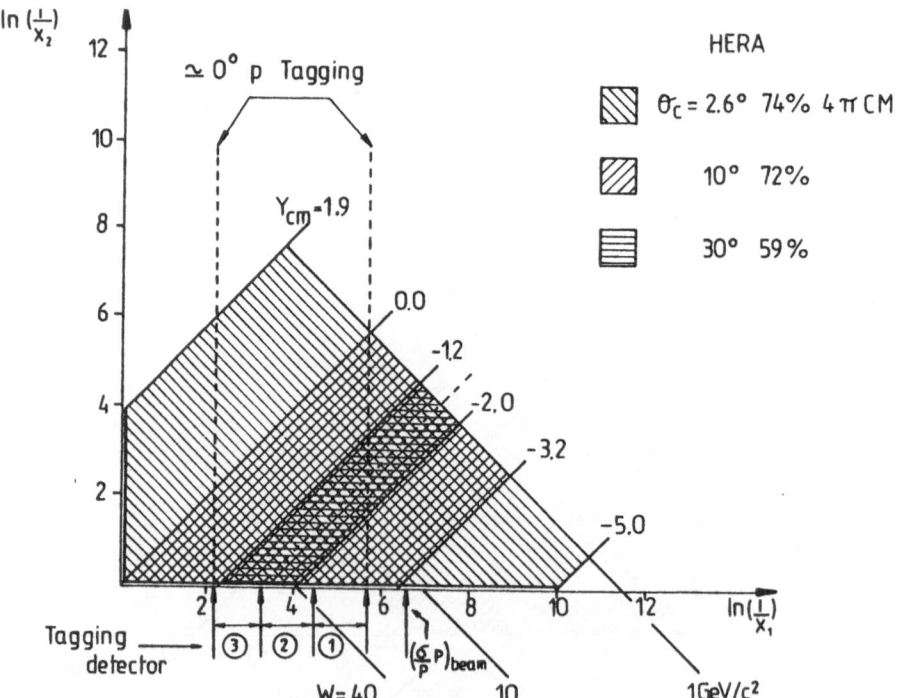

Fig. 10 - Courau plot for HERA showing the effect on the acceptance of different angular cuts, and the kinematical region covered by the proposed ≃0° proton tagger.

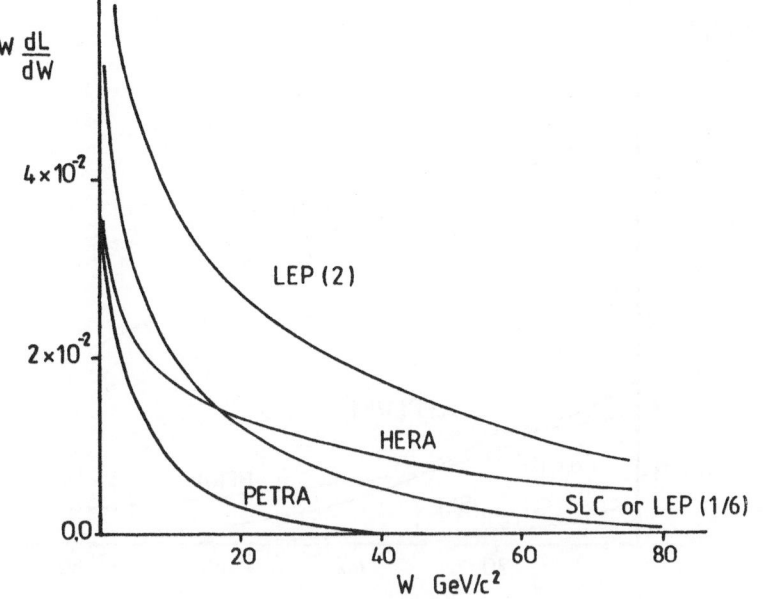

Fig. 11 - *Differential luminosity functions for different colliders.*

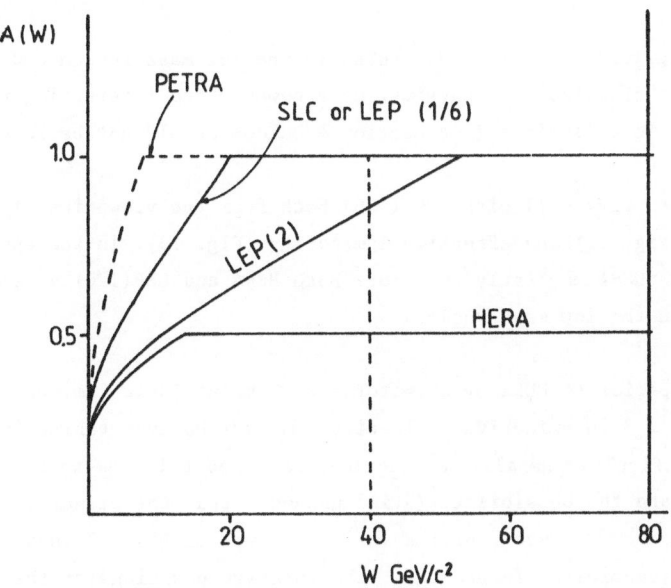

Fig. 12 - *Acceptance function versus W for different colliders.*
$10° < θ < 170°$, $6° < θ^* < 45°$, $y_c = 1.6$.

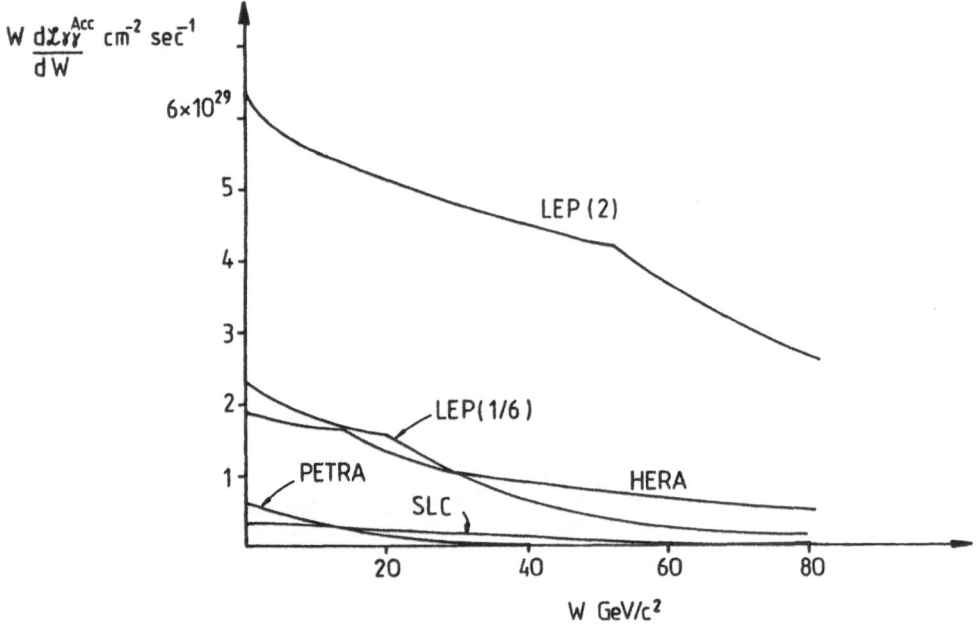

Fig. 13 - Accepted differential luminosity versus W for different colliders.

The following comments can be made on the basis of Fig. 11, 12, 13.

- SLC has a smaller luminosity than PETRA in the low mass region and seems of little
  interest for "classical" 2γ physics. Here however the interesting possibility of
  real γγ or γe collisions (see section 4 below) should not be forgotten.

- HERA competes very well with LEP (1/6) both from the viewpoint of intrinsic 2γ
  luminosity (Fig. 11) and effective luminosity (Fig. 13). In the energy region
  $W > 40$ GeV/c^2 HERA is clearly superior. Both HERA and LEP(1/6) give $\simeq$ 3X PETRA 2γ
  luminosity in the low mass region.

- LEP(2) is superior to HERA by a factor $\simeq$ 2   in intrinsic luminosity (Fig. 11)
  and a factor ∿ 3 in effective luminosity (Fig. 13). However taking into account the
  notorious difficulty that all e^+e^- machines have had till now to reach their design
  luminosity, and the possibility (still unknown) that the situation may be better
  for ep collisions the levels of the various curves in Fig. 13 should be treated
  with   some scepticism. In practice, the relative positions of the LEP (2) and
  HERA curves could easily be reversed. In contrast the curves shown in Figs 11 and
  12 are determined essentially by QED and kinematics and so are subject to a much
  smaller uncertainty.

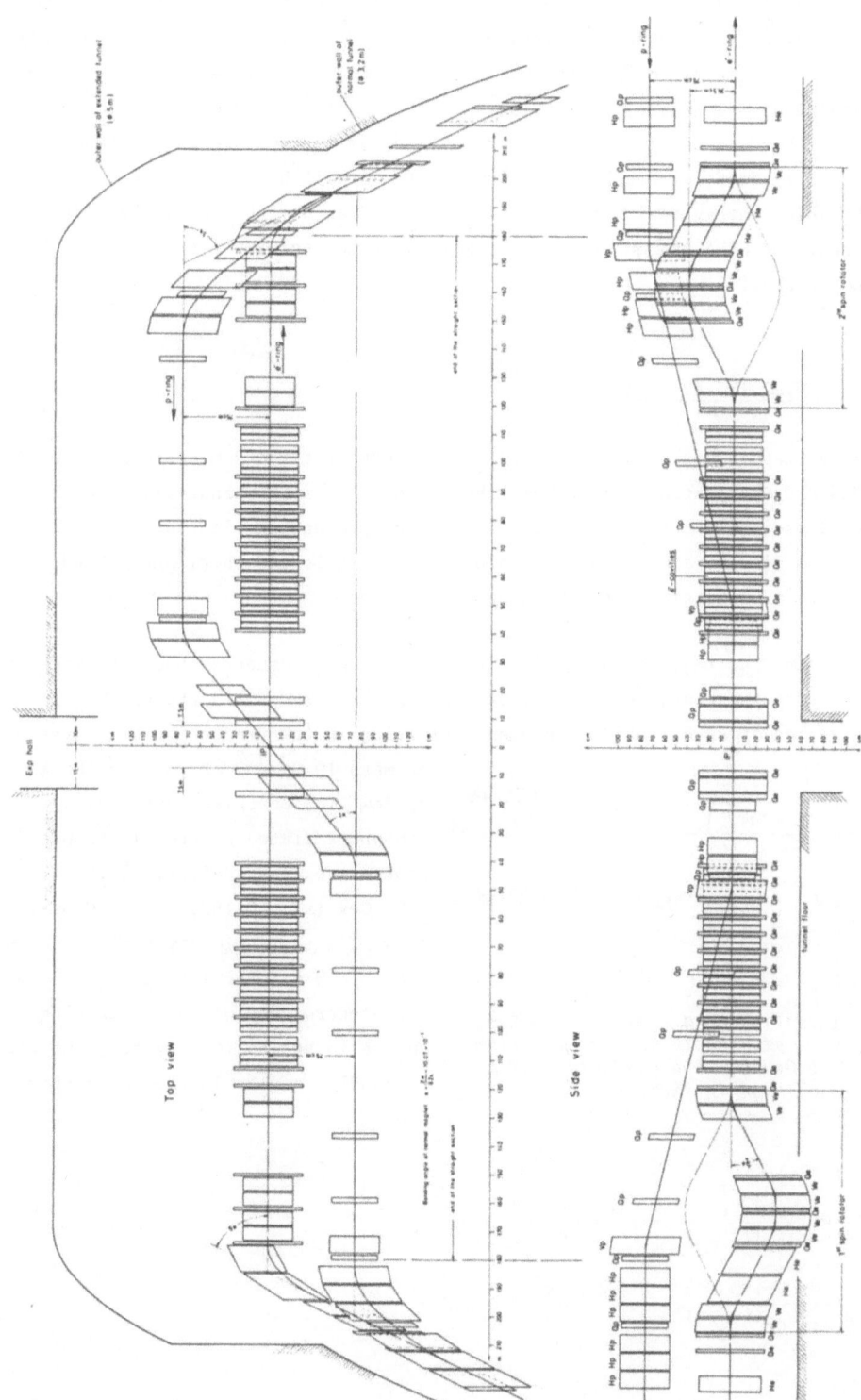

Fig. 14 – Layout of an interaction region at HERA.

The overall conclusion is that both HERA and the first generation LEP machines improve significantly over the existing machines PETRA and PEP (factor ~ 3 in 2γ luminosity at low masses). If LEP(2) reaches its design luminosity it will have almost an order of magnitude higher luminosity than existing machines at low masses. This luminosity will be useful however mainly for untagged physics, as is the case at existing machines. If LEP(2) does not reach design luminosity it will have a very serious competitor in the high energy region in HERA. In any case, as will be discussed in the following section, HERA has quite unique features which facilitate single or double tag two photon physics.

3. SMALL ANGLE TAGGING AT HERA

The beam transport system forseen [27] for a HERA interaction region is shown in Fig. 14. The 820 GeV proton beam crosses the 30 GeV $e^\pm$ beam horizontally at an angle of 20 mrad. Immediately after the beam crossing the proton beam is bent back parallel to the electron line by a horizontal bending magnet. It is this horizontal bend which makes possible $\simeq 0°$ tagging for the scattered protons.

Before discussing the tagging system in more detail some comments should be made on the kinematics of two photon collisions at HERA. Fig. 15 shows, in the ep CM system the configuration of an 2γ collision event resulting in the production of a system of mass 10 GeV/c^2 at rest in the lab system. The energies of the virtual photons radiated by the electron and proton are, respectively, 26.2, 0.96 GeV ($x_2 = 0.167$, $x_1 = 0.0061$). If y_{CM} is the rapididty difference between ep CM and the lab and θ_e, θ_p are the electron and proton scattering angles in the ep CM system, the corresponding angles in the lab system are :

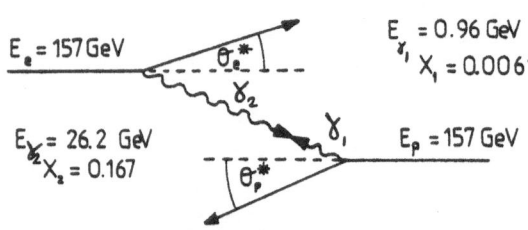

Fig. 15 - *Kinematics of a typical 2 photon event at HERA in the ep CM system. W=10 GeV/c^2. The produced system is at rest in the lab.*

$$\theta_e = \frac{\theta_e^*}{\gamma_{CM}(1-\beta_{CM})} = e^{y_{CM}} \theta_e^* = 5.0 \times \theta_e^*$$

$$\theta_p = \frac{\theta_p^*}{\gamma_{CM}(1+\beta_{CM})} = e^{-y_{CM}} \theta_p^* = 0.2 \times \theta_p^*$$

(12)

Since, as pointed out above, $y_{CM} = 1.6$.

Eqns (12) imply that both small angle proton tagging (where protons are detected at angles < θ_{MAX}) and finite angle electron tagging (where electrons are detected at angles > θ_{MIN}) are more efficient than would be the case if the ep CM system were at rest in the lab :

	$y_{CM} = 0$	$y_{CM} = 1.6$
proton angular acceptance :	$0 < \theta < \theta_{MAX}$	$0 < \theta < 5\theta_{MAX}$
electron angular acceptance :	$\theta_{MIN} < \theta$	$0.2\theta_{MIN} < \theta$

The kinematical effects of the 20 mrad crossing angle are very small and so are neglected in Eqns [12].

If the protons are scattered at angles which are small compared to the angular divergence of the beam at the interaction point, which is given by [27]

$$\alpha_x = 3.9 \times 10^{-5} \text{ rad} \quad \text{(horizontal)}$$

$$\alpha_y = 9.0 \times 10^{-5} \text{ rad} \quad \text{(vertical)}$$

the trajectory of a proton with a given fractional energy loss x_1 can be simply calculated. The horizontal displacement at a given position in the machine is given by :

$$d = D_x \, x_1 \tag{13}$$

where D_x is the horizontal dispersion function. D_x in the proton beam line downstream of a HERA interaction region is shown in Fig. 16.

To have the greatest sensitivity to small values of x_1 the detector should be placed near to the maximum of D_x, which is about 100 m downstream of the interaction point. Other detectors, closer to the interaction point, where D_x is smaller will then be sensitive to larger values of x_1. The smallest value of x_1 which can be detected is limited by :

(i)   The intrinsic energy spread in the proton beam :

$$(\frac{\sigma_p}{p})_{beam} = 1.4 \times 10^{-3} \qquad [27]$$

(ii)   The transverse beam dimension at the position of the detector .

If the detector is placed too close to the beam the lifetime of the latter will be

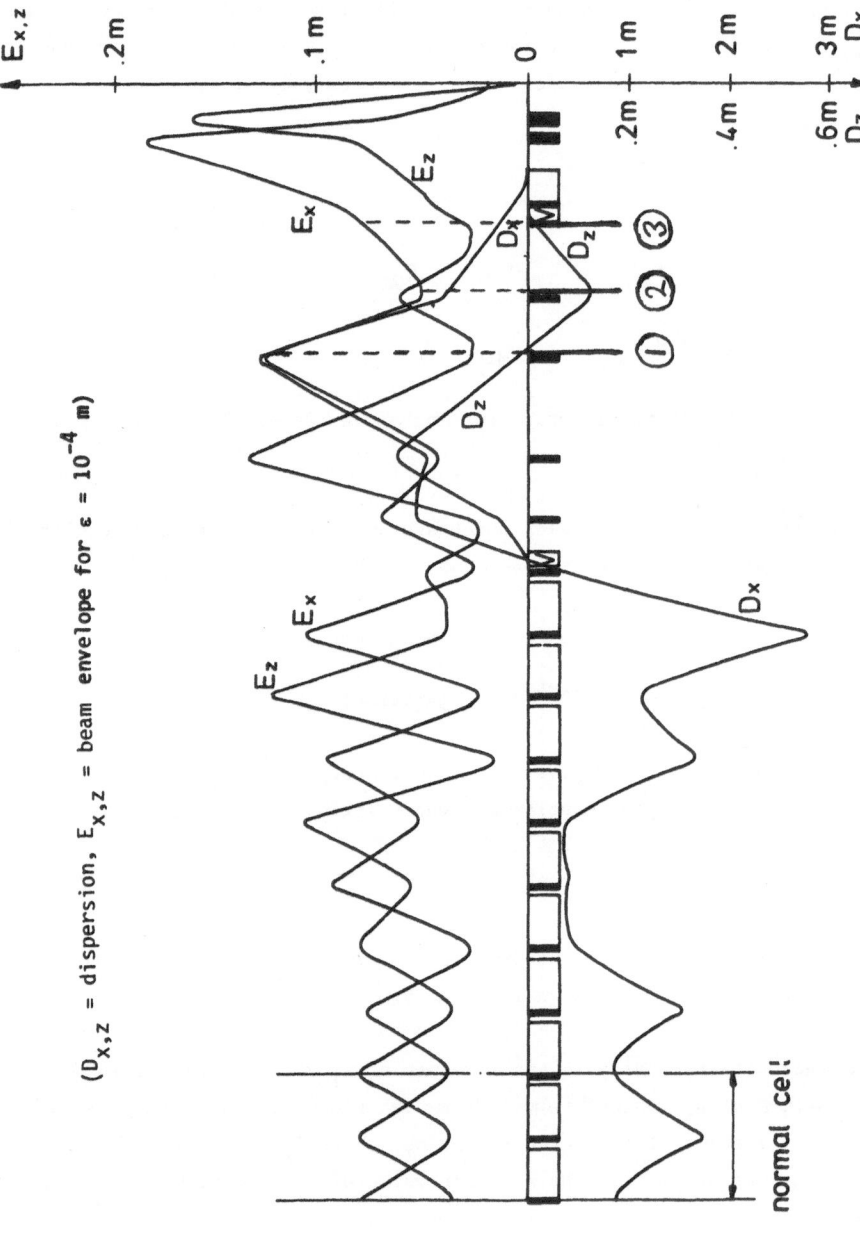

$(D_{x,z}$ = dispersion, $E_{x,z}$ = beam envelope for $\epsilon = 10^{-4}$ m)

Fig. 16 - Beam optics of the proton line downstream of an interaction region at HERA. (1), (2), (3) indicate the locations of tagging detectors.

limited by interactions of the tails of the beam distribution with the detector. In practice the detector should not be closer to the beam than :

$$D_x (\frac{\sigma_p}{p})_{beam} + n\sigma_x \qquad (14)$$

where σ_x = transverse beam size

$$= 0.069 \ \sqrt{E_x} \qquad (\sigma_x \text{ in mm}, E_x \text{ in m})$$

E_x is the horizontal beam envelope function shown in Fig. 16. n should be in the range 5-10 to ensure negligible beam losses.

The largest value of x_1 that can be detected for $\simeq 0°$ scatters is determined by the position of the downstream bending magnet and the radius (3cm) of the vacuum pipe. This last parameter can clearly be made more flexible by installing a special section of beam pipe of larger radius.

By placing 3 detectors at distances of 55 m, 75 m, and 100 m downstream of the interaction point it is possible to cover the range

$$0.0032 < x_1 < 0.11$$

in fractional proton energy loss (or scaled photon energy). Details on the parameters of the tagging detectors are given in TABLE 3.

A possible candidate detector is the Multi Electrode Silicon Detector (MESD) [31] which has the required property of operating easily in high vacuum conditions and could have a spatial resolution in the range 10-300 μ. A telescope of such detectors could mesure both the position and angle of the scattered protons. For protons scattered at 0° a 1mm spatial resolution corresponds to a precision in the fractional energy of the virtual photon of about 5% (see TABLE 3). The positions of the detectors in the lattice are shown in Fig. (16) and the corresponding ranges in the variable $\ln(\frac{1}{x_1})$ are shown in the Courau plot (Fig. 10). Also indicated in Fig. 10 is the value of $\ln(\frac{1}{x_1})$ corresponding to intrinsic beam energy spread $(\frac{\sigma_p}{p})_{beam}$. It is clear from Fig. 10 that the tagging is effective mainly in the high W-region when the angular acceptance for the produced final state particles is properly taken into account.

A realistic calculation of the tagging efficiency to be expected using a system such as that defined in TABLE 3 requires a study of particle tracking through the HERA beam elements.

Here a rough order of magnitude calculation is made by noting that the proposed system covers a range of horizontal scattering angles up to 0.54 mrad in the lab. An optimistic estimate of the tagging efficiency is given by assuming that for a given value of x_1 the whole angular range :

$$0 < \theta < \theta_{MAX}$$

is detected independant of the azimuthal angle of scattering. A more conservative estimate is given by assuming that only protons scattered through angles less than the horizontal beam divergence of 3.9×10^{-5} radians are detected. For comparison the inner edge of the furthest tagging detector and the beam line subtend an angle of 8.7×10^{-5} radians at the interaction point. The first estimate is optimistic because protons can be lost if they scatter in a direction opposite in azimuth to the sense of magnetic deflection, the second is conservative because the overall angular acceptance of the tagging system is underestimated.

The tagging efficiency corresponding to the above angular range in the lab system and fractional energy loss x_1 is given approximately by :

$$\varepsilon(0 < \theta < \theta_{MAX}) = \frac{\ln[r+1] - r/(r+1)}{[2\ln(\frac{2E}{m_p}) - \frac{1}{2}]} \tag{15}$$

where

$$r = \left(\frac{5E\theta_{MAX}}{m_p x_1}\right)^2$$

This formula may be derived from the exact transverse luminosity function [28],[32] in the limit x_1, $\theta_{MAX} \ll 1$. The factor 5 comes from the boost between the ep centre of mass and the lab systems (see Eqns 12). The following values are found for the tagging efficiency :

	"Optimistic"		"Pessimistic"	
	θ_{MAX} = 0.54 mrad		θ_{MAX} = 0.039 mrad	
x_1 = 0.0032	0.8	(0.51)	0.33	(0.08)
0.01	0.59	(0.31)	0.14	(0.005)
0.10	0.19	(0.013)	4×10^{-4}	8×10^{-7}

It can be seen that the small angle tagging becomes ineffective for large values of the fractional energy loss x_1. This is because the peak in the angular distribution

of the scattered protons is at an angle of $\dfrac{x_1 m_p}{E}$ and so, for large x_1, is displaced far from the acceptance region of the tagging system. The figures in brackets indicate the tagging efficiency that would be obtained if the ep CM system were at rest
in the lab i.e. the factor 5 in Eqn(15) is replaced by one. In the physically interesting region where :

$$0.0032 < x_1 < 0.01$$

covered by the tagging detectors 1 and 2 a tagging efficiency of > 40 % can be
expected. In tagging detector 3 protons are detected with a somewhat lower efficiency (probably $\geqslant$ 10%) in the case that they are scattered at angles which are large
as compared to the beam divergence.

With similar approximation to those used in Eqn(15) the electron tagging efficiency
for the angular range :

$$\theta_{MIN} < \theta$$

can be written as

$$\epsilon(\theta > \theta_{MIN}) = \dfrac{2 \ln 5 \left(\dfrac{2x_2}{\theta_{MIN}}\right) + \dfrac{1}{2}}{\left[2 \ln \dfrac{2E}{m_e} - \dfrac{1}{2} \right]} \qquad (16)$$

Some typical values are :

	θ_{MIN} = 45 mrad	θ_{MIN} = 20 mrad
x_2 = 0.037	0.18 (0.06)	0.24
0.167	0.30 (0.17)	0.36
1.0	0.43 (0.31)	0.50

The bracketed figures are found when the factor 5 in Eqn. (16) (coming from the boost
to the ep CM system) is replaced by one.

45 mrad corresponds to the minimum detection angle defined in the HERA proposal [27]
It can be seen that there is little gain in efficiency in reducing θ_{MIN} to 20 mrad.
This is no longer true however when form factor effects are important (see below).

The tagging efficiencies for both protons and electrons are sufficiently large to
make double tag physics viable. The kinematic configuration shown in Fig. 15 for

example has a double tag efficiency, even taking the "pessimistic" figure for the
proton case, of ~ 10%. Depending on the physics, the naïve tagging efficiencies quoted above will be modified by form factor effects [4] .

For a simple VDM type coupling of the photon, for example, a form factor of order

$$F(q^2) \quad = \quad \frac{1}{(1 + \frac{q^2}{m_\rho^2})^2} \qquad\qquad (17)$$

is expected. Because the q^2 values reached by the small angle proton tagging system
are low $\lesssim 0.2 (GeV/c)^2$ the values of F are large at HERA $\gtrsim 0.6$. This may be compared to a value of 4×10^{-3} for tagging at 20 mrad at LEP(2). For electron tagging at
HERA F has maximum values of 0.05, 0.36 for θ_{MIN} = 45,20 mrad. Smaller angle electron tagging than forseen in the current HERA proposal would therefore be advanta-
geous for "soft" single tag and double tag two photon physics. It should be noted
however that the simple VDM suppression factor 5 given above, which corresponds to
the ρ propagator only is ~ 2 times smaller than the prediction of generalised vector meson dominance, where the effect of higher mass vector mesons are taken into
account [4] . When the virtual proton couples to the hadronic system in a point
like manner (as in photon structure function measurements, or processes where jets
are produced at high p_T) no propagator suppression as in Eq. 17 is expected. HERA
would be particularly suitable for a double tag measurement of almost real photons,
using the virtual photons radiated by the proton beam as target, and those radiated
by the electron beam as the probe. The relevant kinematical variables Q^2,x,y can be
determined completely from the double tag measurements, avoiding the $x_{VIS} \rightarrow x$ unfolding problems which arise if only a single tag measurement is made. It is hard to see
how LEP could compete for this type of measurement. The diffractive electroproduction
process which can constitute a serious background for $\gamma\gamma$ physics at low W and q^2 values (see Section 4 below) would be expected to give a negligible contamination in
a structure function experiment.

4. BACKGROUNDS TO TWO PHOTON PROCESSES AT HERA

The most serious background for hadronic states produced in two photon collisions
at HERA (Fig. 17a) is likely to be the diffractive electroproduction process
(Fig.17b).

Using the equivalent photon approximation for the eeγ and ppγ vertices in Fig. 17a and for the eeγ vertex in Fig. 17b the observed cross section ratio, for production of a system X of mass W and rapidity y is estimated to be :

$$\frac{\sigma_{2\gamma \to X}^{obs}}{\sigma_{\gamma p \to X}^{obs}} = \frac{2}{W} \left(\frac{\alpha}{\pi}\right) \left[21n \frac{2E}{m_p} - \frac{1}{2}\right] \frac{\sigma_{\gamma\gamma \to X}(W)}{\frac{d\sigma_{\gamma p \to X}^{diff}}{dW}(W_{\gamma p},W)} \frac{\varepsilon_{2\gamma}(W,y)}{\varepsilon_{\gamma p}(W,y)} \qquad \ldots (18)$$

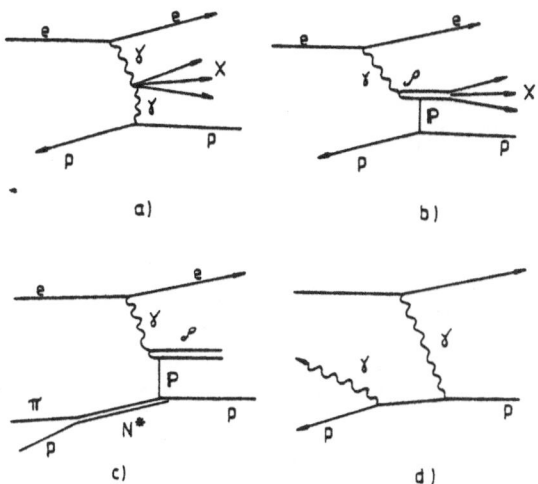

a)   b)

c)   d)

Fig. 17 - Signal and backgrounds in 2γ collisions at HERA. a) 2γ process. b) inelastic diffractive electroproduction. c) diffractive N electroproduction. d) beam beam bremsstrahlung.*

Where $\sigma_{\gamma\gamma \to X}(W)$ is the total γγ cross section to produce system X with effective mass W, and $\frac{d\sigma_{\gamma p \to X}^{diff}}{dW}$ $(W_{\gamma p},W)$ is the differential cross section to produce a system of mass W by diffractive photo production where $W_{\gamma p}$ is the total γp centre of mass energy. $\varepsilon_{2\gamma}$, $\varepsilon_{\gamma p}$ are the proton tagging efficiencies for the two photon and electroproduction process respectively.

One process where measurements exist for both 2γ production [33] and photoproduction [34] is the system $\pi^+\pi^+\pi^-\pi^-$ for masses W of 1.1 – 1.8 GeV/c^2.

It can be seen from Fig. 10 that masses as small as 1.5 GeV/c^2 are detected in the tagging system only for values of x_1 near to the minimum of 0.0032. This implies that $W_{\gamma p}$ = 26.3 GeV/c^2 when W = 1.5 GeV/c^2. Taking the $(E_\gamma^{lab})^{-0.4}$ dependance of the photoproduction cross section found in Ref. [34] it is estimated that :

$$\frac{d\sigma_{\gamma p \to 4\pi}}{dW} = 0.5 \ \mu b/GeV/c^2 (W_{\gamma p} = 26.3 \ GeV/c^2, \ W = 1.5 \ GeV/c^2)$$

while Ref. [33] gives

$$\sigma_{\gamma\gamma \to 4\pi} = 0.2 \ \mu b \qquad (W = 1.5 \ GeV/c^2)$$

Taking θ_{MAX} = 0.039 mrad, then as found above $\varepsilon_{\gamma 2}$ = 0.33. To calculate $\varepsilon_{\gamma p}$ it is noted that, in this case $\begin{bmatrix} 34 \end{bmatrix}$

$$\frac{d\sigma_{\gamma p \to 4\pi}}{dq^2} \simeq e^{-7q^2}$$

If $q^2_{MAX} \simeq 25 \, E^2 \theta^2_{MAX} \ll 1$ then

$$\epsilon_{\gamma p} = \frac{\int_o^{q^2_{MAX}} e^{-7q^2} dq^2}{\int_o^\infty e^{-7q^2} dq^2} \simeq 7 \, q^2_{MAX}$$

giving $\epsilon_{\gamma p} = 6.6 \times 10^{-3}$

for $\theta_{MAX} = 0.039$ mrad.

So from Eqn(18)

$$\frac{\sigma^{obs}_{2\gamma \to 4\pi}}{\sigma^{obs}_{\gamma p \to 4\pi}} = 0.70 \qquad\qquad W = 1.5 \text{ GeV/c}^2$$

Because of the dependence of $\epsilon_{\gamma p}$ on θ^2_{MAX} this ratio would be much smaller if a larger value of θ_{MAX} were used. Conversely a smaller value of θ_{MAX} would result in a sharp reduction in the electro production background, at some cost in tagging efficiency. It is clearly of importance to be able to measure accurately both the angle and the position of the scattered proton.

It is more difficult to give an estimate of the diffractive electroproduction background to be expected at higher masses, where the main physics interest at HERA lies. To make an estimate of the level of such background the kinematical configuration shown in Fig. 15 is considered.

It is assumed that the total inelastic diffractive cross section is the same as the elastic one (i.e. ρ,ω,ϕ production) $\simeq 10$ µb and that $\sigma_{\gamma\gamma \to X}$ is 0.3 µb as given by VDM and factorisation arguements.

The distribution in W is taken to be flat :

$$\frac{d\sigma_{\gamma p \to X}}{dW} = \frac{10 \ \mu b}{W_{\gamma p}} = \frac{10 \ \mu b}{125 \ GeV/c^2}$$

while the q^2 distribution is taken to be less steep [35] than in the low mass region above.

$$\frac{d\sigma_{\gamma p \to X}}{dq^2} \simeq e^{-5q^2}$$

giving $\epsilon_{\gamma p} = 4.7 \ x10^{-3}$ with $\theta_{MAX} = 0.039$ mrad.

From Eqn (15) if is found that :

$$\epsilon_{2\gamma} = 0.22$$

or using Eqn (18).

$$\frac{\sigma^{obs}_{2\gamma \to X}}{\sigma^{obs}_{\gamma p \to X}} = 0.91 \qquad\qquad W = 10 \ GeV/c^2$$

Considering the conservative nature of the above assumptions, particularly the large assumed value for the inelastic diffractive cross section, and the flat W distribution (which surely overestimates the high mass region) this result is quite encouraging.

It should also be pointed out that for the most interesting final states in 2γ collisions accessible at high W values, for example deep inelastic scattering on an almost real photon, high p_T jet production, or exclusive production of high p_T meson pairs, the diffractive electro-production background is expected to be absent. In the high p_T processes there is another type of background coming from direct γ quark or γ gluon scattering where the quark or gluon is a constituent of the incident proton. Such events however (unlike the 2γ collisions where the same hard scattering pro processes may contribute) are always expected, as in the Drell Yan process, to be accompanied by a target fragmentation jet.

Requiring the detection of a scattered proton with very small scattering angle and energy loss should effectively suppress this background. Further rejection , can be obtained requiring the absence of fragmentation products accompanying the detected proton, either in the forward hadronic calorimeters of the detector, or in special veto detectors with an acceptance extending down to smaller angles.

Another possible source of background are events where the proton is diffractively excited : $p \rightarrow N^{*}$ where N^{*} is a resonance with the same quantum numbers as the proton (Fig. 17c). It is not difficult to see, from simple kinematical arguments, that such processes are not expected to give significant backgrounds in the small angle tagging system. Considering for example : the lightest such resonance, the N^{*} (1410) the maximum possible energy for the decay proton is 154.8 GeV in the ep CM system giving x_1 = 0.014. Such events are therefore kinematically forbidden for tagging detector 1 . The rate in the other detectors should be low, as the typical (decay at $\theta^{*} = \frac{\pi}{2}$) decay angle and energy of the proton in the lab system are 0.72 mrad and 580 GeV (x_1 = 0.29) whereas the acceptance range of the tagging system is $\theta < 0.54$ mrad and $x_1 < 0.11$.

A potentially large source of background in the small angle tagging system is that due to the beam-beam bremsstrahlung process where a photon is radiated from the proton line. One graph contributing to this process is shown in Fig. 17d . The cross section, differential in the fractional energy loss x_1 is given by [36] .

$$\frac{d\sigma}{dx_1} = \frac{2\alpha^3}{m_p^2 x_1} \left\{ 2\left[1 + (1 - x_1)^2\right] - \frac{4}{3}(1-x_1) \right\} \left[\ln \frac{4(1-x_1)E^2}{m_p^2 x_1} - \frac{1}{2} \right] \quad (19)$$

Taking the worst case, the minimum value of x_1 = 0.0032 and Δx_1 = 0.001 gives a cross section of :

$$5.2 \times 10^{-33} \text{ cm}^2$$

With a luminosity of $6 \times 10^{31} \text{cm}^{-2} \text{sec}^{-1}$ this corresponds to a total count rate of 0.3 hz. A more relevant parameter is the number of counts per bunch crossing which is 3×10^{-8}. Clearly there is no problem of occupation of the tagging detectors from this source. It is interesting to note, from Eqn (19) that if $e^{+}e^{-}$ collisions under the same kinematic conditions are considered the rate would be a factor 6×10^6 higher, so there would be an occupation problem (0.18 hits per buch cross). At LEP with its lower duty cycle, (4 bunches instead of 210) the situation is even worse and it has been concluded [37] that there small angle tagging is not possible for angles below 1 mrad.

In summary, the rough calculations, presented in this section indicate that two photon physics at HERA should be distinguishable from backgrounds when the small angle tagging system is used. In this connection it is important to measure both the position and the angle of the scattered photon with good precision, as the diffractive electroproduction background drops sharply at small scattering angles. For processes which have a characteristic two photon signature, less stringent angular cuts are needed. Beam beam bremsstrahlung backgrounds (in sharp contrast to LEP and other e^+e^- machines) are negligible.

5. WEAK BOSON PRODUCTION AT HERA.

Before describing the technical realisation and the physics interest of real γe and $\gamma\gamma$ collisions it is interesting to note that a glimpse (or perhaps more) of the physics of the γe collision processes :

$$\gamma e^{\pm} \rightarrow W^{\pm} \nu \qquad\qquad (i)$$

$$\gamma e \rightarrow Z^0 e \qquad\qquad (ii)$$

can already be obtained at HERA. The energy and luminosity of HERA is such that there should be a sufficient flux of high energy Weisacker-Williams photons radiated by the proton beam to observe both (i) [38] and (ii) [39] . In addition $W^{\pm}$ bosons can be produced via the process, specific to ep collisions [40] .

$$\gamma^* p \rightarrow W^{\pm} X^{0+} \qquad\qquad (iii)$$

Where the γ^* is a virtual photon radiated by the electron beam.

At HERA with beam energies and effective luminosity as given in TABLE 2 the following cross sections and numbers of events can be expected for these processes :

	$\sigma_{tot} (cm^2)$	Ref.	Number of events (2 years at $\bar{L}$)
$e^{\pm}p \rightarrow \nu W^{\pm} X^+$	9.0×10^{-38}	[38]	110
$e^{\pm}p \rightarrow e^{\pm} Z^0 X$	1.0×10^{-36}	[39]	1300
$e^{\pm}p \quad e^{\pm} W^{\pm} X^{0+}$	1.2×10^{-36}	[40]	1500

The cross sections quoted correspond to the standard Glashow-Weinberg-Salam electroweak theory. The estimate of Ref. [40] which uses a simple Weisacker-Williams formula to estimate the flux of virtual photons radiated by the $e^{\pm}$ is probably somewhat optimistic. There seems to be an inconsistency between ref. [38], [39] concerning the relative contributions of the elastic and inelastic proton form factors to the virtual photon flux. For a total ep CM energy of order 300 GeV, Ref. [38] found roughly equal contributions, whereas Ref. [34] found, at the same energy that the elastic contribution was more than an order of magnitude smaller then the inelastic one. Until this discrepancy is resolved, the last two results quoted above should be treated with some caution. Putting aside this difficulty the number of events for these processes of $\gtrsim 1000$ should be adequate to identity the Z^o and $W^\pm$ by the clean, purely leptonic, decay signatures :

$$Z^o \rightarrow e^+ e^-, \quad \mu^+ \mu^-$$

$$W^\pm \rightarrow e^\pm \nu, \quad \mu^\pm \nu$$

As will be discussed further below the processes (i) and (iii) are of special interest because they are sensitive to the direct $\gamma W^+ W^-$ coupling, i.e. the magnetic moment of the W boson.

6. REAL γe AND $\gamma\gamma$ COLLISIONS IN ONE PASS LINEAR COLLIDERS.

Linear Colliders

In order to reach significantly higher energies in $e^+ e^-$ collisions than will be provided by LEP, a new type of machine known as the one pass linear collider has been proposed by groups at SLAC [6] , [41] , [42] and Novosibirsk [7]. The idea is to make a head on collision of the beams of two electron-linear accelerators or of two bunches from the same linac . The cost of such a machine and its energy are roughly proportional to its length. In contrast the cost of a $e^+ e^-$ storage ring is roughly proportional to the square of the desired centre of mass energy. It has been pointed out that to achieve significantly higher energies than LEP a linear collider is a more economical proposition. The luminosity of such a machine is given by the expression :

$$\mathcal{L} = \frac{N^2 \nu_{REP}}{4\pi \sigma_x \sigma_y} \tag{20}$$

where N is the number of particles per bunch, ν_{REP} the repetition frequency of the bunch collisions and σ_x, σ_y are the horizontal and vertical beam widths (gaussian profiles are assumed). ν_{REP} is much smaller in a linear collider than in an $e^+ e^-$ storage ring (ν_{REP} = 250 kHz, 50 kHz, 180 Hz for respectively PETRA, LEP, and S.L.A.C.). In compensation N is made very large and σ_x, σ_y as small as technically possible. For the SLC project [6] , [41] ,[42] it is planned to increase N by a factor ⪴ 50 and values of σ_x, σ_y as small as 1 - 2 μm with a corresponding beam emittance of 3×10^{-5} rad.m are proposed. The enormous mutual focussing forces destroy the bunches during collision, but, in so doing, actually increase the luminosity by the so called 'pinch effect' [6] , [42] . The SLC projet at SLAC has two aims, first to build a collider capable of studying Z^o physics at relativly low cost and secondly to act as a prototype for a more ambitious collider giving centre of mass energies of up to 1 TeV.

Fig. 18 which shows a schematic diagram of the VLEPP project at Novosibirsk gives an idea of how such a high energy collider might look. Only one half of the machine is shown. The operating cycle is as follows : After colliding the e^+e^- bunches in one or more interaction regions (11), a pulsed deflection magnet (9) and a spectrometer (18) deflect the beam into a helical undulator (12)where of the order of 1% of the total beam energy is radiated in the form of circularly polarised photons with ≃ 10 MeV energy. After passing through the undulator the beam is deflected by bending magnets and is used either in a stationary target experiment (16) or is absorbed in a beam dump. The polarised photons strike a conversion target (14) producing a beam of longitudinally polarised electrons and positrons. After charge selection these particles are accelerated up to an energy of 1 GeV in an intermediate accelerator (2). The bunch length of 1 cm is stretched by a factor of ≃ 10 in the debuncher (3) and the longitudinal polarisation in rotated into the transverse direction. After collecting the beam in the large acceptance storage ring (4) it is cooled down to the required low emittance value by radiation damping in the cooling ring (5). The transverse polarisation is maintained in (4) and (5) by the Sokolov-Ternov mechanism [43]. After cooling the beam passes into the buncher (6) where the bunch length is reduced to ~ 1 cm and the polarisation is rotated back into the longitudinal direction. To complete the cycle the beam is accelerated to high energy in the linear accelerator (7), (8) and the whole process is repeated. An injector (1) provides the initial beam.

Such a machine can provide collisions between longitudinally polarised $e^\pm$ at energies up to and beyond 1 TeV with luminosities of 10^{32} $cm^{-2}sec^{-1}$. Many technical problems remain, however to be resolved [6],[7],[42]. Experience with the SLC, whose principles of operation are essentially the same as described above for VLEPP should indicate whether these problems can be overcome.

Another feature of such a machine is the possibility to convert, with high efficiency the polarised electron beams of one or both of the colliding linacs into a beam of polarised real photons [8] ,[9],[10] . The principle of operation of such a "photon accelerator" will now be brefly described.

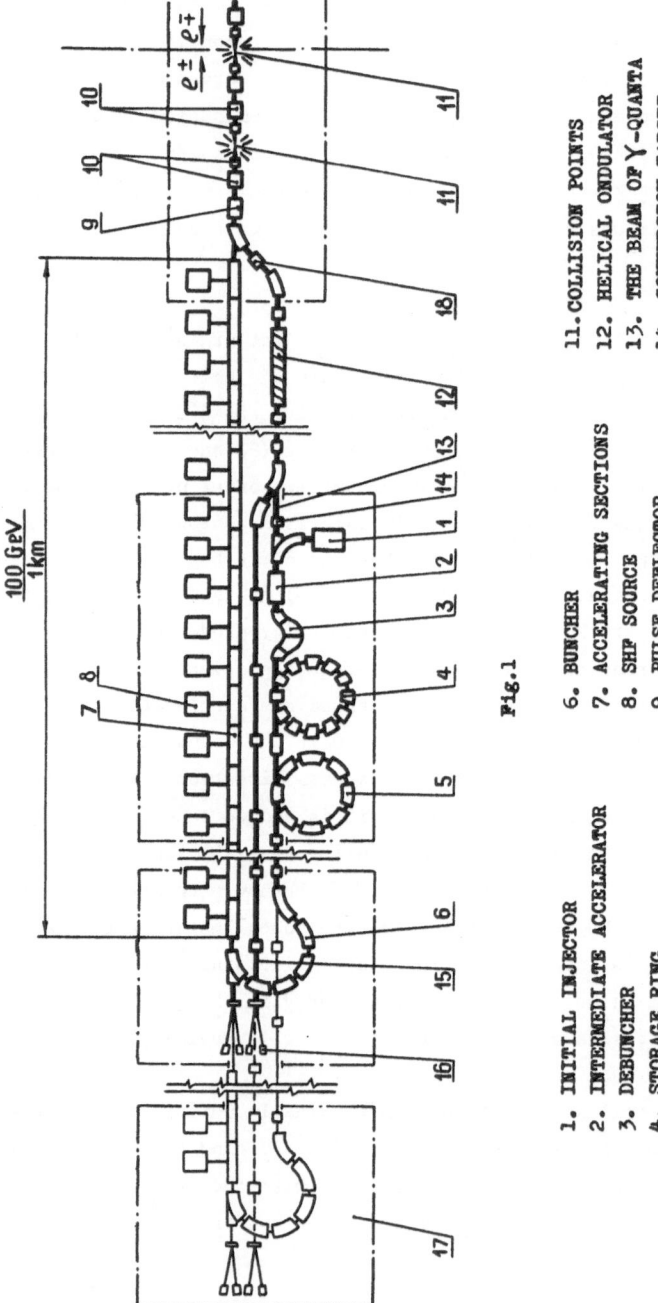

$\dfrac{e^{\pm}}{}\;\dfrac{e^{\mp}}{}$

$\dfrac{100\ GeV}{1km}$

Fig.1

1. INITIAL INJECTOR
2. INTERMEDIATE ACCELERATOR
3. DEBUNCHER
4. STORAGE RING
5. COOLER – INJECTOR

6. BUNCHER
7. ACCELERATING SECTIONS
8. SHF SOURCE
9. PULSE DEFLECTOR
10. FOCUSING LENSES

11. COLLISION POINTS
12. HELICAL ONDULATOR
13. THE BEAM OF γ-QUANTA
14. CONVERSION TARGET
15. RESIDUAL ELECTRON BEAM

16. ELECTRON (POSITRON) BEAM EXPERIMENTS
    WITH STATIONARY TARGET
17. THE SECOND STEP
18. SPECTROMETER

Fig. 18 – The VLEPP project.

Compton Back Scattering

The production of high energy photons by collisions of highly relativistic particles with light quanta of optical wave lengths was first calculated by Feenberg and Prima-koff [44] in an astronomical context. They considered the interaction of high energy cosmic electrons with starlight quanta.

The interest of such a process for high energy physics experimentation was pointed out, after the advent of the laser, by Milburn [45] and Arutyunian and Tumanian [46] One experiment, in which a polarised photon beam produced by Compton back scattering of a ruby laser was incident on a hydrogen bubble chamber was carried out some 15 years ago [47] .

The definition of kinematic quantities is given in Fig. 19. In addition the following dimensionless variables are introduced (the notation of Ref. [10] is followed)

$$y = \omega/E \qquad\qquad y_o = \omega_o/E$$

$$x = \frac{4y_o}{\gamma^2} \cos^2 \frac{\alpha_o}{2} \quad = \quad \frac{4E\omega_o}{m_e^2} \cos^2 \frac{\alpha_o}{2}$$

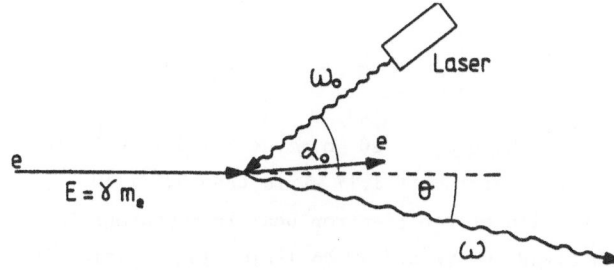

Fig. 19 - Kinematics of Compton back-scattering.

Taking typical values E = 50 GeV, ω_o = 1.17 eV (neodymium glass laser) and α_o = 0 x is found to be 0.90. For this case the energy of the incident photon in the electron rest frame is low $\simeq 0.12$ MeV and the total scattering cross section is close to the classical (Thompson) limit of 2.5×10^{-25} cm^2.

The maximum energy of scattered photon is given by :

$$y_{MAX} = x/(1+x) \qquad\qquad\qquad (21)$$

while the energy of the scattered photon is related to its scattering angle θ by the relation ($\theta \ll 1$)

$$y = \frac{y_{MAX}}{1 + (\theta/\theta_o)^2} \qquad\qquad\qquad (22)$$

where $\qquad \theta_o = \sqrt{1+x}/\gamma$

High energy photons are scattered at angles smaller than $\theta_0 \simeq \frac{1}{\gamma}$. The parameters a_e, a_γ and b are defined in Fig.20. a_e is the half width of the collision region of the electron beams. a_γ is the corresponding parameter for the laser and electron beams, while b is the distance from the electron electron beam collision point to the laser electron beam collision point. For both SLC and VLEPP $a_e \simeq$ 1 to 2 μm, $a_\gamma \simeq$ 20 μm

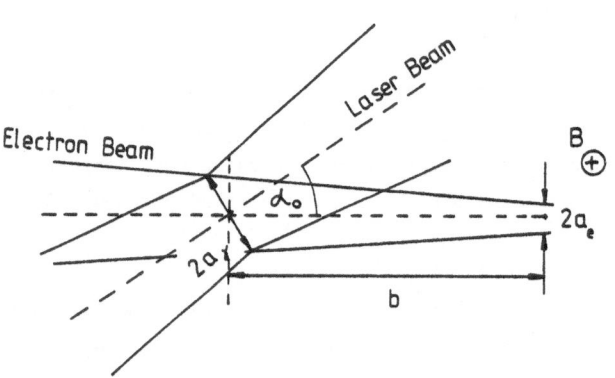

and b is 5 to 10 cm. A transverse magnetic field B of order 1 T is need to deflect the electron beam after the laser crossing region to avoid direct electron electron interactions of the residual beam.

The order of magnitude of the laser power needed to convert ~ 100 % of the electron beam into photons can be readily calculated. If S is the cross sectional

Fig. 20 - Definition of beam parameters for Compton backscattering of a laser beam.

area of the interaction region of the electron and laser beams, and n the number of photons in the laser pulse, saturation of the Compton scattering process will be reached if

$$n\sigma_c = S \tag{23}$$

Taking $\sigma_c = 2.5 \times 10^{-25}$ cm^2 and S = πa_γ^2 where a_γ = 20 μm gives n = 5 x 10^{19}. The corresponding energy in the laser pulse, taking ω_0 = 1.17 eV is then 5.9 x 10^{19} eV or 9.4 joules. The efficiency of conversion of the electron beam into photons is measured by the coefficient k, which gives the γe and $\gamma\gamma$ collision luminosities in terms of the electron electron luminosity :

$$\mathcal{L}_{\gamma e} = k\mathcal{L}_{ee} \; , \qquad \mathcal{L}_{\gamma\gamma} = k^2 \mathcal{L}_{ee}$$

The energy spectrum of the scattered photons for various values of E is shown in Fig. 21 for ω_0 = 1.17. As can be seen from Eqn 21, the higher the beam energy the closer is the end point of the scattered photon spectrum to the electron beam energy. The differential luminosity for $\gamma\gamma$ collisions, as a function of the mass of the produced system for the same series of E values is shown in Fig. 22. In Figs 23 and 24 the differential luminosity for eγ and $\gamma\gamma$ collisions respectively for E = 150 E = 150 GeV, ω_0 = 1.17 eV are compared with the corresponding luminosity expected from the virtual Weisacker-Williams photons radiated at the same energy. Even with k values as low as 30 %, the luminosity available in the real $\gamma\gamma$ collisions near the

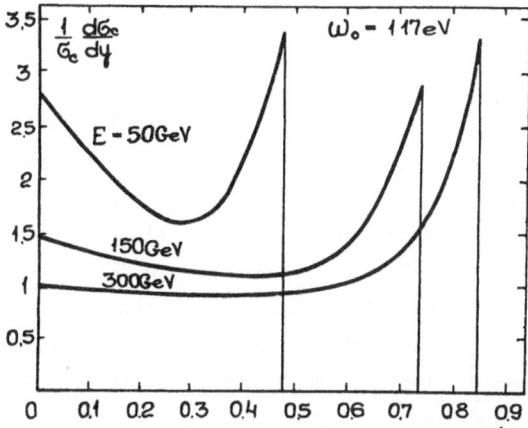

Fig. 21 - *Energy spectrum of back scattered photons.*

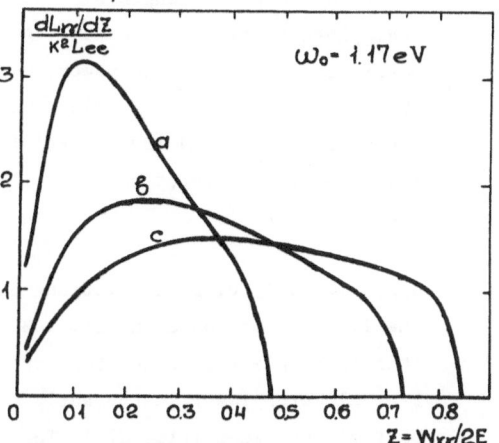

Fig. 22 - *Differential luminosity of $\gamma\gamma$ collisions for $\rho^2 \ll 1$. Curves a,b,c correspond to E = 50,150, 300 GeV.*

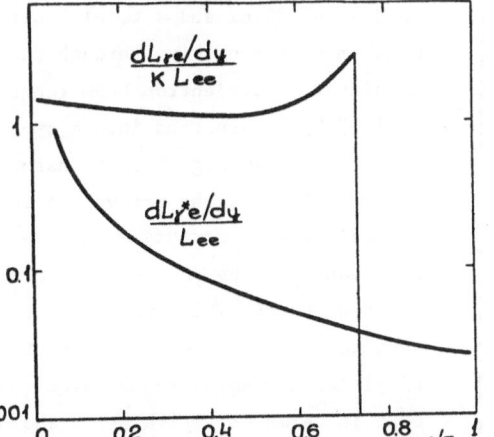

Fig. 23 - *Differential luminosity for γe and $\gamma^* e$ collisions E=150 GeV, ω_o = 1.17 eV, $\rho^2 \ll 1$.*

kinematical end point is some two orders of magnitude larger than in the virtual $\gamma\gamma$ collisions ! It can be seen from Eqn 22, that if photons are selected at small scattering angles to the incident electron beam, the energy spectrum will be concentrated more in the region of y_{MAX}, giving a photon beam with a narrower energy spread. Such a "monochromatisation" of the colliding beams can be most easily obtained by increasing b (Fig.20). If however the ratio :

$$\rho = \left(\frac{b\theta_o}{a_e}\right)^2 \tag{24}$$

becomes large, the typical scattering angle of the photon (θ_o) is greater than the angular extent of the ee collision region as seen from the e laser interaction point (a_e/b) and a loss of luminosity results, even for the high mass region. This effect is demonstrated in Figs 25 and 26 where the differential luminosity functions for $e\gamma$ and $\gamma\gamma$ collisions respectively are shown for different values of ρ . It can be seen that monochromaticity is bought only at the cost of luminosity. If the laser photons have polarisation (circular or linear) this polarisation is largely retained by the scattered photons. If the laser light is unpolarised and the electrons are longitudinally polarised, the scattered photons have a high degree of circular polarisation. Details of these polarisation effects may be found in Ref. [10] .

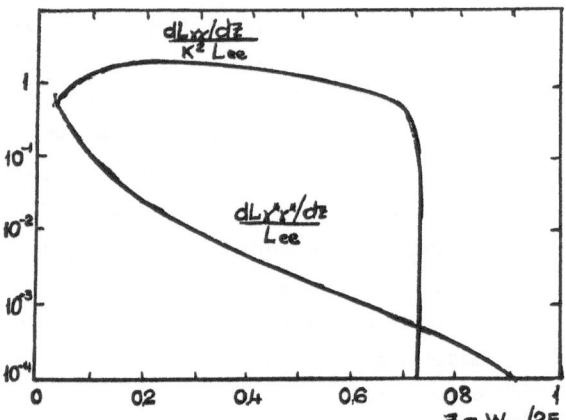

Fig. 24 - Differential luminosity for $\gamma\gamma$ and $\gamma^*\gamma^*$ collisions. E = 150 GeV, ω_o = 1.17 eV, $\rho^2 \ll 1$.

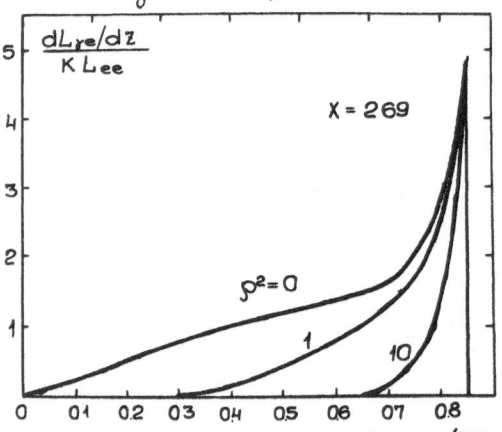

Fig. 25 - Differential luminosity for γe and $\gamma^* e$ collisions (E=150 GeV, ω_o=1.17 eV) for different ρ^2 values.

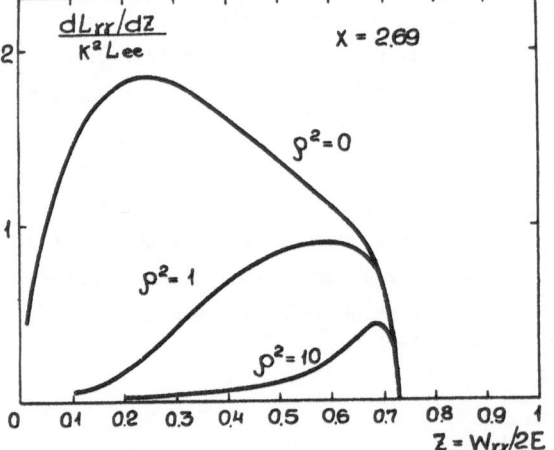

Fig. 26 - Differential luminosity for $\gamma\gamma$ and collisions (E=150 GeV, ω_o=1.17 eV) for different ρ^2 values.

The required laser power seems to be within the reach of existing technology. The outstanding problem is the high power output which is required in combination with a relatively high repetition rate. (180 Hz for SLC, 10 Hz for VLEPP). To date considerably higher power outputs have been achieved but at low repetition rates, and the required repetition rates or greater have been reached, but with somewhat lower power [10]. The required pulse length of 100 ps seems less difficult to obtain. A possible solution to the power and repetition rate problem would be to use a battery of 10 or 100 lasers of lower power, triggered in co-incidence.

A particularly elegant solution for the laser is proposed in Ref. 48. This is to place an undulator, or free electron laser in the early part of the linac (Fig. 27.). To provide sufficient laser power this needs to have a field of 2T, a period λ_o of 20 cm and a total length of 40 m. After passing through the undulator, the electron beam (energy 10 GeV) is diverted into a by-pass line (B in Fig. 27) so that the pulse of laser light arrives at the correct time, in advance of the electron bunch, to be focussed on to the opposing bunch. This system has the advantage that synchronisation of the laser beam collision is garanteed both in terms of arrival time and of pulse overlap. It is unfortunately not applicable to a one linac colli-

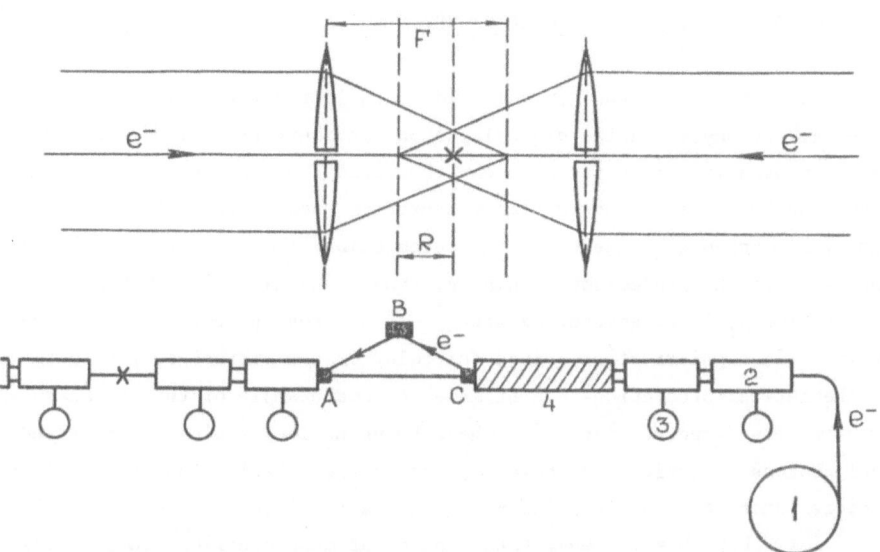

Fig. 27 - Free electron laser (undulator) for generating high energy photon beams.
1. - injector, 2. - accelerating system. 3. - High frequency generator.
4. - undulator.

der der such as the SLC, where magnetic bending is needed to bring the bunches into collision.

The problem of calibration of the luminosity of real γγ collisions has been considered at Novosibirsk by Kuraev, Schiller and Serbo [49]. At the very high energies of VLEPP traditional QED processes such as :

$$\gamma\gamma \rightarrow \mu^+\mu^-$$

$$\gamma\gamma \rightarrow e^+e^-$$

are not suitable, both because of the low value of the cross section and its rapid $(1/W^2)$ energy dependance. It is proposed to use instead the reaction :

$$\gamma\gamma \rightarrow e^+e^-\mu^+\mu^-$$

which has a larger, energy independant, cross section of 5.7×10^{-39} cm^2 and relatively clean signature due to the presence of a low mass muon pair which carries almost the full energy of one of the colliding photons. Special muon pair detectors in the forward direction will be required to observe this process.

7. THE PHYSICS OF REAL γe AND γγ COLLISIONS.

The advantage of real γe and γγ collisions, in luminosity terms, as compared to the "classical" 2γ physics using virtual photons is clear from Figs 23 and 24. Many of the most interesting processes from the theoretical viewpoint - those where QCD predictions can be tested - require large centre of mass energies in order to be in the kinematical region where perturbative calculations should be valid. Examples of such processes are the production of high p_T quark and gluon jets [50, 51] or of pairs of mesons at high p_T [52]. Another advantage of the real γ beams is that they can be produced in various polarisation states (circular or transverse) so that more refined tests of theoretical predictions can be made. A good example of this is measurement of the real photon structure function, where by using longitudinally polarised elec-trons, and photons in arbitary circular or transverse polarisation states all the pho-ton structure functions can, in principle, be measured. Even at the lowest beam ener-gy proposed for VLEPP, E = 150 GeV, total centre of mass energies, in eγ collisions, comparable to those that will be attained at HERA in ep collisions should be obtain-ed. In this case the contributions of Z^O exchange to the photon structure function Fig. 28a and the corresponding W exchange process with ν production Fig. 28b should

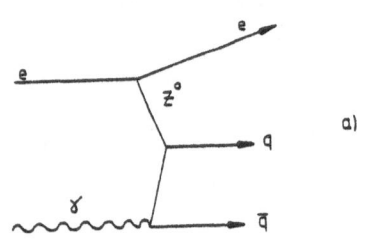

be measurable. At SLC with E = 50 GeV, the maximum q^2 in real eγ collisions is $\simeq 10^4 (GeV/c)^2$ where γ and Z^O exchange contributions are already comparable but it is doubtful if the luminosity will be adequate to exploit this possibility.

A particularly important field of physics that will be opened to experimental study by real γe and γγ collisions at high energy is the production of the gauge bosons Z^O, $W^{\pm}$.

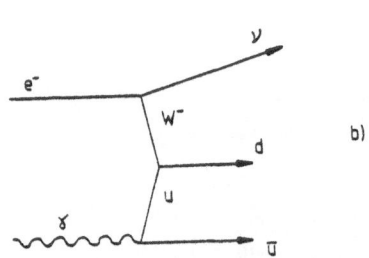

Fig. 28 - a) Z^O exchange contri-bution to the photon struc-ture function.
b) Charge current inter-action on a photon target.

The process $e^+e^- \to W^+W^-$ has been studied in some detail in connection with the LEP project [53]. Although, in principle, sensitive to the fundamental gauge couplings γW^+W^-, $Z^O W^+W^-$, it is in fact diffi-cult to observe these contributions because the cross section is dominated by the neutrino exchange graph. In contrast the processes γe → Wν and γγ → W^+W^- enable the couplings γW^+W^- and $\gamma\gamma W^+W^-$ to be directly measured. A brief discussion of gauge boson production in γe and γγ collisions follows.

$\gamma e \rightarrow Z^o e$

The differential cross section for this process is given in Ref. [13]. The total cross section as a function of the γe centre of mass energy, and the lowest order diagrams which contribute are shown in Fig. 29. The total cross section is [14]:

$$\sigma(\gamma e \rightarrow Z^o e) = \frac{\tilde{\sigma}}{x} \left[(1 - \frac{2}{x} + \frac{2}{x^2}) \, L + \frac{1}{2}(1 - \frac{1}{x})(1 + \frac{7}{x}) \right]$$

where $x = (W_{\gamma e}/M_Z)^2$

$$\tilde{\sigma} = \frac{\pi \alpha^2}{2M_Z^2 \sin^2 2\theta_W} \left[1 + (4\sin^2\theta_W - 1)^2 \right] = 5.9 \text{ pb}$$

$$L = 2\ln \left(\frac{W_{\gamma e}^2 - M_Z^2}{m_e W_{\gamma e}} \right) \tag{25}$$

No fundamentally new couplings are measured in this process. The cross section is however large just above threshold. The dominant graph at high energy is u channel electron exchange, which results in emission of the Z^o preferentially antiparallel to the incoming photon direction.

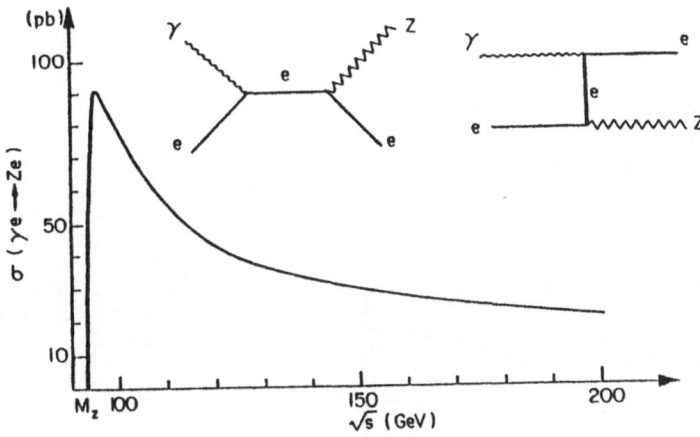

Fig. 29 - *Cross section for the process* $\gamma e \rightarrow Z^o e$ *as a function of* $W_{e\gamma} = \sqrt{s}_{e\gamma}$ *and lowest order contributing graphs.*

$\gamma e \to W^{\pm} \nu$

This process is discussed in Ref. [13], [14]. The lowest order diagrams and the total cross section versus $W_{e\gamma}$ are shown in Fig. 30. t channel W exchange gives an important contribution which is sensitive to the gauge coupling $\gamma W^{+}W^{-}$. At high energies the W exchange graph is predominant and the produced W boson is emitted mostly parallel to the incoming photon. The total cross section can be written in terms of the degrees of longitudinal polarisation of the electron and photon beams $P^{\ell}_{e^{\mp}}$, P^{ℓ}_{γ} as [14]:

$$\sigma(\gamma e^{\mp} \to W^{\mp}\nu) = (1 \mp P^{\ell}_{e^{\mp}})(\sigma^{np} + P^{\ell}_{\gamma} \tau)$$

$$\sigma^{np} = \tilde{\sigma}\left[\left(1 - \frac{1}{x}\right)\left(1 + \frac{5}{4x} + \frac{7}{4x^2}\right) - \frac{1}{x}\left(2 + \frac{1}{x} + \frac{1}{x^2}\right) \ln x\right]$$

$$\tau = \frac{\tilde{\sigma}}{x}\left[\frac{x-1}{4x}\left(13 + \frac{3}{x}\right) - \left(1 + \frac{3}{x}\right) \ln x\right]$$

$$\tilde{\sigma} = \frac{\pi\alpha^2}{M_W^2 \sin^2\theta_W} = 47 \text{ pb} , \quad x = (W_{\gamma e}/M_W)^2 \tag{26}$$

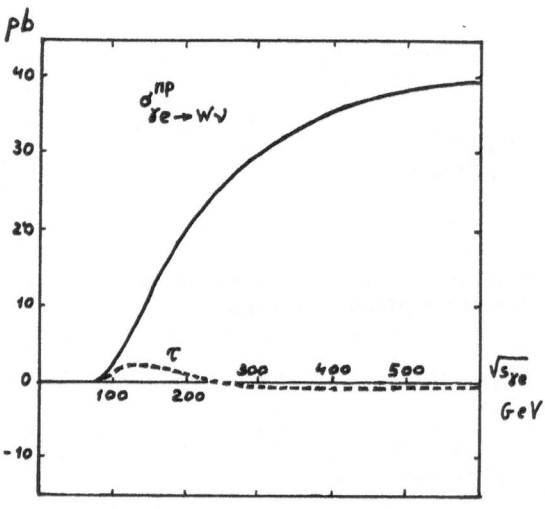

Fig. 30 - Cross section for the process $\gamma e \to W\nu$ as a function of $W_{e\gamma} = \sqrt{s}_{e\gamma}$ and lowest order contributing graphs.

The cross section vanishes for right handed electrons and left handed positrons. τ is half the difference between the cross sections for photons with right handed and left handed polarisation. This is perhaps measurable in the region near threshold (see Fig. 30).

$\gamma\gamma \rightarrow W^+W^-$

Several theoretical calculations of this process have been made [14] , [54] , [55] [56]. The total cross section can be written as [14] :

$$\sigma(\gamma_1\gamma_2 \rightarrow W^+W^-) = \sigma^{np} + P^\ell_{\gamma_1} P^\ell_{\gamma_2} \tau^a + \frac{1}{2} P^t_{\gamma_1} P^t_{\gamma_2} \tau\cos 2 \Delta\phi$$

$$\sigma^{np} = \tilde{\sigma}v \left[1 + \frac{3}{16x} + \frac{3}{16x^2} - \frac{3}{16x^2} (1 - \frac{1}{2x}) L \right]$$

$$\tau^a = \frac{\tilde{\sigma}v}{16x} \left[- 19 + (8' - \frac{5}{x}) L \right]$$

$$\tau = \frac{3}{16} \frac{\tilde{\sigma}v}{x^2} \left[1 + \frac{1}{2x} L \right]$$

where $\tilde{\sigma} = \dfrac{8\pi\alpha^2}{M_W^2} = 86$ pb $\qquad x = (W_{\gamma\gamma}/2M_W)^2$

$$v = \sqrt{1 - \frac{1}{x}} \qquad L = \frac{1}{v} \ln \frac{1 + v}{1 - v} \tag{27}$$

Here $P^\ell_{\gamma_i}$, $P^t_{\gamma_i}$ are the degrees of longitudinal and transverse polarisation respectively of the photon i. $\Delta\phi$ is the azimuthal angle between the planes of transverse polarisation of the two photons. τ and τ^a can also be written in the form :

$$\tau = \sigma_\parallel - \sigma_\perp$$

$$\tau^a = \frac{1}{2} (\sigma_o - \sigma_2) \tag{28}$$

where $\sigma_\parallel$, $\sigma_\perp$ are the cross sections for photons with parallel, perpendicular planes of transverse polarisation and σ_o, σ_2 are the cross sections for states of total helicity 0,2 for the two photons. Fig. 31 shows the dependance of $\sigma^{np}, \tau, \tau^a$ on $W_{\gamma\gamma}$ as well as the 3 lowest order diagrams (all involving fundamental couplings of gauge bosons) which contribute to the cross section. It can be seen from Fig. 31 that the effects of longitudinal polarisation can be more easily measured than these of transverse polarisation. Such a measurement is sensitive to the different contributions of the γW^+W^- and the $\gamma\gamma W^+W^-$ couplings (Fig. 31).

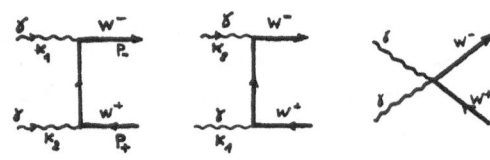

$$\gamma\gamma \rightarrow H \rightarrow Z^{o}Z^{o}$$

In the case that the Higgs scalar meson of the standard electroweak theory is very heavy, the above process with a cross section of $\simeq 10^{-2}$pb is expected to be the dominant one for the process $\gamma\gamma \rightarrow Z^{o}Z^{o}$ [14]. This reaction would be a particularly clean way to detect a the Higgs scalar in this region, since unlike the reaction $\gamma\gamma \rightarrow H \rightarrow W^{+}W^{-}$ there is no "born term" corresponding to $W^{\pm}$ exchange.

The total cross section to be expected for the processes discussed above, when account is taken of the energy spectrum of the real photons produced by Compton back-scattering, are shown in Fig. 32. The cross sections shown are averaged over the energy distribution of the scattered electrons :

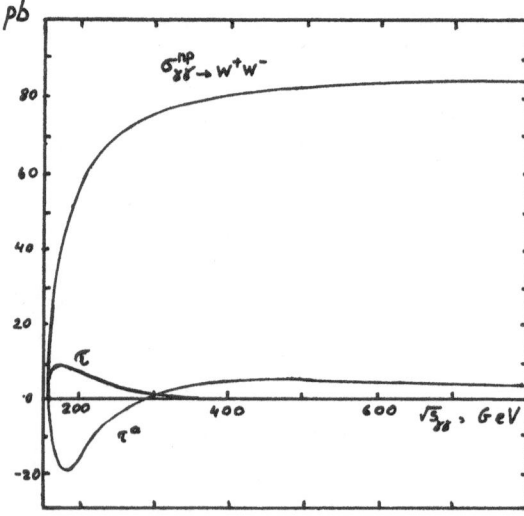

Fig. 31 - Cross section for the process $\gamma\gamma \rightarrow W^{+}W^{-}$ as a function of $W_{\gamma\gamma} = \sqrt{S}_{\gamma\gamma}$ and lowest order contributing graphs. For definitions of τ, τ^{a} see text.

$$\frac{1}{N_{\gamma}} \frac{dN_{\gamma}}{dy} = \frac{2\pi\alpha^{2}}{xm_{e}^{2}\sigma_{c}(x)} \left[\frac{1}{1-y} + 1 - y - \frac{4y}{x(1-y)} + \frac{4y^{2}}{x^{2}(1-y)^{2}} \right] \tag{29}$$

where x, y are defined in section 6 above, $\sigma_{c}(x)$ is the Compton cross section, and a laser photon energy of 1.17 eV is chosen. x is $\simeq 1$ for SLC energies and $\simeq 2-5$ for VLEPP. In the curves corresponding to γe and $\gamma\gamma$ collisions a conversion efficiency k of 0.5 is assumed. Some counting rate estimates using Fig. 32 for SLC and VLEPP taking in each case 2 years of "effective luminosity" are presented in TABLE 4. Notice that for SLC the effective luminosity is now a factor 3 smaller than in TABLE 2, since the "pinch effect" enhancement is no longer operative. It should be remarked in general however that the parameters for optimum luminosity in γe and $\gamma\gamma$ collisions are not the same as in $e^{+}e^{-}$ [7], implying that the estimates of TABLE 2 are perhaps conservative. For SLC the maximum [6] beam energy of 70 GeV is taken.

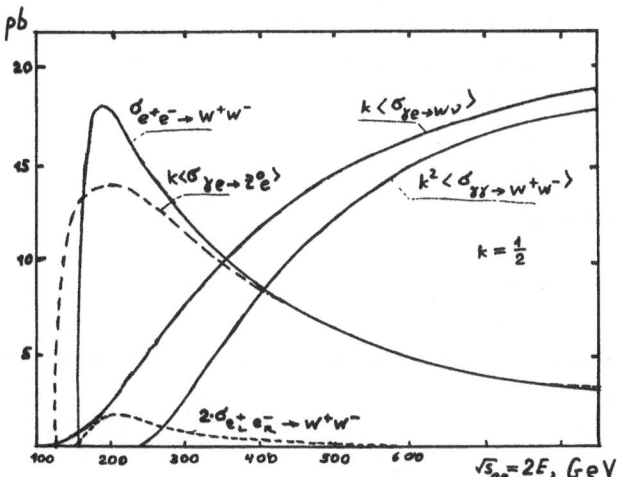

Fig. 32 - Cross sections for various Z^o, $W^{\pm}$ production processes, weighted by the photon energy distribution of Eqn(29) as a function of $2E = \sqrt{S}_{ee}$.

8. SUMMARY

A comparative study of the 2γ physics potential of PETRA, SLC, LEP(1/6), LEP(2) and HERA has been made. On the basis of the quoted design luminosities, both HERA and LEP (1/6) (the initial version of LEP with 50 GeV beams) offer a significant advantage in terms of accepted luminosity over the current (PETRA, PEP) generation of e^+e^- storage rings. LEP(2) (the ultimate version of LEP with superconducting R.F. and 130 GeV beams) will permit an order of magnitude increase in accepted luminosity over PETRA and PEP, in the low energy region as well as giving useful counting rates at 2γ masses as high as 100 GeV/c^2. Because of technical limitations on the acceptance of tagging detectors and background problems LEP (2) will be best used for untagged physics.

The SLC seems to be of little interest for virtual 2γ physics because of its rather low luminosity. The above statements are conditional on the respective machines reaching their design luminosities. Taking into account the difficulty all e^+e^- machines have experienced in approaching their design luminosity, it may well be that HERA will approach or even surpass LEP(2) for untagged two photon physics. In any case HERA seems to offer quite unique possibilities for single tag physics when the scattered proton is detected at very small angles. A first look at the background from diffractive electroproduction indicates that it should be less than or of the order of the rate from 2γ collisions. Even at 0° the expected background level from beam-beam bremsstrahlung is negligible. The production of Z^o and $W^{\pm}$ bosons in γ e and γ p collisions should be observable at HERA.

In the more distant future a single pass linear collider such as VLEPP or a higher energy successor to the SLC could provide intense real photon beams by Compton back scattering of laser light. Such photon beams (which can be circularly or transversely polarised) will enable real $\gamma\gamma$ and γe collisions to be observed at centre of mass energies as high as 1 TeV. Electroweak interference in the photon structure function as well as copious production of Z^O and $W^{\pm}$ via the reactions :

$$\gamma e \rightarrow Z^O e$$
$$\gamma e^{\pm} \rightarrow W^{\pm}\nu$$
$$\gamma\gamma \rightarrow W^+W^-$$

is to be expected. The last two reactions are sensitive to the fundamental gauge couplings γWW, $\gamma\gamma WW$.

ACKNOWLEDGEMENTS

I should like to thank A. Courau for a clarifying discussion on the calculation of tagging efficiency and J.M. Levy for valuable comments on the experimental results on diffractive photoproduction.

The material in sections (6) and (7) is drawn almost entirely from material provided directly to me by our colleagues in Novosibirsk. I thank them for their promptness in transmitting the information to me, and apologise to them for any errors or omissions in the account given above of their work.

HIGH ENERGY COLLIDERS

Machine	Colliding Beams (Energy in Gev)	Design Luminosity $(cm^{-2} sec^{-1})$
CERN $\bar{p}p$	$270\ p + 270\ \bar{p}$	10^{29}
SLC	$50\ e^{\pm} + 50\ e^{-}$	6×10^{30} (*)
	possibly $\quad e^{\pm} + 30\ \gamma$	3×10^{29} (†)
	$30\ \gamma + 30\ \gamma$	4.5×10^{28} (†)
LEP $(\frac{1}{6})$ (**)	$50\ e^{+} + 50\ e^{-}$	4×10^{31}
LEP (2) (††)	$130\ e^{+} + 130\ e^{-}$	10^{32}
HERA	$30\ e^{\pm} + 820\ p$	6×10^{31}
FNAL $\bar{p}p$	$1000\ p + 1000\ \bar{p}$	10^{30}
ISABELLE pp	$400\ p + 400\ p$	10^{33}
VLEPP	$e^{\pm} e^{-}$ $e^{\pm} \gamma$ $\gamma\ \gamma$ All beams polarised Energy range $150 \rightarrow 500$	10^{32}

(*) A factor 3 enhancement from the "pinch effect" assumed.

(†) Monochromatic γ beam with $\Delta E/E \simeq 0.1$ Ref. [8]

(**) 1st Stage of LEP

(††) Superconducting R.F. required.

Table 2

ENERGIES AND EFFECTIVE LUMINOSITIES FOR PETRA, LEP SLC AND HERA

Machine	PETRA	LEP $(\frac{1}{6})$	LEP (2)	SLC	HERA
Colliding beam Energies (GeV)	$20\ e^{+}$ $20\ e^{-}$	$50\ e^{+}$ $50\ e^{-}$	$130\ e^{+}$ $130\ e^{-}$	$50\ e^{+}(e^{-})$ $50\ e^{-}$	$820\ p$ $30\ e^{\pm}$
W_{MAX} (GeV/c)	40	100	260	100	314
$\bar{\mathcal{L}}$ $(cm^{-2} sec^{-1})$	4×10^{30}	1.3×10^{31}	3×10^{31}	2×10^{30}	2×10^{31}

Table 3

PARAMETERS OF SMALL ANGLE TAGGING DETECTORS

Detector	1	2	3
Distance from interaction point (m)	100	75	55
Distance of detector from beam line (mm)	$8.7 < r < 30$	$8.9 < r < 30$	$10.5 < r < 30$
Effective distance of inner edge from beam	$D_x(\frac{\sigma_p}{p_{beam}}) + 6\sigma_x$	" " $+ 28\sigma_x$	" " $+ 18\sigma_x$
D_x (m)	2.7	0.8	0.28
σ_x (mm)	0.86	0.28	0.5
Acceptance in x_1	$0.0032 \to 0.011$	$0.011 \to 0.0375$	$0.0375 \to 0.11$
$\frac{<\sigma_{x_1}>}{x_1}$ for $\sigma_r = 1$ mm	0.057	0.053	0.048

Table 4

COUNTING RATE ESTIMATES FOR Z°, $W^\pm$

PRODUCTION IN γe, $\gamma\gamma$ COLLISIONS AT SLC AND VLEPP

SLC   E = 70 GeV $\bar{\mathcal{L}}^{(\dagger)} = 6.7 \times 10^{29}$ cm^{-2} sec^{-1}

VLEPP   E = 150 GeV $\bar{\mathcal{L}}^{(\dagger)} = 3.3 \times 10^{31}$ cm^{-2} sec^{-1}

Numbers of events for 2 years at $\bar{\mathcal{L}}$

	SLC	VLEPP
$\gamma e \to Z^\circ e$	504	23,000
$\gamma e^\pm \to W^\pm \nu$	24	15,000
$\gamma\gamma \to W^+ W^-$	–	4,000

(†) The effective luminosity $\bar{\mathcal{L}}$ is $\frac{1}{3}$ the quoted design luminosity. No "pinch effect" enhancement is included for the SLC.

DISCUSSION

Q. - Vermaseren

In a reaction ep → eμμ p vs ee → eeμμ   most of the HERA advantage for tagging
goes away when the observation of both μ's is demanded. See for instance
NIKHEF preprint H/82-15.

A. - J.H.F.

This remark is true only for electron tagging. If a hadronic system is produced rather than muon pairs, the smaller form factor suppression will favour HERA
over LEP. LEP cannot compete with HERA for the $\simeq 0°$ tagging as both technical
acceptance limitations and beam-beam bremsstrahlung background forbid tagging
at angles less than $\simeq$ 2 mrad at LEP.

Q. - Brodsky

There is an interesting hadronic diffractive process which would be a potential
background to γγ physics in ep collisions ; specifically γ + pomeron → X or
γ + (gg) → X where the proton scatters elastically and forward at large x_F.
This is an interesting process to study in itself ; careful studies are needed to see if the γγ reactions are separable.

A. - J.H.F.

There is some data for the corresponding reaction in pp collisions (Fig. 2)
which indicates, perhaps rather surprisingly, that "pomeron pomeron" collisions
are rather ineffective in the production of low mass, low multiplicity systems.
For the ep collisions however, I agree, diffractive electroproduction is a potentially important background. In the written version of the talk I shall try
to make some estimate of its magnitude relative to the 2 photon process.

Q. - Jönsson

I would like to point out that the possibility of making 0°-tagging in principle
also exists at DORIS, where the beams are vertically bent at some distance from
the interaction point ; however, if you investigate the background situation
you find that you will have problems with the beam beam single bremsstrahlung
processes, which even if they don't contribute to the double tag rather directly will give a high single tag rate. To get around this problem you have to
introduce an angular cut at about 1 mrad, which reduces the background by more
than a factor 100.

If we wouldn't have these background problems the double tag efficiency at
DORIS would be of the order of 25% but with a cut at 1 mrad it comes down to
about 8% which still is quite impressive.

I suppose you have to worry about the same background problems at HERA as at
DORIS.

A. - J.H.F.

Again, in the written version of the talk I shall give an estimate of the beam-
beam bremsstrahlung background for tagging at HERA.

Q. - Wacker

I found in a study of 0° tagging at DORIS, that a measurement of position and
angle of the scattered particle behind the bending magnet does not always give
a unique solution, there are sometimes two possible tracks with different momenta and scattering angles. This is due to the quadrupoles in the beam line.
In case of e beams, a shower counter can resolve this, but that is obviously
impossible with protons.

A. - J.H.F.

Such ambiguities may also exist at HERA. If there is a sufficiently large difference between the proton momenta in the two solutions, the ambiguity can perhaps be resolved by observation of the produced final state. For exclusive
final states measurement of the visible energy and rapidity define the energies
of both virtual photons.

Q. - Kessler

I come back to Stan Brodsky's objection concerning $\gamma\gamma$ collisions at HERA.
Actually we gave up studying $\gamma\gamma$ physics at HERA, because it came out from
discussions with strong interaction specialists that - even tagging the proton
at 0° - $\gamma\gamma$ reactions would be dominated by an overwhelming background from γ
Pomeron collisions. As I understand, the experiment you mention is showing that
the Pomeron is not there in double peripheral reactions ?

A. - J.H.F.

The ISR experiment shown in Fig. 2 can be interpreted in terms of conventional
Regge exchanges only. There is certainly no "dominance" of double pomeron
exchange in this experiment, in which the total centre of mass energy is 23 or
30.7 GeV. This has already been pointed out by H. Sens at the 1981 Paris
photon-photon colloquium.

REFERENCES

1) L. Camilleri, J.H. Field, E. Gabathuler and G. Preparata, CERN 76-18 P169 Geneva 1976.

2) J.H. Field and P. Landshoff. Proceedings of the LEP Summer Study, CERN 79-01 P553 Geneva 1979.

3) ECFA-LEP Working Group Progress Report 1979, ECFA/79/39 P145.

4) J.H. Field in Proceedings of the International Workshop on γγ collisions. Amiens, France 1980.
Lecture Notes in Physics 134 Springer Verlag 1980 P248.

5) M. Davier in Proceedings of the Fourth International Colloquium on Photon Photon Interactions Paris 1981. World Scientific Singapore (1981) P411.

6) B. Richter SLAC-PUB-2854 November 1981.

7) A. Siderov in Proceedings of 1981. International Conference on Lepton and Photon Interactions at High Energies. Bonn 1981, Ed. W. Pfeil, P944.

8) C. Akerlof. University of Michigan, pre-print UM HE 81-59 1981.

9) I.F. Ginzburg, G.L. Kotkin, V.G. Serbo and V.I. Telnov.
Novosibirsk pre-print 81-50 1981.

10) I.F. Ginzburg, G.L. Kotkin, V.G. Serbo and V.I. Telnov.
Novosibirsk pre-print 81-102 1981. Nuclear Instruments and Methods (in Print).

11) A.M. Kondratenko, E.V. Pakhtusova and E.L. Saldin.
Novosibirsk pre-print 81-85 1981.

12) P. Kessler. Seminar on Gamma-Gamma Physics Paris 1982. Collège de France pre-print LPC/82-14 P61.

13) F.M. Renard. Z. Phys. C14(1982) 209.

14) I.F. Ginzburg. G.L. Kotkin, S.L. Panfil. and V.G. Serbo.
Novosibirsk pre-print   TP-3(127) 1982.

15) A. Donnachie (As 5) above P303.

16) C. Tao. As 5) above P281.

17) C. Carimalo, P. Kessler and J. Parisi. Phys. Rev. D18(1978) 2443.

18) R. Moore Z. Phys. C14 (1980) 351.

19) B. Schrempp and F. Schrempp, Nucl. Phys. B182 (1981) 343.

20) S. Drell and T.M. Yan Phys. Rev. Lett. 25 (1970) 316.

21) F. Vannucci as 4) above P238.

22) J.A.M. Vermaseren. NIKHEF (Amsterdam) preprint, NIKHEF-H/82-15.

23) C. Carimalo, P. Kessler and J. Parisi, Orsay 2 Photon Seminar 1981.
    Orsay pre-print LAL 82/03 and College de France pre-print LPC 81-30.

24) Comment by H. Sens (see Ref. 5) P301.

25) J.C.M. Armitage et al. Phys. Lett. $\underline{82B}$ (1979) 149.

26) G. Coignet, As 4) above P399.

27) DESY-HERA 81/10 July 1981

28) J.H. Field Nucl. Phys. $\underline{B168}$ (1980) 477. Erratum $\underline{B176}$ (1980) 545.

29) A. Courau. as 4) above P19.

30) The LEP "Pink book" CERN/ISR-LEP/79-33.
    August 1979 and Ref. 4) above.

31) S.R. Amendolia et al. Physica Scripta $\underline{23}$ (1981) 674.

32) Ch. Berger and J.H. Field. Nucl. Phys. $\underline{B187}$ (1981) 585.

33) See talk of H. Kolanoski at this workshop.

34) D. Aston et al. Nucl. Phys. $\underline{B189}$ (1981) 15.

35) D. Aston et al. Nucl. Phys. $\underline{B166}$ (1981) 1.

36) G. Altarelli and B. Stella.Lettre al Nuovo Cimento 9(1974) 416.

37) ECFA-LEP Working Group 1979. Progress Report. Ed. A. Zichichi ECFA/79/39 P161.

38) H. Neufeld Z. Phys. $\underline{17C}$ (1983) 145.

39) P. Salati and J.C. Wallet. Z. Phys. $\underline{16\ C}$(1982) 155.

40) A.N. Kamal, J.N. Ng. and H.C. Lee. Phys. Rev. $\underline{24}$ (1981) 2842.

41) W. Panofsky as 7) P957.

42) H. Wiedemann SLAC-PUB-2849 November 1981.
    A.A. Sokolov and I.M. Ternov, Sov. Phys. TETP 4(1957) 369.

43) and Sov. Phys. Doklady $\underline{8}$ (1964) 1203.

44) E. Feenberg and H. Primakoff.Phys. Rev. $\underline{73}$ (1948) 449.

45) R.H. Milburn.Phys. Rev. Lett. $\underline{10}$ (1963) 75.

46) F.R. Arutyunian and V.A. Tumanian. Phys. Lett. $\underline{4}$(1963) 76.

47) J. Ballam et al. Phys. Rev. Lett. $\underline{23}$ (1969) 498.

48) A.M. Kondratenko, E.V. Pakhtusova and E.L. Saldin.
    Novosibirsk preprint 81-85 1981.

49) E.A. Kuraev, A. Schiller and V.G. Serbo.
    Novosibirsk pre-print 82-107 1982.

50) S.J. Brodsky, T. de Grand, J. Gunion and J. Weis. Phys. Rev. $\underline{D19}$ (1979) 1418.

51) K. Kajantie and R. Raito. Nucl. Phys. $\underline{B159}$ (1979) 528.

J.H. Field, E. Pietarinen and K. Kajantie, Nucl. Phys. B171 (1980) 377.

52) S.J. Brodsky and G.P. Lepage. Phys. Rev. D24 (1981) 1808.

53) W. Alles, C. Boyer and A.J. Buras. Nucl. Phys. B119 (1977) 125.

54) K.J. Kim and Yung-Su Tsai. Phys. Rev. D8 (1973) 3109.

55) G. Tupper and M.A. Samuel. Phys. Rev. D23 (1981) 1933.

56) M. Katuya. Phys. Lett. 124B (1983) 421.

SUMMARY

OF THE EXPERIMENTAL DISCUSSION SESSION

H. Spitzer

Universität Hamburg
Notkestr. 85
D 2000 Hamburg - 52

Abstract A total of 26 papers have been contributed to the experimental discussion
session. I will discuss some of the tools needed for 2γ experimentation and then
point to highlights and regions of particular progress in the last two years. The
process $\gamma\gamma \rightarrow \eta \rightarrow \gamma\gamma$ has been measured for the first time as well as the cross section
for $\gamma\gamma \rightarrow \rho^+\rho^-$. The latter rules out a resonance interpretation for σ ($\gamma\gamma\rightarrow\rho^0\rho^0$).
Progress has been made in understanding the f^0 shape by an interference mechanism.
Previous determinations of σ_{tot} ($\gamma\gamma \rightarrow$ hadrons) have suffered from limited detector
acceptance. The ongoing analyses of PLUTO, JADE and PEP4 + PEP9 data look promising.
A separation of σ_{TT} and σ_{TL} seems possible. The first measurement of the Q^2 dependence of high p_T jet production was presented. Both high p_T and high Q^2 help in extracting the Born process $\gamma\gamma \rightarrow q\bar{q}$ by suppressing competing processes. Progress
has been made in measuring the photon structure function F_2. The data available have
increased from the original 110 PLUTO events in 1981 to well above 2000 events from
five experiments. The Q^2 limit accessible has grown from 15 GeV^2 to about 200 GeV^2. A
proper unfolding method which converts the measured event distribution into a structure function F_2 (x, Q^2) has been presented.

1. Introduction

Two years ago at the Paris International Colloquium on Photon-Photon Interactions
seven papers were contributed to the experimental discussion session. We had reports
on η', f^0 and A_2^0 production and first qualitative results on the total $\gamma\gamma$ cross
section, on high p_T jets and the photon structure function. This year 26 speakers
presented their data. The field has grown both in quality and quantity. Table 1 gives
a summary of the topics covered. Most of the data and analyses have

Table 1   Contributions to the experimental discussion session

field	speaker	topic
resonance production	A. Weinstein (Crystal Ball)	$\gamma\gamma \to \eta$
	R. Mir (TASSO)	$\gamma\gamma \to \eta'$
	M. Zachara (PLUTO)	$\gamma\gamma \to \eta'$
		$\gamma\gamma \to f^0$
	F. Kovacz (CELLO)	$\gamma\gamma \to f^0$
	J. Olsson (JADE)	$\gamma\gamma \to A_2$
	L. Köpke (TASSO)	$\gamma\gamma \to f'$
	U. Karshon (TASSO)	$\gamma\gamma \to \iota/\theta/\eta_c$ limits
exclusive processes	C. Williams (PEP9)	$\gamma\gamma \to \mu^+\mu^-$
	R. Kellogg (PLUTO)	$\gamma\gamma \to h^+h^-$
	H. Kueck (TASSO)	$\gamma\gamma \to p\bar{p}$
	M. Wollstadt (TASSO)	$\gamma\gamma \to \pi^+\pi^+\pi^-\pi^-$
	H.J. Behrend (CELLO)	$\gamma\gamma \to \pi^+\pi^+\pi^-\pi^-$
	J. Olsson (JADE)	$\gamma\gamma \to \rho^+\rho^-$
	A. Shapira (TASSO)	$\gamma\gamma \to \rho^0\rho^0$ analysis method
σ_{tot}	N. Wermes (SLAC)	problems in σ_{tot} determination
high p_T processes	E. Duchovny (TASSO)	high p_T inclusive
	H. Lierl (CELLO)	high p_T incl./jets
	S. Cartwright (PLUTO)	high p_T jets
structure functions	F. Kirschfink (TASSO)	F_2, endcap tag
	G. Carnesecchi (CELLO)	F_2, endcap tag
	T. Nozaki (JADE)	F_2, EC + barrel tag
	G. Knies (PLUTO)	F_2, endcap tag
	B. King (PLUTO)	F_2, LAT tag
	F. Raupach (PLUTO)	F_2, LAT tag
	S. Maxfield (PLUTO)	F_2, double tag
	V. Blobel (PLUTO)	unfolding methods
other	D. Miller (U. C. London)	tagging at LEP

been shown at this conference for the first time.

Faced with the task of summarizing the discussion session I realized that J. Olsson, N. Wermes, H. Kolanoski, W. Wagner and M. Pohl have given excellent summaries of the experimental results contributed to this conference. Instead of repetition I decided to concentrate rather on the status and future potential of 2γ experiments and to present a few highlights which illustrate the experimental progress we have made in the last two years. The presentation will follow the list of topics given in table 1.

For more detailed information on the 26 contributions I refer the reader to the abstracts, which are included in the symposium proceedings, and to the authors themselves.

2. Tools for 2γ experimentation

For 2γ experimentation you need first sufficient luminosity from an e^+e^- storage ring which produces the flux of virtual colliding photons.

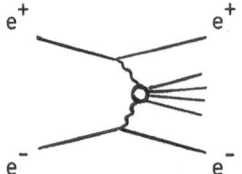

More importantly you need a detector with very good particle acceptance in both the central and forward regions.   Finally 2γ experiments are not easy to analyze. Detection efficiencies can be quite low. Particles are produced preferably near the forward direction ($\theta=10\text{-}20°$, see fig. 10), where your detector might be insensitive. So you need a lot of skill and endurance to determine a cross section.

Table 2 shows the integrated luminosities available from detectors involved in 2γ physics measurements. As is obvious from the table most of the present data come from PETRA experiments. However this trend might reverse in the future because of a dramatic luminosity increase at PEP since the beginning of 1983. In contrast the luminosity of PETRA will be relatively low during the top search with beam energies above 20 GeV.

Fig. 1 shows the acceptance of various detectors for measuring charged hadrons (full lines) and neutral showering hadrons (dashed lines). It is mandatory to have optimum coverage since in most experiments the mass of the $\gamma\gamma$ system, $W_{\gamma\gamma}$, has to be determined from the measured hadrons. The PLUTO and PEP9 experiments are best in this respect followed by JADE, MAC, CELLO and TASSO. The endcap counters of CELLO and TASSO have been not used in the past for neutral hadron measurement.

Table 2   Luminosity taken til April 1983 by e^+e^- storage ring detectors involved in 2γ measurements.

	detector	integrated luminosity analysed(pb^{-1})	lumi taken, not analysed (pb^{-1})
multi-purpose detectors	TASSO	∿ 75	-
	JADE	∿ 75	-
	CELLO	∿ 10	-
	MARK II		∿ 50
	MAC		∿ 40 *
specialized 2γ detectors	PLUTO II	45	-
	PEP 4 + 9	3	∿ 30
calorimeters	MARK J	∿ 75	-
	Crystal Ball	3 (γγ → η)	
		21 (γγ → f)	13
		SPEAR	DORIS
present data taking rate : (June 1983)	DORIS (E_b = 5 GeV)		1 - 2 pb^{-1}/week
	PEP (E_b = 14.5 GeV)		8 pb^{-1}/week
	PETRA (E_b = 20 GeV)		0.5 - 1 pb^{-1}/week

Another requirement is a potential for tagging and Q^2 measurement of electrons scattered at moderate angles (θ<100 mrad). This is vital for analyzing the Q^2 dependence of 2γ reactions and for selecting reactions with one quasireal photon (the so called target photon) by "antitagging" on the second electron. PLUTO, PEP4/9 and JADE have used such devices extensively. CELLO, TASSO and MARK II will have them for future analyses (see table 3).

Table 3   Range of small angle tagging devices with potential for Q^2 measurement

detector	tagging range (mrad)
JADE	43 - 78
PLUTO 81/82	30 - 60, 85 - 250
PEP 4 + 9	22 - 180
MARK II	20 - 85
CELLO	50 - 90 (since 1983)
TASSO	25 - 115 (since 1983)

---

* A first measurement of the structure function F_2 has been shown at this conference.

hadron acceptance of 2γ detectors

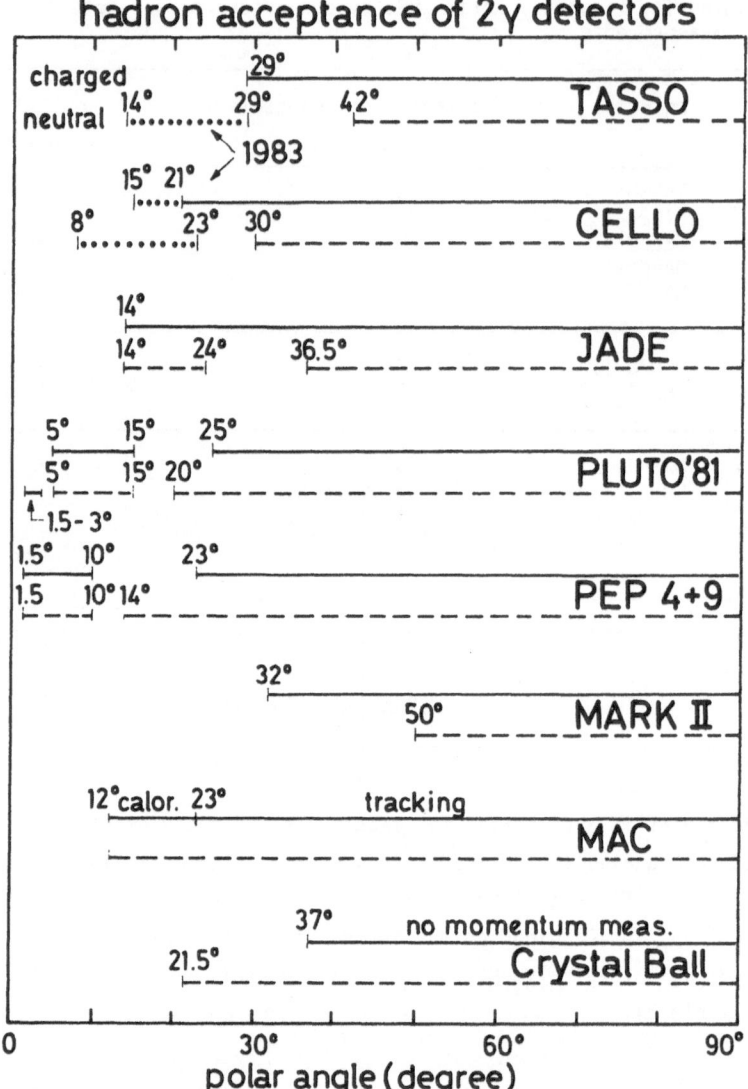

polar angle (degree)

Fig. 1 Acceptance for measurement of charged hadrons (full line) and neutral showering hadrons (dashed line) in existing 2γ detectors at DESY and SLAC.

Two experiments have been equipped with specialized detectors for 2γ physics. Fig. 2 shows a view of the PLUTO detector as used in 1981/82. The central PLUTO detector, which has been a well known work horse since DORIS and early PETRA times has been supplemented by two forward spectrometers. These devices consist each of a magnetic spectrometer ($\sigma_p = 0.025 \cdot p$), a small angle and a large angle shower counter (SAT, LAT) and a muon detector. A similar setup is available at PEP. Fig. 3 shows a view of the PEP4 and PEP9 detectors which look like a big brother of PLUTO. The magnetic forward spectrometers cover the range $24 < \theta < 180$ mrad with a resolution $\sigma_p = 0.01 \cdot p$ (p in GeV). In the heart of the PEP9 setup is a small angle NaI tagger with a resolution of $\sigma_E/E = 1$ % at 14 GeV. This will allow double tagging measurements with an independent bias free determination of $W_{\gamma\gamma}$. The PEP4 and PEP9 experiments started combined data taking in early 1982. Whereas PLUTO is no longer in the beam, PEP4 and PEP9 will hopefully continue for some years.

3. Resonance production

In the last four years two photon experiments have led to a renaissance of C=+1 meson spectroscopy. The excitation of meson resonances like η, η', f^0, A_2, f' occurs with ample rate in the reaction

$$\gamma\gamma \to \text{meson},$$

where both photons are quasireal (no tag mode). The final state can be selected easily as long as all particles from the resonance decay are detected. The measurements yield the radiative decay widths $\Gamma(M \to \gamma\gamma)$ which give us insight into the quark structure of meson resonances. In addition one can set limits on the production of exotic states like glueballs or 4 quark states.

Since particle momenta from resonance decay are low (usually below 1 GeV), one needs efficient low multiplicity triggers. The boost of the $\gamma\gamma$ system in the beam direction tends to push a sizeable fraction of the decay products out of the acceptance region of existing detectors. Table 4 lists the channels which have been studied in contributions to this conference. Charged final states can be measured with detection efficiencies between 3 and 15 %, whereas for final states involving π^0' or photons the detection efficiencies drop below 1 % (with the exception of $\gamma\gamma \to \eta \to \gamma\gamma$).

In order to illustrate the difficulties encountered in the analysis I show in fig.4(a) a typical photon energy spectrum from the decay $\eta' \to \rho\gamma$, as measured by TASSO. The average photon energy is around 200 MeV, but the measurement ends at 100 MeV. Here the photon detection has dropped to 40 % for TASSO and 65 % for JADE (see fig. 4(b)).

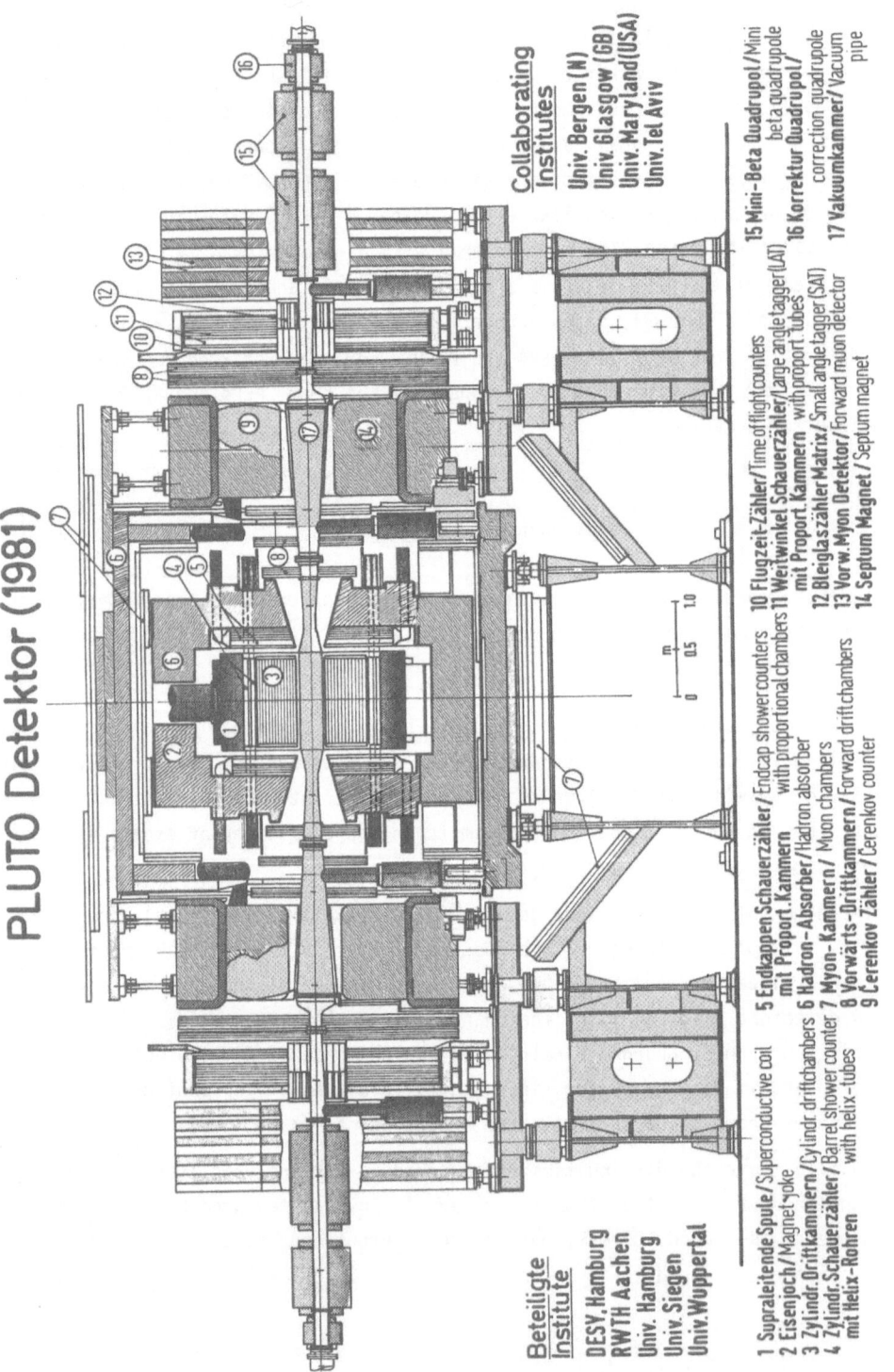

PLUTO Detektor (1981)

Beteiligte Institute

DESY, Hamburg
RWTH Aachen
Univ. Hamburg
Univ. Siegen
Univ. Wuppertal

1 Supraleitende Spule / Superconductive coil
2 Eisenjoch / Magnet yoke
3 Zylindr. Driftkammern / Cylindr. driftchambers
4 Zylindr. Schauerzähler / Barrel shower counter
  mit Helix-Rohren   with helix-tubes
5 Endkappen Schauerzähler / Endcap shower counters
  mit Proport. Kammern with proportional chambers
6 Hadron-Absorber / Hadron absorber
7 Myon-Kammern / Muon chambers
8 Vorwärts-Driftkammern / Forward driftchambers
9 Čerenkov Zähler / Čerenkov counter

10 Flugzeit-Zähler / Time of flight counters
11 Weitwinkel Schauerzähler / Large angle tagger (LAT)
  mit Proport. Kammern with proport. tubes
12 Bleiglaszähler Matrix / Small angle tagger (SAT)
13 Vorw. Myon Detektor / Forward muon detector
14 Septum Magnet / Septum magnet

15 Mini-Beta Quadrupol / Mini
               beta quadrupole
16 Korrektur Quadrupol /
  correction quadrupole
17 Vakuumkammer / Vacuum
          pipe

Collaborating Institutes

Univ. Bergen (N)
Univ. Glasgow (GB)
Univ. Maryland (USA)
Univ. Tel Aviv

Fig. 2 Side view of the PLUTO detector as operated from 1981 to 1982.

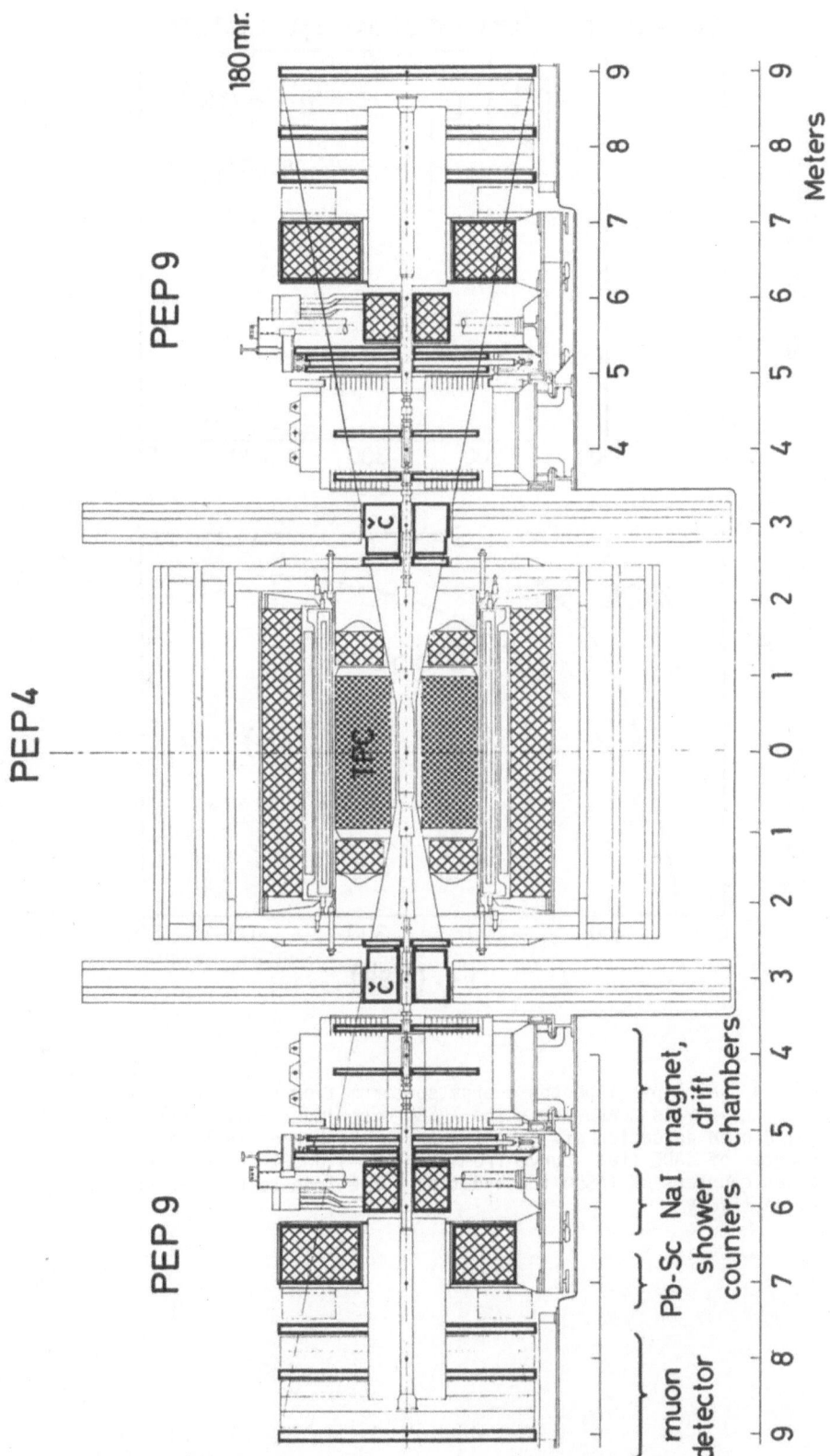

Fig. 3 Side view of the PEP4 and PEP9 detectors at SLAC. The shower counters surrounding the TPC and in the forward spectrometer are shown cross-hatched.

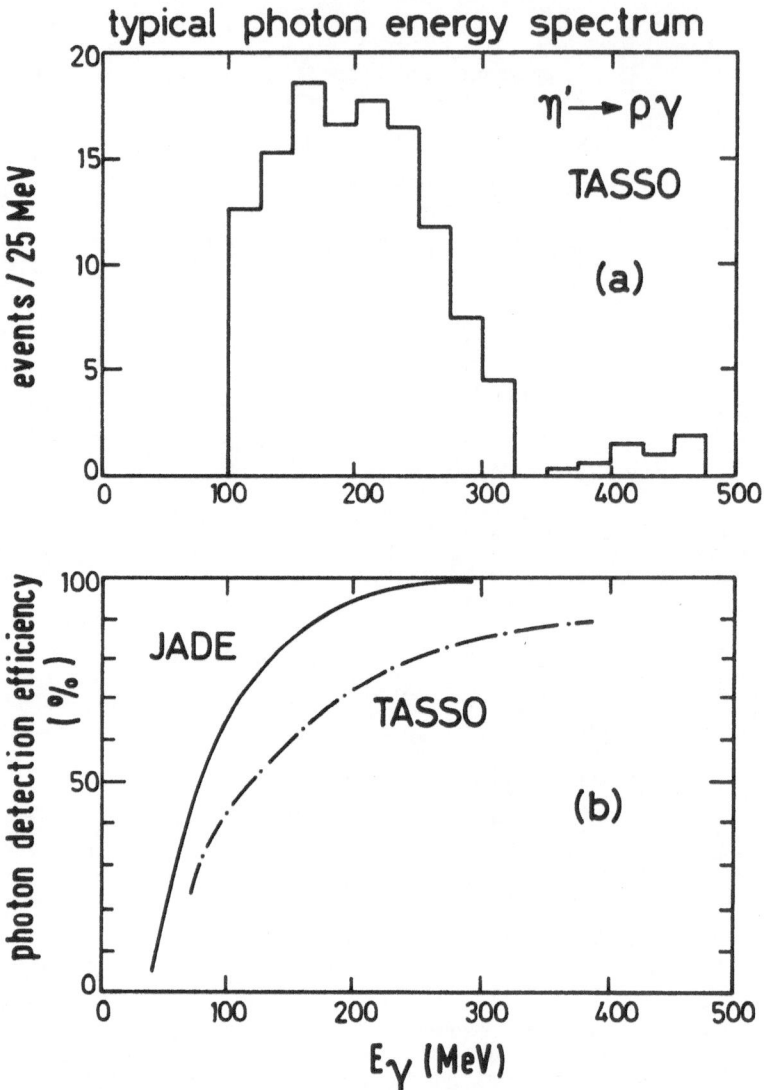

Fig. 4 (a) Typical photon energy spectrum from resonance decay: decay η → ργ as measured in the TASSO experiment (talk of R.Mir). (b) Photon detection efficiencies using the barrel lead glass counter of JADE (full curve from ref. 10) and the barrel liquid argon counters of TASSO (dash-dotted curve from talk of R. Mir), respectively.

Table 4   Production of meson resonances and four pion final states in 2γ reactions as reported at this conference

final state	detection efficiency	detector	number of events
$\eta \to \gamma\gamma$	15 %	Crystal Ball	56 ± 12
$\eta' \to \pi^+\pi^-\gamma$	0.5 %	TASSO	∿ 210
	< 1 %	PLUTO	165
$f^0 \to \pi^+\pi^-$	14 %	CELLO	∿2000 $\pi^+\pi^-$ events
$A_2 \to \rho^\pm \pi^\mp$	∿ 1 %	JADE	200
$f^0 \to K^+K^-$	∿ 3.5 %	TASSO	∿ 100
$\to K^0K^0$	∿ 5 %		∿ 25
$\pi^+\pi^+\pi^-\pi^-$	∿ 3 %	TASSO	2400
	∿ 10 %	CELLO	850
$\pi^+\pi^-\pi^0\pi^0$	0.25-0.5 %	JADE	235

In spite of these difficulties considerable progress has been made in the analysis of resonance final states involving neutrals. The reactions $\gamma\gamma \to \eta \to \gamma\gamma$ and $\gamma\gamma \to \pi^+\pi^-\pi^0\pi^0$ have been measured for the first time. Fig. 5 shows as an example the two photon mass spectrum of the Crystal Ball group. A clear peak at the mass of the η (550) is observed, leading to a radiative width of

$$\Gamma_{\eta\gamma\gamma} = (0.56 \pm 0.09) \text{ keV}.$$

This has to be compared to the table value of $\Gamma_{\eta\gamma\gamma} = 0.324 \pm 0.046$ keV determined from a measurement at Cornell involving the Primakoff effect via $\gamma N \to \eta N$ [1]. More data on $\gamma\gamma \to \eta \to \gamma\gamma$ can be expected soon, e.g. from JADE.

One of the first resonances observed in 2γ reactions has been the f^0 meson [2]. Last year a breakthrough has been made in understanding the shape of the f^0 meson peak. Previous experiments have reported a downward shift of the mass peak from the table value of 1273 MeV by about 40 - 60 MeV [3]. G. Mennesier has recently provided a theoretical framework which explains the mass shift [4]. The model describes the production of $\pi\pi$ and $K\bar{K}$ pairs by real photons using an unitary and analytic coupled channel formalism. The unitary corrections (final state rescattering effects) are taken from fits to previously measured $\pi\pi$ and KK phase shift data. The CELLO Collaboration has applied the model successfully to their data as shown in fig. 6(a) [5]. The curve contains only the Born term for $\gamma\gamma \to \pi^+\pi^-$ and a helicity 2 f^0 amplitude both with unitary corrections. The only free parameter is $\Gamma_{f\gamma\gamma}$. The f^0 meson is assumed to have the standard mass. Fig. 6(b) illustrates how the mass shift is produced by the interference between the Born and f^0 amplitudes (curve A). If the interference did not occur the mass shift would disappear (curve B).

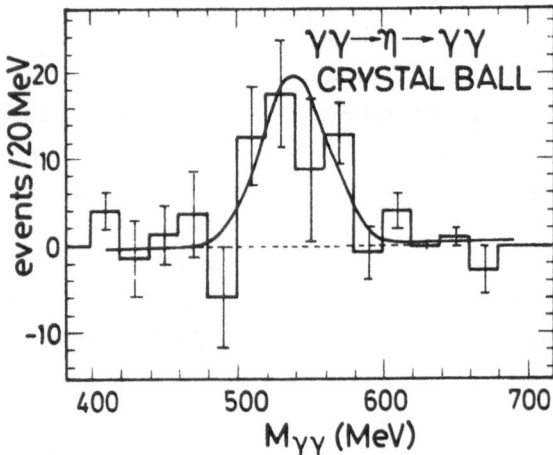

Fig. 5 Two photon mass spectrum of the reaction $\gamma\gamma \rightarrow \eta \rightarrow \gamma\gamma$ measured by the Crystal Ball detector at SPEAR. Background from cosmic rays etc. has been subtracted. The data are from a run with $\int Ldt = 2.7$ pb^{-1} at 2.5 GeV $< E_{beam} <$ 3.625 GeV (from the talk of A. Weinstein).

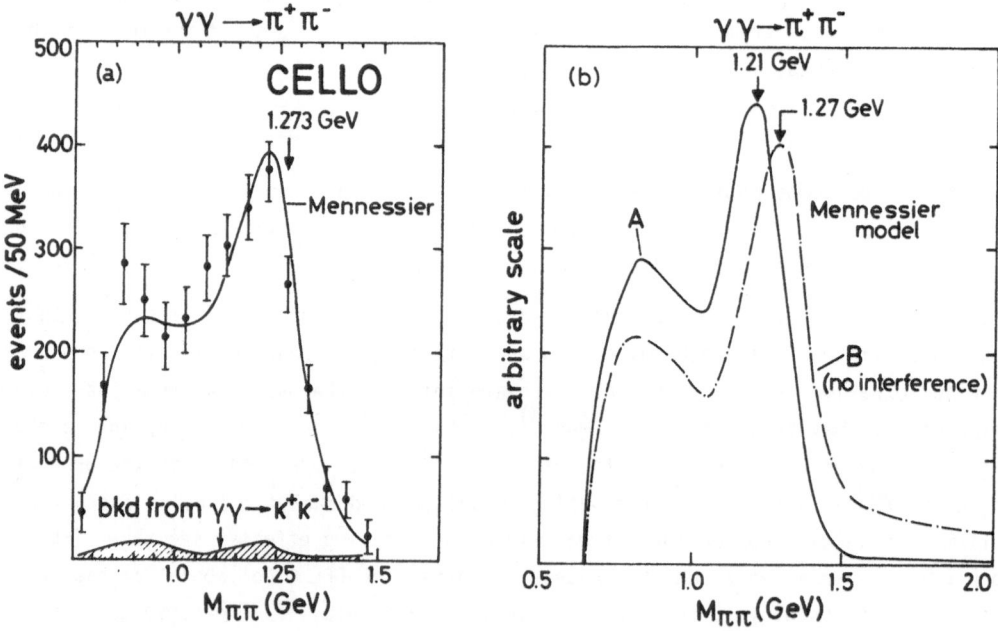

Fig. 6 (a) Dipion mass spectrum of reaction $\gamma\gamma \rightarrow \pi^+\pi^-$ from CELLO fitted to the Menessier model (no tag data with $<Q^2> = 0.0016$ GeV^2) (b) Curve A is the prediction of the Menessier model for interference of $\gamma\gamma \rightarrow f^0 \rightarrow \pi^+\pi^-$ (helicity 2, $\Gamma_{f\gamma\gamma} = 3$ keV) and an unitarized Born term amplitude. Curve B would result, if the f^0 and Born amplitudes would add incoherently (from talk of F. Kovacz).

The PLUTO Collaboration has taken a look at the Q^2 dependence of f^0 production. The histogram in fig. 7 shows the two particle mass distribution (assuming pion masses) measured in the no tag mode ($<Q^2>$ = 0.007 GeV^2). The open points were taken in the single tag mode with an average Q^2 of 0.4 GeV^2. The latter event distribution was multiplied by a factor of 10. Both spectra still contain a smooth background from e^+e^- and $\mu^+\mu^-$ pairs. Whereas a clear f^0 peak is observed at $<Q^2>$ = 0.007 GeV^2, little is left at $<Q^2>$ = 0.4 GeV^2. The PLUTO Collaboration will have more results on the Q^2 dependence of f^0 and η' production in the future.

A beautiful example of 2 particle final state analysis has been presented by the TASSO Collaboration. Fig. 8 shows a three dimensional scatter plot for the reaction $\gamma\gamma \rightarrow 2$ charged particles, where both particle masses have been determined by a time of flight measurement. Apart from the dominant e^+e^-, $\mu^+\mu^-$ and $\pi^+\pi^-$ production clear enhancements from the reactions $\gamma\gamma \rightarrow f'$ (1515) $\rightarrow K^+K^-$ and $\gamma\gamma \rightarrow p\bar{p}$ are observed.

Finally we turn to the production of 4 pion final states. The CELLO Collaboration has presented a new measurement of the cross section for the reaction $\gamma\gamma \rightarrow \pi^+\pi^+\pi^-\pi^-$. The cross section reaches values close to 200 nb around 1.6 GeV (fig. 9(a)). Slightly more than half of this cross section is made up by the subprocess $\gamma\gamma \rightarrow \rho^0\rho^0$. The full points in fig. 9(b) show respectively the new results from CELLO and TASSO.

One of the highlights of this conference has been the first measurement of the reaction

$$\gamma\gamma \rightarrow \pi^+\pi^- \pi^0\pi^0$$

as reported by J. Olsson (JADE). Here four γ's had to be measured in order to reconstruct the two π^0's. The resulting upper limits for the cross section σ ($\gamma\gamma \rightarrow \rho^+\rho^-$) are shown in fig. 9(b) (open points). The $\rho^+\rho^-$ cross section is much smaller than the $\rho^0\rho^0$ cross section below 1.8 GeV. As discussed in the plenary talk of H. Kolanoski this result rules out most of the previous resonance interpretations of the peak in $\sigma(\gamma\gamma \rightarrow \rho^0\rho^0)$ near 1.6 GeV.

I cannot cover the reports on exclusive two particle final states in detail. But I would like to emphasize that two groups have exploited the potential for particle identification of their detectors resulting in measurements of the reactions $\gamma\gamma \rightarrow h^+h^-$ (PLUTO) and $\gamma\gamma \rightarrow p\bar{p}$ (TASSO). TASSO has now 10 times more events than originally presented. It is remarkable that QCD calculations for the exclusive final states give the correct order of magnitude for $W_{\gamma\gamma} > 2$ GeV and $<p_T> \simeq 1$ GeV (talks of R. Kellogg and H. Kueck).

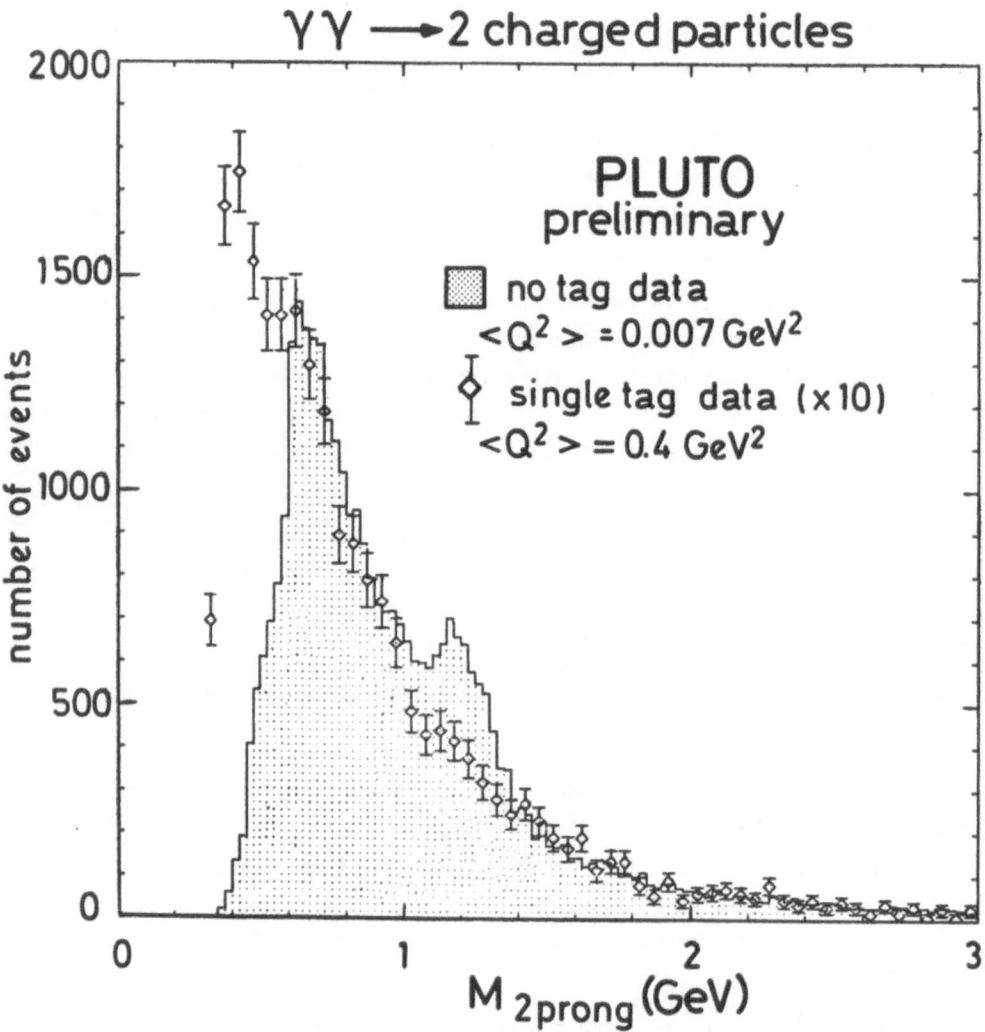

Fig. 7 Invariant two particle mass distribution (assuming pion masses) of the reaction γγ → 2 charged particles. The data still contain the dominant background contribution from e⁺e⁻ and μ⁺μ⁻ final states. Data from PLUTO taken in the no tag mode (histogram, $<Q^2>$=0.007GeV^2) and in the tag mode multiplied by a factor 10 (open points, $<Q^2>$=0.4 GeV^2) (from the talk of M. Zachara).

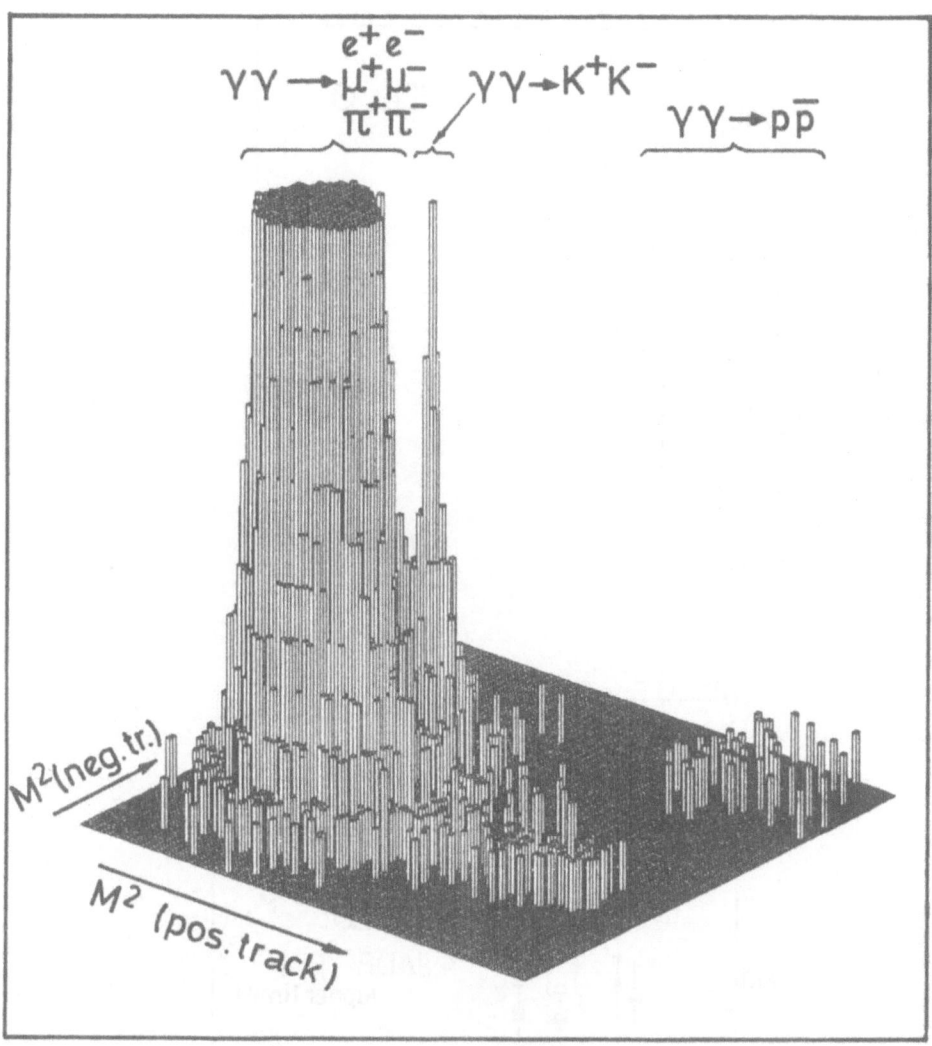

Fig. 8 Scatter plot of M^2 (positive track) vs. M^2 (negative track) from the reaction $\gamma\gamma \to 2$ charged particles, where both particle masses have been determined by a time of flight measurement. Data from TASSO as reported by L. Köpke.

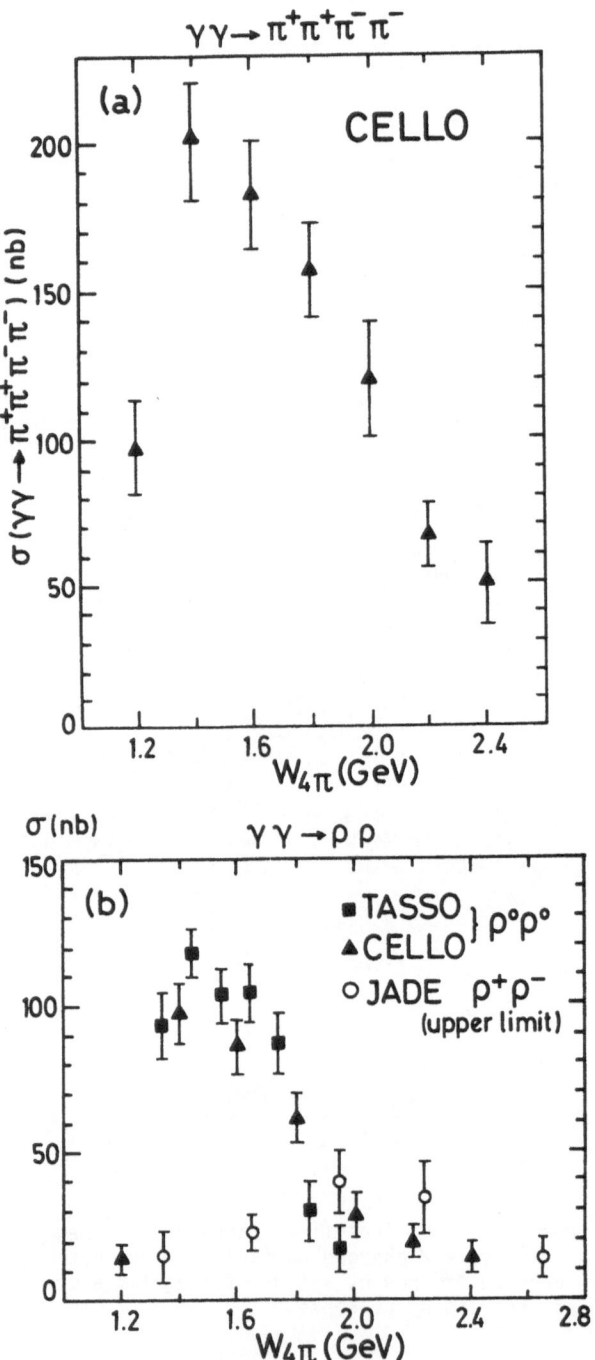

Fig. 9 (a) The cross section for reaction $\gamma\gamma \to \pi^+\pi^+\pi^-\pi^-$ as measured in the no tag mode by the CELLO collaboration. (b) The cross section for $\gamma\gamma \to \rho^0\rho^0$ from TASSO and CELLO and upper limits for $\sigma(\gamma\gamma \to \rho^+\rho^-)$ from JADE. TASSO assumed isotropic production and decay of the ρ's. CELLO assumed isotropic production and decay angular distributions as determined from the data. (from the talks of H.J. Behrend, M. Wollstadt and J. Olsson).

3. The total hadronic $\gamma\gamma$ cross section

3.1 What quantities can be measured?

The total hadronic $\gamma\gamma$ cross section is a fundamental quantity. The cross section for two real photons is thought to be dominated by hadron-like interactions. Hard scattering processes with pointlike coupling of the photon to quarks take over at high Q^2 and p_T[6].

When discussing the terms contributing to the total cross section we have to consider three cases:

a) No tag mode (both electrons at angles $\lesssim$ 25 mrad). Here one approaches a measurement of the collision of two real (transverse) photons, which is described by a single cross section term:

$$\sigma(W) = \sigma_{TT}(W)$$

b) Single tag mode (one electron scattered at angles $\gtrsim$ 25 mrad). The target photon is treated as a real photon. The other photon has finite Q^2 and hence longitudinal and transverse polarization components. The cross section can be written as a sum of two terms:

$$\sigma(Q^2,W) = \sigma_{TT}(Q^2,W) + \varepsilon\sigma_{TL}(Q^2,W),$$

where σ_{TL} describes the collision of a longitudinal (virtual) photon with a transverse target photon; ε is the polarization parameter (see below, 3.3).

c) Double tag mode (both electron at angles $\theta \gtrsim$ 25 mrad). Here five terms contribute (σ_i, τ_i). For details see e.g. the talk of W. Wagner at this conference.

A full program of total cross section measurements has the following objectives (in order of increasing complexity):

1) Measure of the Q^2 and W dependence of $\sigma_{TT} + \varepsilon\sigma_{TL}$ at finite Q^2.

2) Determine $\sigma_{TT}(W,Q^2=0)$ either by extrapolation to $Q^2=0$ or by "0^0 tagging" see below, 3.3).

3) Separate $\sigma_{TT}(Q^2,W)$ from $\sigma_{TL}(Q^2,W)$ by measurements at different values of ε.

4) Determine all σ_i's and τ_i's.

3.2 Difficulties with previous cross section determinations

So far only steps 1) and 2) have been attempted. The PLUTO Collaboration has published the Q^2 and W dependence of $\sigma_{TT} + \varepsilon\sigma_{TL}$ and an extrapolation to $Q^2 = 0$[7].

The low energy part of this extrapolation has been debated (see e.g. the talk of H. Kolanoski at this conference). No other group has published total cross section measurements so far. N. Wermes has reanalyzed the TASSO total cross section data taken up to 1980. He concludes that "given the TASSO detector of 1980 they can't determine the W dependence of $\sigma_{\gamma\gamma}$" [8].

Why are measurements of the total cross section so difficult? I see two reasons: (a) the mismatch between the final state angular distributions and the acceptance of available detectors and (b) the lack of clearcut models for the final state particles. Model parameters have to be optimized by comparison with the data, but are sensitive to effects from limited acceptance.

In order to illustrate the first point the histogram in fig. 10 shows the polar angular distribution of final state hadrons with respect to the incoming e^+e^- axis as expected from a model, which simulates hadronic events at small Q^2 (single tagging at 30-60 mrad). The distribution is strongly peaked near the forward and backward directions. The peaking becomes less pronounced for events generated according to a point-like process $\gamma\gamma \rightarrow q\bar{q}$ with subsequent fragmentation (solid curve for $Q^2 > 1$ GeV^2). The reason is twofold: (a) the finite Q^2 of the scattered electron (corresponding to $\theta > 5$ degrees) pushes the event axis away from the extreme forward/backward region (b). Since the model used constituent quark masses ($m_q = 300$ MeV), the extreme forward/backward region is depopulated. A finite momentum transfer is necessary for creating the quark mass.

Cross section determinations at $Q^2 < 1$ GeV^2 have to cope with particle distributions as given by the histogram. I have therefore renormalized the scale in order to get constant particle flux per cm of abscissa. Fig. 11 shows the hadron acceptance available for total cross section measurements of different detectors using this scale. The TASSO detector as of 1980 provided less than 50 % charged particle coverage and no measurement of neutral hadrons. The average detection efficiency for the hadronic final state (given a tagged event) was about 10 %. This might explain some of the particular difficulties encountered by TASSO. The PLUTO detector as of 1979 had about 50 % coverage for charged and neutral showering hadrons.

In the single tag mode the $\gamma\gamma$ cms energy has to be reconstructed from the measured hadrons. The measured energy W_{vis} is usually smaller than the true energy W. The unfolding from W_{vis} to W becomes increasingly difficult with decreasing particle acceptance.

3.3 Future potential for σ_{tot} measurements at small Q^2

The bottom part of fig. 11 shows the particle acceptance of JADE, of the 1981/82 version of PLUTO and the PEP4 + PEP9 experiment. Clearly the acceptance has improved.

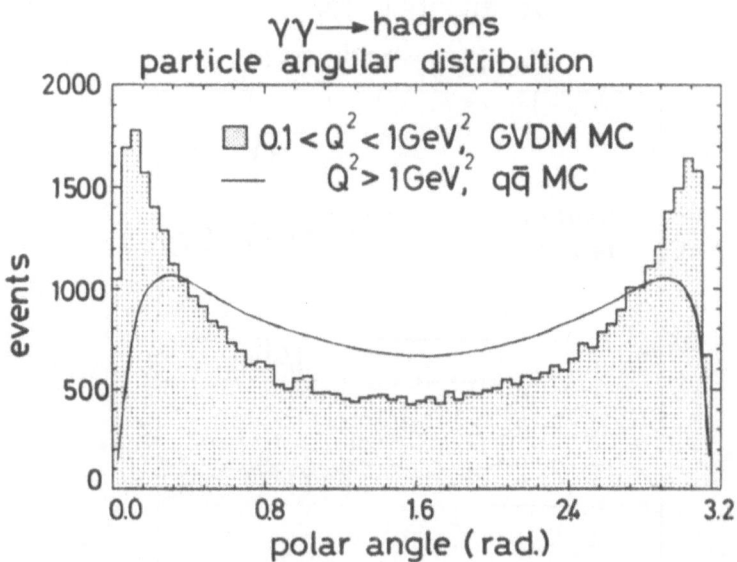

Fig. 10 Generated polar angular distribution of charged and neutral hadrons for the process γγ→hadrons at small Q^2 (histogram). The spectrum has been generated by a GVDM Monte Carlo program with Lund fragmentation (for details see "Discussion") for E_{beam} = 17.5 GeV and the SAT tagging range of the PLUTO detector. No detector effects are included. The curve shows a corresponding distribution for the large angle tagging range of PLUTO ($Q^2>1GeV^2$) and $W>3GeV$. The distribution was generated by a γγ→q$\bar{q}$ Monte Carlo program using constituent quark masses and Feynman-Field fragmentation.

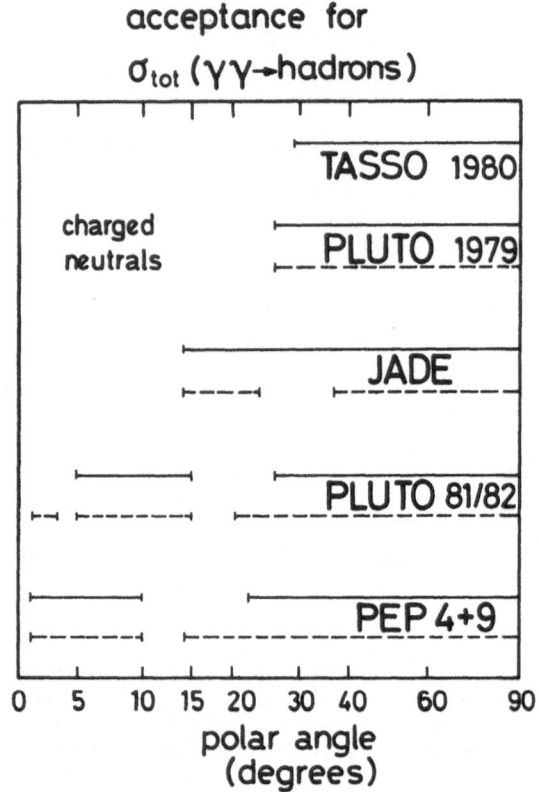

Fig. 11 Acceptance for σ_{tot} ($\gamma\gamma \to$ hadrons) available in experiments at PETRA and PEP. The abscissa has been renormalized in order to get constant flux per cm of abscissa using the histogram of fig. 10.

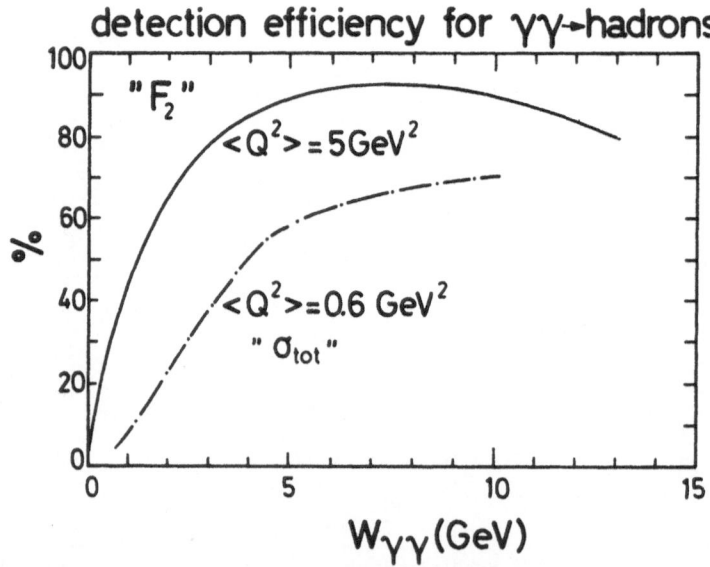

Fig. 12 Detection efficiency for multihadron final states as a function of $W_{\gamma\gamma}$ (true). Full curve: PLUTO structure function analysis with a $\gamma\gamma$ LAT tag ($<Q^2>=5GeV^2$). Dash-dotted curve: JADE total cross section analysis (in progress) with a small angle tag ($<Q^2>=0.6$ GeV^2).

In the following I will discuss two cases:

a) <u>Single tagging experiments</u> ($Q^2 < 1$ GeV^2)

Here we are looking forward to analyses which exploit the improved acceptance. The dash-dotted curve in fig. 12 shows the efficiency for detecting a hadronic final state if the event was tagged at an average Q^2 of 0.6 GeV^2. The efficiency has an average of 30% as compared to 10% in the 1980 TASSO data. Below 2 GeV the efficiency drops below 20%.

The PEP4 + PEP9 experiment has also a potential for separating σ_{TT} and σ_{TL}. At finite Q^2 the total cross section is given by

$$\sigma_{\gamma\gamma} (W,Q^2) = \sigma_{TT} (W,Q^2) + \varepsilon\sigma_{LT} (W,Q^2)$$

with $\qquad \varepsilon = \dfrac{2 (1-y)}{2-2y+y^2}$

$$y \simeq 1 - \frac{E_{tag}}{E_{beam}} \text{ (at small } \theta_{tag})$$

Using the forward shower counters of PEP9 clean tagging down to energies E_{tag} = 3.5 GeV should be possible. With E_{beam} = 14.5 GeV one then obtains for

$$E_{tag} = 0.9 \ E_{beam} \qquad : y = 0.1 \text{ and } \varepsilon = 0.994$$
$$E_{tag} = 0.24 \ E_{beam} \qquad : y = 0.76 \text{ and } \varepsilon = 0.45.$$

At ε = 0.45 the contribution of σ_{LT} is suppressed by about 50 %. By combining data from different ε values a separation of σ_{TT} and σ_{TL} should be possible.

b) <u>Double tagging experiments</u>

When both scattered electrons are measured with sufficient accuracy the energy $W_{\gamma\gamma}$ can be determined in an unbiassed way. This is clearly an advantage. However in the double tagging mode the dominant QED processes $\gamma\gamma \rightarrow l^+l^-$ are tagged also: one either has to rely on a large QED subtraction or on particle identification (lepton rejection), which again confronts one with the problem of limited acceptance.

Two groups have existing devices or future plans for double tagging. The PEP9 group will use their NaI tagging counters in the region from 24 to 90 mrad. With a measured resolution of σ_E/E = 1 % at 14.5 GeV one expects from

$$W_{\gamma\gamma} \simeq 4 \ (E_{beam} - E_{e1}') \ (E_{beam} - E_{e2}')$$

a resolution of

$W_{\gamma\gamma}$ (GeV)	$\sigma (W_{\gamma\gamma})$ (GeV)
2	0.27
4	0.25

This will be adequate for σ_{tot} determination above $W_{\gamma\gamma}$ = 2 GeV.

The ARGUS Collaboration which works at DORIS has proposed the addition of a 0° double tagging system [9]. This system will exploit the vertical bends in DORIS close to the interaction point. Momentum degraded electrons from 2γ interactions which emerge at very small angles (between 1 and 17 mrad) will be deflected into 2 counter arrays placed close to the beampipe. Using Bismuth-Germanate (BGO) shower counters with an energy resolution of $\sigma_E/E \leq 3$ % in the energy range of E_{tag} = 3-5 GeV one expects a resolution of

$W_{\gamma\gamma}$ (GeV)	$\sigma (W_{\gamma\gamma})$ (GeV)
1	0.27
2	0.24
4	0.18

The question of backgrounds is not yet fully settled.

4. High p_T processes

We now turn to processes where one particle or a jet of particles is emitted at high transverse momentum(p_T) relative to the photon direction in the $\gamma\gamma$ cms. High p_T reactions have the promise of unravelling the basic process $\gamma\gamma \rightarrow q\bar{q}$ and competing hard scattering processes. Both the single particle and the jet p_T distributions from hard scattering processes like $\gamma\gamma \rightarrow q\bar{q}$ are predicted to yield the characteristic p_T^{-4} behaviour [11]. Processes where one or both of the photons interact like a ρ should give higher powers of p_T (p_T^{-6}, p_T^{-8} ...).

Data on high p_T processes first became available three years ago [12]. The inclusive single particle p_T spectra showed a hard component at large p_T ($p_T \gtrsim 2$ GeV), which could not be attributed to hadron-like $\gamma\gamma$ interactions [6], [13]. Data on jet production were encouraging but limited in statistics. At this conference a measurement of the Q^2 dependence of high p_T jet production was reported for the first time. We also have two new analyses of high p_T production in notag data.

Any analysis of high p_T jet production in $\gamma\gamma$ processes has to cope with a large background from incompletely detected annihilation events which mimic a $\gamma\gamma$ event. Table 5 shows the no tag event samples from TASSO and CELLO. After appropriate cuts the number of $\gamma\gamma$ events with $p_T^{jet} > 2$ GeV is 410 for TASSO and 60 for CELLO, but the annihilation background is of similar magnitude. It is possible to determine cross sections; but it will be difficult to analyse event shapes.

Table 5 Samples of $\gamma\gamma$ events with $p_T^{jet} > 2$ GeV presented at this conference. The TASSO analysis uses charged particles only for computing W_{vis}. Their limit at 6.8 GeV corresponds to the 10 GeV limit of CELLO.

	TASSO no tag	CELLO no tag	PLUTO single tag
W_{vis} (GeV)	3.4 - 6.8	4 - 10	4 - 13
background in final event sample	~ 40 %	51 %	3 %
events after bkd subtraction	~ 410	~ 60	181

The PLUTO group has avoided this problem by using single tag data only. The track chambers in front of the LAT help in reducing the annihilation background by a factor of 10 (by requiring $p > 3$ GeV and the correct sign for the electron track). Similar reductions are obtained for the SAT tag data. Table 6 shows the resulting event samples for $4 < W_{vis} < 13$ GeV before p_T cuts. The remaining backgrounds are on the 3% level. Above $W = 10$ GeV annihilation events with hard photon radiation become an increasingly important source of background in the tagged data sample.

Table 6 PLUTO event samples used for the high p_T analysis (before p_T cuts). Cuts: $4 < W_{vis} < 13$ GeV, $n_{ch} \geq 4$, $E_{beam} = 17.5$ GeV, $E_{tag} > 8$ GeV.

tagging device Q^2 range (GeV^2) $\int L dt$ (pb^{-1})	SAT 0.1 - 1 28	LAT 1 - 18 39
events observed	918	491
backgrounds: $\gamma\gamma \to \tau\tau$ $e^+e^- \to$ hadrons beamgas $e^+e^- \to \tau\tau$ inelastic Compton	12 9 ~1 } negligible	20 5 ~2 negligible
events after bkd subtraction	869	464

As a next step a jet analysis has been performed by determining the thrust axis in the hadronic c.m.s. Particles detected in the forward spectrometers of PLUTO have been included. Fig. 13 shows the event distribution as a function of the p_T^2 value of the thrust axis for (a) SAT tags and (b) LAT tags. The full curve is from a Monte Carlo program [14] for the process $\gamma\gamma \to q\bar{q}$ with subsequent fragmentation [15]. The dashed curve is from a generalized vector dominance model for hadron production. The data show an excess over the $\gamma\gamma \to q\bar{q}$ model, which decreases with increasing p_T and increasing Q^2. This is seen more clearly in fig. 14, where the ratio

$$\tilde{R}_{\gamma\gamma} = \frac{d\sigma/dt \ (\gamma\gamma \to \text{jets})}{d\sigma/dt \ (\gamma\gamma \to q\bar{q})}$$

is plotted as a function of p_T^{jet}. The dashed line at $\tilde{R}_{\gamma\gamma} = 1$ corresponds to production via $\gamma\gamma \to q\bar{q}$ only. Whereas the SAT data (fig. 14(a)) approach $\tilde{R}_{\gamma\gamma} \approx 1$ only for $p_T > 3$ GeV, the LAT data (fig. 14 (b)) level off at $\tilde{R}_{\gamma\gamma} \approx 1.5$ already between 0 and 3 GeV. This value of $\tilde{R}_{\gamma\gamma}$ is expected if other hard processes with a p_T^{-4} behaviour (apart from $\gamma\gamma \to q\bar{q}$) are included [11].

In summary: Considerable progress has been made in extracting evidence for the process $\gamma\gamma \to q\bar{q}$ by Q^2 suppression of competing processes. There is no need for a sizeable p_T^{-6} or p_T^{-8} component for $Q^2 > 5$ GeV^2 (fig. 14 (b)). It will be a challenge to study the jet topology in more detail by looking for the predicted 3- and 4-jet events [11].

5. Photon structure function

5.1 Introduction

Two photon reactions give access to deep inelastic electron photon scattering via

$$e\gamma \to eX.$$

Here one of the electrons is scattered at high Q^2, whereas the other one is the source of a quasireal photon (4 momentum squared $P^2 \approx 0$). The structure of the low P^2 photon is probed by the high Q^2 electron. The double differential cross section for this reaction is given by standard factors from QED and by the structure function $F_2^{\gamma}(x,Q^2)$ of the photon

$$\frac{d\sigma}{dxdy} \sim F_2^{\gamma} \ (x, \ Q^2),$$

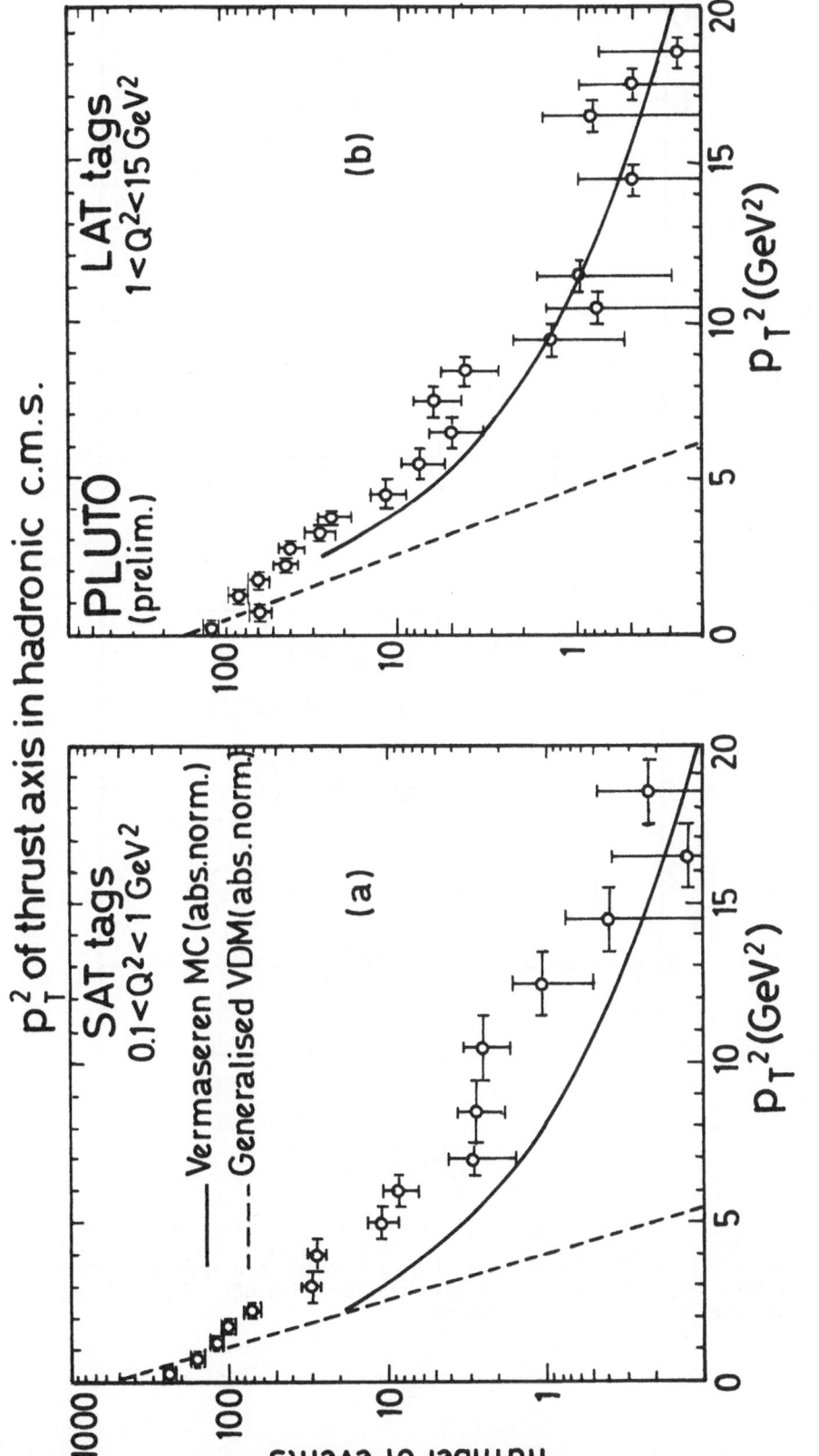

Fig. 13 The p_T^2 distribution of the thrust axis in the hadronic cms of events $\gamma\gamma\rightarrow$hadrons from PLUTO: (a) events tagged at small Q^2, (b) events tagged at large Q^2.

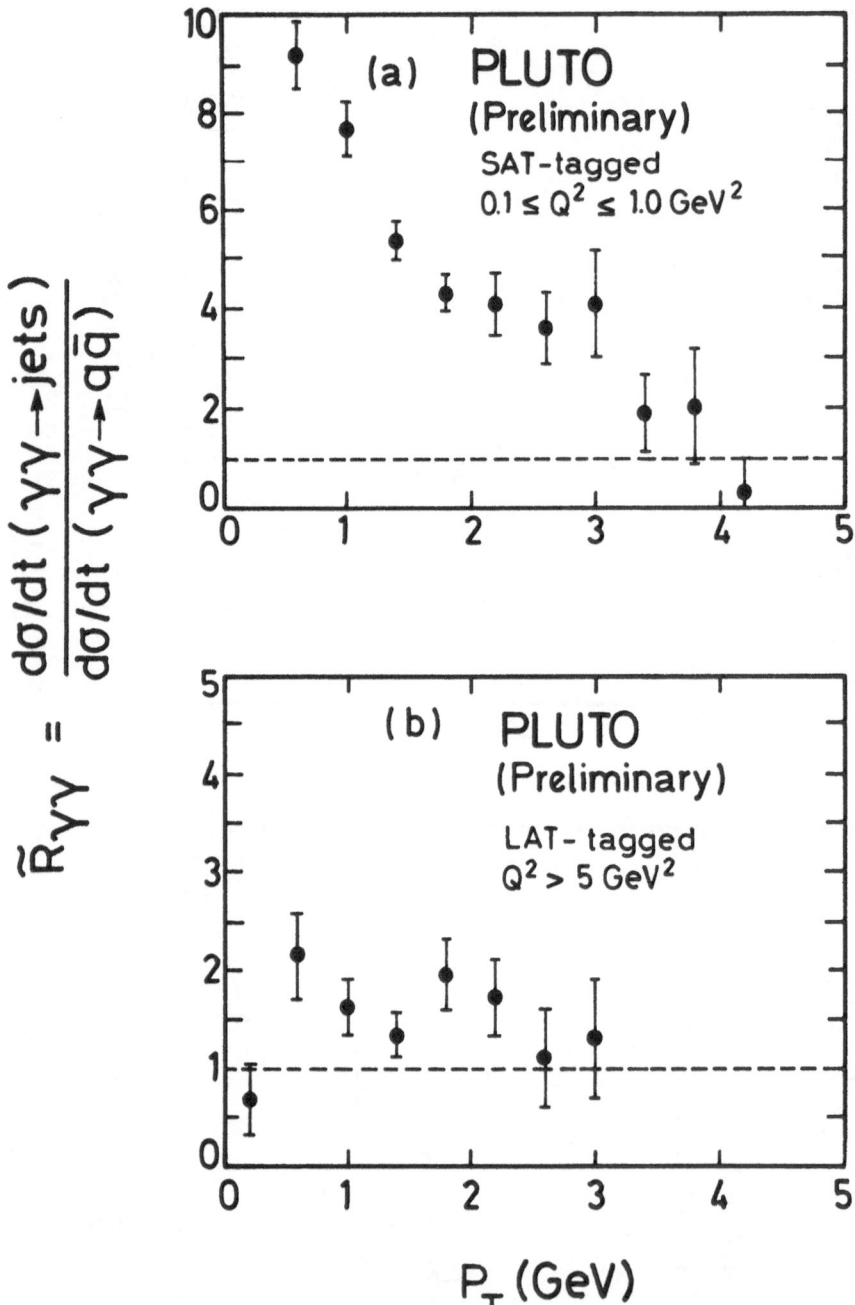

Fig. 14 The ratio $\tilde{R}_{\gamma\gamma}$ of the measured cross section for $\gamma\gamma{\rightarrow}$hadrons and the expected cross section for $\gamma\gamma{\rightarrow}q\bar{q}$ (u, d, s, c quarks) plotted vs. the p_T of the thrust axis. I.e. all events are analyzed assuming a 2jet shape. Dotted line: prediction for σ ($\gamma\gamma{\rightarrow}q\bar{q}$). (Figs. 13 and 14 from the talk of S. Cartwright).

where x is the Bjorken variable $x = Q^2/(Q^2 + W^2)$ and $y \approx (E_{beam} - E_{tag})/E_{beam}$. The formula holds in the limit $xy^2 \ll 1-y$ which is valid for all data presented so far (see e.g. ref. 6a for more details on the formalism). The structure function has an intuitive meaning. In the quark model F_2^γ is proportional to the momentum weighted probability distribution $xG_{q|\gamma}(x,Q^2)$ of finding a quark with momentum fraction x inside the target photon, when the photon is probed in a deep inelastic collision.

The photon structure function can be calculated in leading and higher orders of QCD (at least in the region $0.2 \lesssim x \lesssim 0.8$) [17]. Detailed measurements of the structure function then permit tests of QCD.

Experimentally the determination of the photon structure function confronts one with two difficulties: (a) Due to the incomplete detection of the final state hadrons the measured quantities x_{vis} and W_{vis} differ from the true ones:

$$W_{vis} < W$$
$$x_{vis} > x.$$

Data from different experiments should be unfolded in x and W in order to be comparable. But unfolding has not yet been done in most cases. Note that the Q^2 of the scattered electron is measured with a resolution of 6 - 10% directly from the electron track direction and the energy deposited in the shower counters. No systematic shift of Q^2 occurs. (b) The Monte Carlo program needs to be "fine tuned" in order to perform adequate acceptance corrections. This is easier than in the case of the total cross section at low Q^2 since the detection efficiency for events with $Q^2 > 1$ GeV^2 is higher (see the full curve in fig. 12).

The first pioneering measurement of F_2^γ based on 110 events published two years ago by the PLUTO Collaboration [18]. At this conference much progress has been reported in four areas:

(a) The PLUTO Collaboration has now much improved statistics at "low" Q^2
    ($<Q^2> = 5$ GeV^2).

(b) We have heard reports on x distributions at previously unaccessible values of
    Q^2 ($<Q^2> = 50 - 100$ GeV^2).

(c) A proper unfolding method has been presented.

(d) The first measurement of deep inelastic $\gamma\gamma$ scattering in double tagging mode
    has been made.

In the following I will discuss the progress in (a) - (d).

5.2. Q^2 range of different experiments

The variety of new data and the largely extended Q^2 range has become possible by exploiting systematically the barrel and endcap shower counters for the tagging of high Q^2 electrons. Fig. 15 (a) shows the tagging range accessible at PETRA. Full lines refer to endcap shower counters, dash-dotted lines to barrel counters, and the dashed line to the PLUTO LAT counter. Fig. 15 (b) gives the corresponding event numbers after background subtraction.

The CELLO Collaboration has already published their x distributions [19]. The TASSO group presented for the first time x distributions using the liquid argon endcap counters. JADE doubled the event numbers with endcap tag compared to their publication [20] and showed beautiful new data from the barrel counter at an average Q^2 of 110 GeV^2. The latter data are displayed in fig. 16 as an example. The PLUTO Collaboration has ten times more events from the LAT than 2 years ago and also a measurement in the endcap counters. Finally a first measurement of the x distribution from MAC at PEP was shown. For a complete display of the new data I refer to the talk of W. Wagner [16].

One of the corrections necessary for a comparison with QCD calculations is the target mass effect. The target photon will occasionally obtain sizeable values of P^2 unless this can be excluded by antitagging. The last column in fig. 15 (b) shows the antitagging range of the PETRA detectors. Only JADE and PLUTO have performed antitagging below 100 mrad on the data presented. As discussed by W. Wagner an uncertainty of 5% (15%) will be introduced into the α_s determination from F_2^γ if antitagging is performed only above 30 mrad (100 mrad) [16].

5.3. A procedure for unfolding $F_2^\gamma(x,Q^2)$

A complete determination of $F_2^\gamma(x,Q^2)$ requires the unfolding of the measured data for detector effects like track losses and finite resolution. A proper method of unfolding has been presented by V. Blobel at this conference [21]. In the following I will sketch this method. The objective is to determine the structure function $F_2^\gamma(x)$ averaged over a given interval of Q^2 from the measured values of W_{vis} and Q^2 of the events. The effect of detector resolution can be found by a Monte Carlo simulation of the events in the detector. The density of events in the $Q^2 - W_{vis}$ plane for a structure function $F_2(x)$ is then given by

potential for structure function measurement

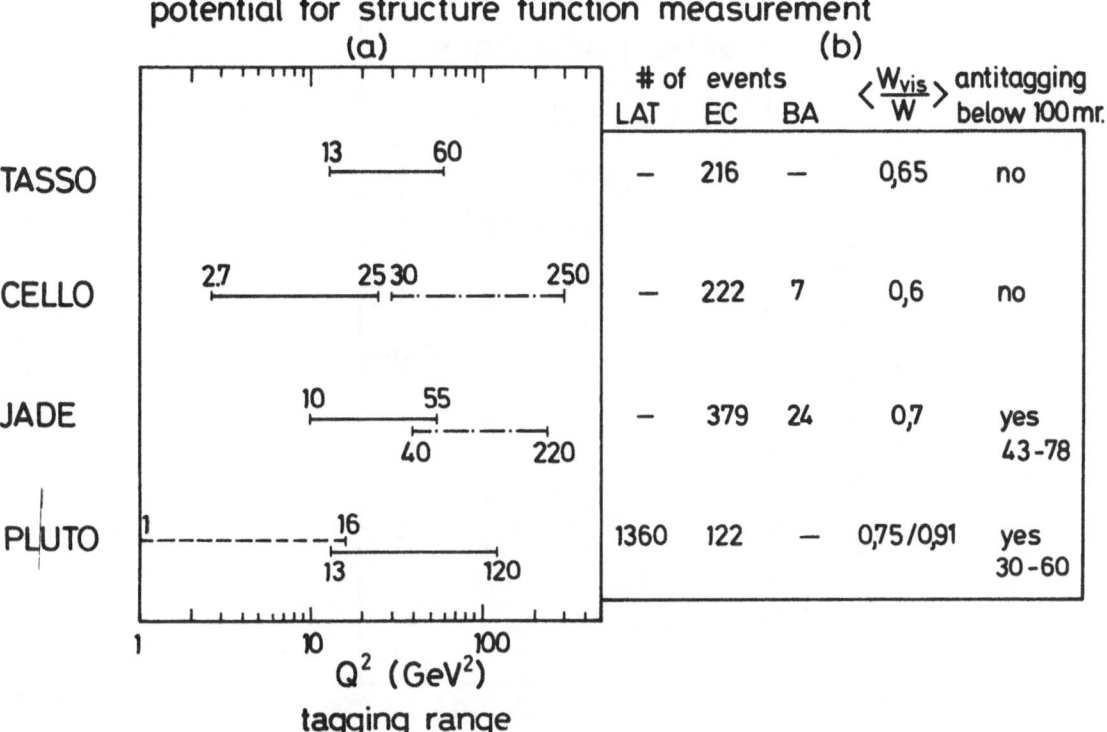

	# of events			$\langle \frac{W_{vis}}{W} \rangle$	antitagging
	LAT	EC	BA		below 100 mr.
TASSO	–	216	–	0,65	no
CELLO	–	222	7	0,6	no
JADE	–	379	24	0,7	yes 43-78
PLUTO	1360	122	–	0,75/0,91	yes 30-60

Fig. 15 (a) Tagging range accessible for measurements of the photon structure function using PETRA detectors. Full lines: endcap
counters (EC), dash-dotted lines: barrel counters (BA), dashed line:
PLUTO large angle tagger (LAT). (b) Numbers of deep inelastic eγ
scattering events found after background subtraction. Also given is
the ratio W_{vis}/W of measured and "true" energy of the hadronic final
state and the potential for antitagging (vetoing) on a second scattered electron below 100 mrad.

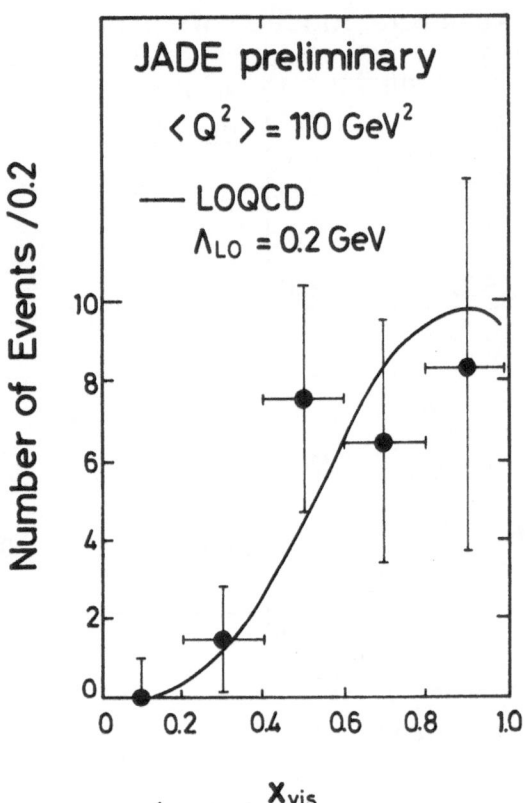

Fig. 16 x_{vis} distribution from deep inelastic $e\gamma$ scattering with an electron tagged in the barrel shower counter of JADE. The full curve is from a leading order QCD calculation for u,d,s,c quarks. Effects of the finite photon target mass are included. For details of the calculation see ref. 20 (from talk of T. Nozaki).

$$\frac{d^2 N^{MC}}{dQ^2 dW_{vis}} = \int R (W_{vis}, Q^2, x) F_2(x) \, dx, \tag{1}$$

where $R(W_{vis}, Q^2, x)$ includes the photon flux, kinematical factors, the fragmentation model and the effect of the detector.

In the method, the structure function $F_2(x)$ is written as a linear superposition of B-spline functions $b_i(x)$,

$$F_2(x) = \Sigma \, a_i \, b_i \, (x) \tag{2}$$

B-splines [22] are piecewise cubic polynomials, continuous up to the second derivative; they are non-negative, and each $b_i(x)$ is non-zero only in a small region of x. They have the property $\Sigma \, b_i \, (x) \equiv 1$. B-splines have been introduced because they have optimal interpolation properties.

Inserting (2) into (1), one obtains

$$\frac{d^2 N^{MC}}{dQ^2 dW_{vis}} = \Sigma \, a_i \left[\int R(W_{vis}, Q^2, x) \, b_i(x) \, dx \right]$$

$$= \Sigma \, a_i \left[\frac{d^2 N^{MC}}{dQ^2 dW_{vis}} \right]_i \tag{3}$$

i.e. the generated distribution in the $Q^2 - W_{vis}$ plane has been decomposed into contributions from the B-splines $b_i(x)$.

Now the coefficients a_i can be obtained by a maximum likelihood fit of (3) to the measured event distribution in the $Q^2 - W_{vis}$ plane. Inserting the fitted coefficients a_i into (2) yields the unfolded structure function $F_2(x)$. Data points are calculated from (2) by integration over bins of x.

Due to the finite resolution of the detector, there are negative correlations between bin contents, resulting in local (bin to bin) oscillations. These oscillations can be reduced by the assumption that $F_2(x)$ is a smooth function. I.e. one assumes that $F_2(x)$ has no narrow structures. This is enforced by requiring that the "curvature" of $F_2(x)$ is small. The curvature is measured by an integral over the square of the second derivative of $F_2(x)$,

$$\int \left[F_2''(x) \right]^2 dx = \Sigma\Sigma\, a_i\, c_{ij}\, a_j$$

For details I refer to ref. 23.

The unfolding procedure has been applied to the PLUTO LAT data. This is a sample of 1360 events in the Q^2 range $1 < Q^2 < 18$ GeV^2.

Fig. 17(a) shows the structure function F_2 as a function of x_{vis}. Fig. 17(b) shows the same data after unfolding. The unfolded $F_2(x)$ has a systematic shift to smaller x values as compared to $F_2(x_{vis})$, however, the shift and the overall difference is small. As a second result it was found that 'systematic' differences in $F_2(x)$ for different fragmentation models are on the level of 1σ of the statistical error.

The unfolding procedure has a relatively small effect for two reasons: (1) in the Q^2 range considered $F_2(x)$ has little structure. (2) The good particle acceptance of PLUTO makes smearing effects small. This is illustrated in fig. 18, which shows the ratio W_{vis}/W as determined from Monte Carlo events which were traced through the PLUTO central detector (shaded histogram) and the combination of central and forward detectors (open histogram). A clear improvement in resolving W_{vis} is observed in the 1981 version of PLUTO. The average ratios of W_{vis}/W from different detectors are listed in fig. 15(b) for comparison. What matters in unfolding is not just the average value of W_{vis}/W but also the width and shape of the distribution. As a combined measure we use the ratio of width to mean. The smaller the ratio, the better. Typical values of σ/mean are shown in table 7.

Table 7   The ratio "standard deviation σ/mean value" of the distribution of W_{vis}/W for different detectors. The last line refers to an analysis where only charge balanced events are included.

detector	mean $= \langle \dfrac{W_{vis}}{W} \rangle$	$\dfrac{\sigma}{mean}$
PLUTO 1979	~ 0.6	0.6
JADE	0.7	0.3
PLUTO 1981 LAT	0.75	0.27
EC	0.91	
PLUTO 1981 LAT $\Sigma q_i = 0$	0.90	0.18

Photon structure function

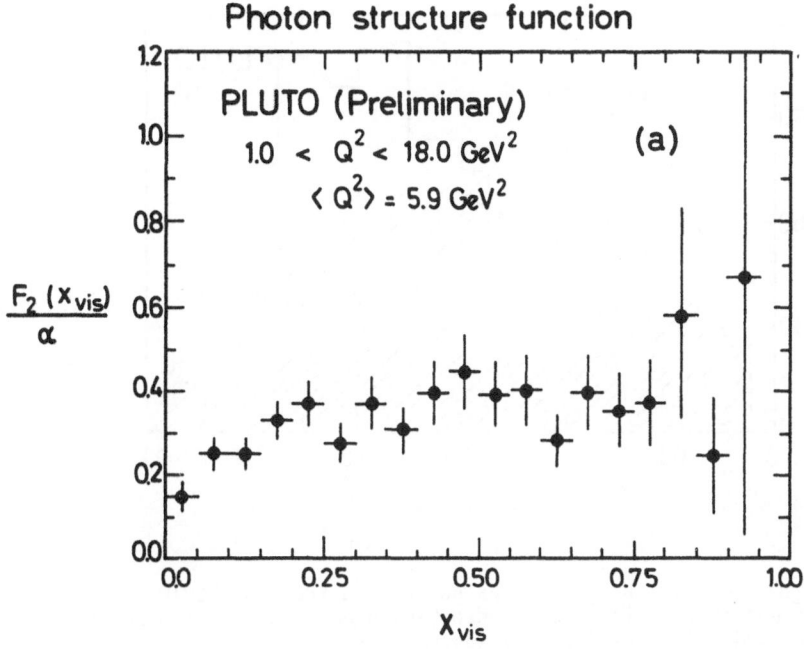

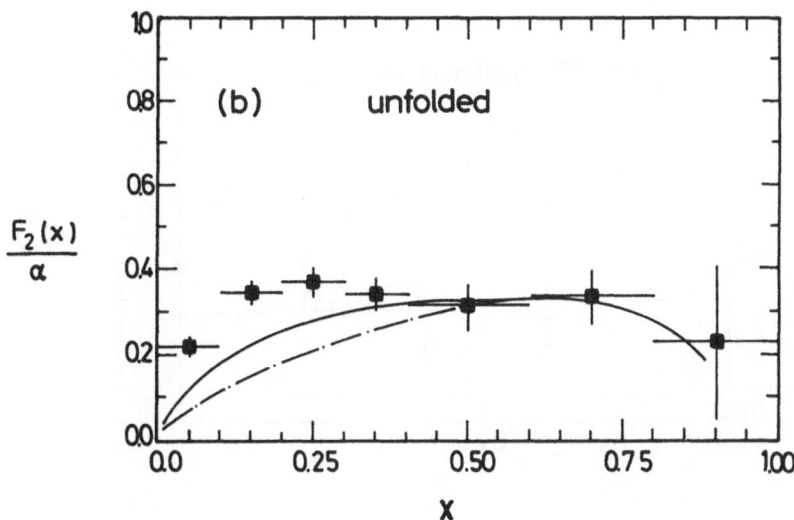

Fig. 17 (a) The structure function $F_2(x_{vis})/\alpha$ averaged over the range $1 < Q^2 < 18$ GeV^2 as a function of x_{vis}. Data from PLUTO. The data points still contain a background of about 12% (mostly from $\gamma\gamma \to \tau\tau$).

(b) The unfolded structure function $F_2(x)/\alpha$. The background has been subtracted in the unfolding procedure. The curves are from a quark parton model calculation for uds quarks (dash-dotted curve) and udsc quarks (full curve) using $m_u = m_d = 0.3$ GeV, $m_s = 0.5$ GeV and $m_c = 1.6$ GeV.

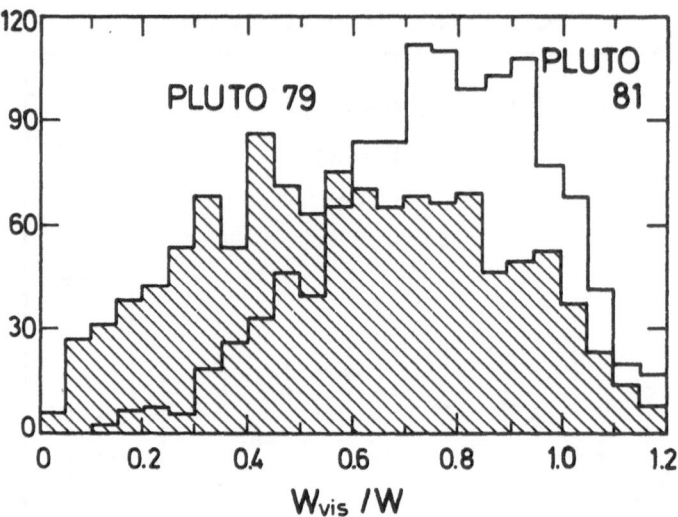

Fig. 18 The ratio W_{vis}/W as determined from the PLUTO central shower and track detectors (shaded histogram) and from the central and forward detectors (open histogram).

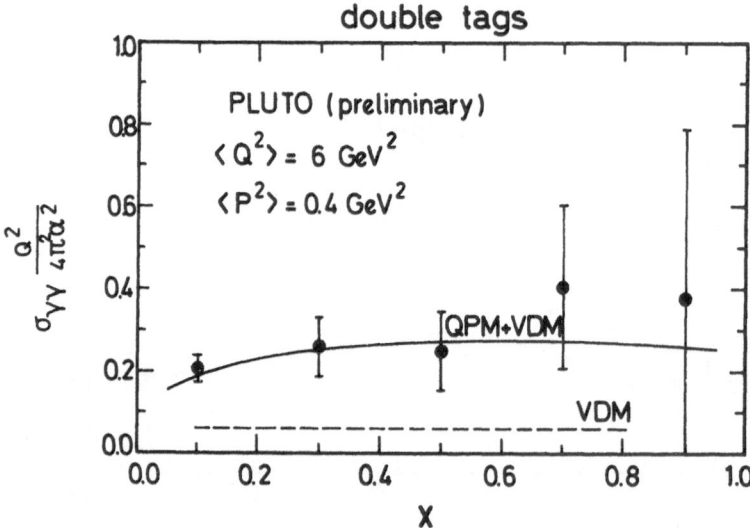

Fig. 19 The hadronic cross section $\sigma_{\gamma\gamma} Q^2/(4\pi^2\alpha^2)$ measured in the double tag mode by the PLUTO Collaboration at $<Q^2> = 6$ GeV^2 and $<P^2> = 0.4$ GeV^2. The dashed curve is from a VDM calculation assuming the interaction of a highly virtual photon with a target ρ meson. The full curve is the sum of the VDM contribution and a quark parton model calculation for u,d,s,c quarks with constituent masses.

A method for minimizing the systematic shift of W_{vis} has been presented by
F. Raupach from the PLUTO collaboration. He reduced the track losses by selecting
charge balanced events with $n_{ch} \geq 4$ only. As a consequence the W_{vis}/W ratio
rises to 0.9 and σ/mean drops to 0.18, which makes unfolding easy. However
a price is paid by loosing two thirds of the events.

In summary, unfolding the measured x_{vis} distribution can and should be done.
It requires an adequate acceptance and resolution of the detector (σ/mean < 0.3)
and sufficient statistics.

5.4. $\gamma\gamma$ Cross section from double tagging

First measurements of hadron production in double tag reactions were presented by the PLUTO Collaboration. The outgoing electrons were tagged by combinations
of the SAT and LAT shower counters. A total of about 200 events was found after
background subtraction. When restricting to LAT - SAT combinations ($<Q^2>$ = 6 GeV²,
P^2 = 0.4 GeV²) about 70 events are left. Fig. 19 shows the resulting cross section
$\sigma_{\gamma\gamma} \cdot Q^2/4\pi^2\alpha^2$ as a function of x. Here x is defined by $x = Q^2/(Q^2 + P^2 + W^2)$. The dashed
curve is from a VDM calculation assuming the interaction of a highly virtual photon
with a target ρ meson [24]. The full curve is the sum of the VDM contribution and a
quark parton model calculation. Good agreement is found.

6. Plans for Gamma - Gamma Tagging at LEP

(contributed by David J. Miller, University College London)

6.1. The Four Forward-Detectors

It was difficult to get any information from the LEP collaborations in time for the
Aachen meeting. The technical proposals were still being written and some features of
the detectors were not settled. Table 8 is a "snapshot" of the plans in April 1983.

For ALEPH there is a note from Heidelberg [25] which discusses the forward detector
entirely in terms of luminosity monitoring. The details in the table are according
to that note, but during the presentation at Aachen it was said [26] that improvements
would be made to turn the forward detector into a proper tagger. For DELPHI the
details given were obtained by word of mouth at Aachen [27]. For L3 the information
came by telephone [28]. The OPAL design is our responsibility [29].

Table 8   Plans for LEP Forward Detectors; April 1983

	ALEPH	DELPHI	L3 small angle	large angle	OPAL
θ_{min} (mr)	45	43	24	<58	45
θ_{max} (mr)	80	$\sim$ 144 (beyond flare)	58	120	117
z front (cm from intersection)	248	233	300	150	216
Beam pipe	flared (trumpet)	flared			parallel Be
Calorimeter Technique	Pb with gas-tubes	Pb/Bi with scin-tillating fibres[*] and VPTs[+]	BGO with Si Photo-diodes		Pb-scintillator with VPTs[+] or GPDs
Longitudinal Division	Yes	No	No		Preradiator + 2 layers[*]
Tracking	Minidrift or drift tubes	Tube chambers	Yes		Radial drift
Spacing of chambers	over 20 cm	over 10 cm	?		2 over 40 cm + 1 after preradia-tor
Luminosity Monitoring					
- Fine	Chambers	Chambers	Chambers		Chambers plus counters
- Through miniβ	Yes	No	Yes		Yes

* see text
+ VPT is vacuum phototriode, GPD is gas filled diode (under development at UCL)

<u>Acceptance:</u> The LEP beampipe fixes the minimum tagging radius. All but L3 have their forward detector at about 2.2 m from the intersection. L3 divides the forward detector into two parts and is able to go to smaller angles by placing the inner part of the tagger further away.

<u>Beampipes:</u> All except OPAL expect to use a flared beampipe. OPAL wants a parallel pipe so that the inner vertex chamber can slide over it.

<u>Calorimetry:</u> The DELPHI calorimeter will use Wood's metal (low melting point Pb-Bi alloy) as a dense matrix around scintillating fibres. L3 have sufficient light from BGO to be able to use silicon photodiodes. DELPHI and OPAL plan to use vacuum phototriodes or gas filled photodiodes.

<u>Tracking:</u> ALEPH had planned only partial ϕ coverage for their tracking chambers but may be changing to full 2π coverage. Only OPAL plans to have sufficient lever-arm to be able to project back to the crossing point with any precision.

6.2. <u>Motivation for the OPAL Design</u>

The OPAL forward detector is probably the most elaborate [30]. It is intended to tag electrons down to the lowest possible energy; to $\sim 10\%$ of beam energy if possible. Only at low tagged energies will it be possible to resolve the contribution from the longitudinal structure function F_ℓ to deep inelastic $e\gamma$ scattering. To recognise these low energy electrons it will be necessary to resolve their tracks clearly before the calorimeter and to reject the pion background which becomes serious at less than 40% of beam energy in the forward direction. This is why OPAL has a large space for tracking and three longitudinal subdivisions to its forward electromagnetic calorimeter.

6.3. <u>Rates</u>

The rates for deep inelastic $e\gamma$ scattering for a given integrated luminosity should be higher than at PETRA/PEP. As an example one expects for $E_{beam} = 50$ GeV, $20 < \theta_{tag} < 200$ mr, $5 < E_{tag} < 50$ GeV, $\int Ldt = 10$ pb^{-1} and full hadron acceptance about 1400 events with $Q^2 > 5$ GeV^2 and $W^2 > 10$ GeV, as compared to 240 events at PETRA (with same cuts except $E_{beam} = 18$ GeV, $3 < E_{tag} < 18$ GeV) [31]. In summary, the prospects for 2γ physics at LEP are good.

7. Final remark

We have seen an impressive and growing variety of attempts to master the art of
2γ experimentation. The results have mainly come from PETRA. I hope that the
colleagues from PEP and elsewhere will join in soon with their results.

Acknowledgement

I thank all speakers of the experimental discussion session for their cooperation.
I also thank Prof. Christoph Berger, who worked hard on making this stimulating,
high level conference possible. I am indebted to Prof. V. Blobel and Dr. J. Dainton
for their help in preparing this manuscript.

References

1. Particle Data Group, Phys. Lett. 111B (1982)

2. PLUTO Collaboration, Ch. Berger et al., Phys. Lett 94B (1980) 254

3. S. Cooper, talk at the 2nd International Conference on Physics in Collision,
    Stockholm, 2. - 4. June 1982, DESY report 82-050 (1982)

4. G. Mennessier, Zeits. Phys. C16 (1983) 241

5. CELLO Collaboration, to be published and talk of F. Kovacz at this conference

6. Ch. Berger, Proc. Fourth International Colloquium on Photon-Photon
    Interactions, Paris, 1981, ed. G.W. London. World Scientific, Singapore (1981),
    D. Cords, ibidem.

7. PLUTO Collaboration, Ch. Berger et al., Phys. Lett. 99B (1981) 287

8. N. Wermes, thesis, Bonn University, report BONN-IR-82-27 (July 1982)
    and talk in the Experimental Discussion Session

9. ARGUS Collaboration, A proposal to study $\gamma\gamma$ interaction with the detector
    ARGUS at DORIS, 1983, PRC 83/06 (unpublished)

10. JADE Collaboration, Phys. Lett. 113B (1982) 190

11. S. Brodsky, T. De Grand, J. Gunion, J. Weis, Phys. Rev. D19 (1979) 1418

12. W. Wagner, Proc. XX. Intern. Conf. on High Energy Physics Madison, Wisc. 1980,
    ed. L. Durand and L.G. Pondrom, American Institute of Physics, New York 1981

13. N. Wermes, talk at this conference

14. J.A.M. Vermaseren, program writeup (1978), unpublished

15. The Feynman-Field option inside the Lund program was used;
    T. Sjostrand, Lund University, Theor. Phys. report 80-3 (1980)

16. W. Wagner, talk at this conference

17. see eg. D. Duke, W. Frazer, talks at this conference

18. PLUTO Collaboration, Ch. Berger et al., Phys. Lett. 107B (1981) 168

19. CELLO Collaboration, Phys. Lett. 118B (1982) 211
    and DESY report 83-018 (1983), to be published

20. JADE Collaboration, Phys. Lett. 121B (1983) 203

21. The method has already been applied to ν structure function data in:
    Charm Collaboration, Phys. Lett. 102B (1981) 67 and 109B (1982) 133.
    See also S. Provender, Comp. Phys. Comm. 27 (1982) 213

22. C. de Boor, A practical guide to splines, New York 1978

23. V. Blobel, transparencies of talk in the experimental discussion session,
    unpublished

24. J. Dainton, S. Maxfield, private communication

25. "Luminosity Monitor for ALEPH", B. Falkenburg et al., Heidelberg;
    ALEPH Note 93, February 1983

26. Dr. Grupen, Siegen, private communication

27. Dr. Lillestøl, Bergen, private communication

28. Prof. R. Kraemer, Carnegie Mellon University

29. U. C. London, Maryland, Bologna

30. Technical Proposal for the OPAL Experiment, submitted to the LEP Committee
    at CERN, May 1983

31. M. Davier, $\gamma\gamma$ Physics at LEP, Proc. Fourth International Colloquium on
    Photon-Photon Interactions, Paris 1981, World Scientific, Singapore; 1981

Discussion

J. Nozaki (Lancaster)

Are there any reasons why you did not comment on the ln Q^2 dependence of F_2 which was observed in the Q^2 range from 10 - 100 GeV^2?

H. Spitzer (Hamburg)

I consider the new data on F_2 to be an important achievement. The ln Q^2 dependence of F_2 has been summarized by W. Wagner.

J. Nozaki (Lancaster)

A comment: The structure function F_2 increases by about 50 % from 10 to 100 GeV^2 of Q^2 value according to the ln (Q^2/Λ^2) prediction. On the other hand the contribution of c quark production increases the F_2 function by about 30 - 40 % for the same Q^2 range. These two factors altogether give a factor 2 increase of F_2 in this Q^2 range. This is in good agreement with the data.

N. Wermes (SLAC)

Comment concerning 1γ background in high p_t jet processes: As pointed out in the experimental parallel session by E. Duchovni the 1γ background to 2γ jet events is due to initial state radiation only in the TAG case. For NOTAG data 1γ events are accepted dominantly due to acceptance losses.

H. Spitzer (Hamburg)

My statement about the dominance of 1γ background accompanied by a hard radiation in fact referred to the TAG case. It becomes important for $W_{vis} > 10$ GeV.

N. Wermes

I personally think that the experimental situation concerning a possible measurement of 3 or more jet processes is not that hopeless. I agree with you that a positive identification of the beam pipe jet is probably very difficult if not impossible. But applying an "ANTI-BEAM-PIPE-JET"-cut, requiring no hadronic particles in forward/backward directions and comparing to present data which do not apply this cut could possibly improve our knowledge about these processes a lot. Would you like to comment on this?

H. Spitzer

This should be tried, but it will be hard. Recall that the average width of jets is expected to be large, e.g. 50 % of the energy from $e^+e^- \to q\bar{q}$ jets at $E_{cm} = 9.4$ GeV comes outside a cone of 54^0 full opening angle.

H. Kolanoski (Bonn)

You showed plots where the jet cross section has been described by $\gamma\gamma \to q\bar{q}$ + GVDM. What do you mean with "GVDM"?

H. Spitzer

The GVDM model of the PLUTO group assumes that the cross section has a Q^2 dependence as given by the GVDM model of Ginzburg and Serbo (Phys. Lett. 109 B (1982) 121) and a W dependence of (240 + 270/W) nb, (W in GeV). Particles are produced in two clusters along the photon direction in the $\gamma\gamma$ cms and fragmented according the Lund model.

S. Barshay (Aachen)

Is there any enhanced or resonant-like behavior in $\gamma\gamma \to \pi\pi$ in the experimental data in the mass region of 500 MeV - 1 GeV?

H. Kolanoski

There is a measurement of $\gamma\gamma \to \pi^+\pi^-$ for $m_{\pi\pi}$ from threshold to ~ 0.7 GeV from DCI. They observe about twice as large a cross $^{\pi\pi}$ section as expected from the Born cross section. However, the statistical significance is only 2.3 standard deviations.

S. Barshay

The reason that this is interesting is that there are at present some small discrepancies in muonic X-rays from muons close to a nucleus. It would be nice to know whether the 2π field at low mass can connect significantly, via 2γ, with the muon.

J. Olsson (DESY)

I would like to answer concerning $\gamma\gamma \to \pi\pi$. There are limits from JADE on the product of $B_{\pi\pi} \cdot \Gamma_{\gamma\gamma}$ between 600 MeV and 1 GeV for narrow resonances ($\Gamma \leq 50$ MeV). These limits are ~ 0.5 KeV. We are working on similar limits for broad resonances.

F. Kovacs (Paris VI)

Answer to S. Barshay about $\pi\pi$ events under 1 GeV/c^2 mass: We are starting an analysis at CELLO on $ee \to (ee) \gamma\gamma \to (ee) \pi\pi$ in the mass range of 500 MeV/c^2 to 1 GeV/c^2. We have there a lot of data. For the moment our problem is to subtract tails coming from inclusive events with p_t^{total} conserved like:

$\pi\pi$ from 4π events

$\pi\pi$ from η' (where the γ is soft)

$\pi\pi$ from $A_2 \to \rho^\pm\pi^\pm$ (where the π^0 is hard with low p_t)

etc.

Than, if we have an excess of events we will try to fit our data with Menessier's model which includes broad resonances in that mass region and try to do the junction with DCI results at very low W. It seems indeed of great interest to have results of $\pi\pi$ production by $\gamma\gamma$ in the low mass region.

J.H. Field (Paris VI)

Two comments on the CELLO contributions:

1. The number of $\pi\pi$ events used in the f^0 analysis for $1.0 < M_{\pi\pi} < 1.4$ GeV/c^2 was 2031 not 50 (the latter figure corresponds to $\eta' \to \pi^+\pi^-\gamma$ production).

2. In the untagged high p_t jet analysis for a cut $p_t^{Jet} > 2$ GeV we see ~ 2 times the box contribution which is not inconsistent with the PLUTO findings.

SUMMARY OF THEORY PARALLEL SESSIONS

D.M. Scott

Department of Applied Mathematics and Theoretical Physics,
University of Cambridge, Silver Street, Cambridge CB3 9EW
U.K.

Introduction

The theory parallel sessions contained 16 talks, each of about 15 minutes:

1.   I. Schmitt, Structure functions for heavy quarks,
2.   M. Gorn, Structure functions for massless quarks,
3.   E. Masso, $\gamma\gamma$ total cross sections,
4.   E. Gotsman, Generalised vector dominance,
5.   S.L. Grayson, Wide-angle production of meson pairs,
6.   P.H. Damgaard, $\gamma\gamma \rightarrow p\bar{p}$,
7.   P. Singer, Radiative decays of 2^+ mesons,
8.   J. Kühn, $\gamma\gamma$ collisions and radiative quarkonium decays,
9.   E. Reya, Supersymmetry in two-photon processes,
10. J.A. Grifols, $\gamma\gamma \rightarrow$ squarks and sleptons,
11. J.A.M. Vermaseren, Radiative decays,
12. P.H. Daverveldt, Monte Carlo simulations of radiative corrections,
13. G. Missonnier, Corrections to jet production,
14. P. Kessler, Direct photon-pair production,
15. A. Janah, Jet production and quark charges,
16. S.J. Brodsky, Closing talk.

Clearly there was a broad range of topics, preventing a general overview. Consequently a thumbnail sketch of each presentation will be given.

Structure functions

In the range $0.4 \lesssim x \lesssim 0.8$, where two-loop corrections to the photon structure function F_2 are smallest, the charm threshold is crossed as Q^2 varies over the experimentally accessible range $Q^2 = 2 - 50$ GeV^2 . Because $e_c = 2/3$, the charm contribution can be important. It can (i) provide a QCD test in itself, and (ii) provide Q^2 dependence of F_2 through the threshold turn-on, which could be confused with the sought after QCD $\ln Q^2$ behaviour. Schmitt presented results of a calculation[1] of the structure functions F_2 and F_L, involving the box diagram together with $0(\alpha_s)$ corrections. Although QCD corrections are expected to be unreliable for $W^2/m_c^2 \gtrsim 10$ (because of large logarithms) and for $W^2/m_c^2 \lesssim 5$ (because the squared matrix element contains a factor $1/v$ (v = velocity) and the effective expansion parameter α_s/v is required to be small, say $\lesssim 1/2$) there is a window where the calculations may be trusted. Some results are shown in fig. 1. The $0(\alpha_s)$ corrections do not qualitatively change the box diagram result for charm production: at $Q^2 \simeq 20$ GeV^2, $x \simeq 0.5$ charm contributes to $F_{2,L}$ at the 20% level, and its contribution increases with Q^2.

Gorn[2] was interested in the other extreme, namely $m_q = 0$. In the box diagram for the photon's structure functions m_q appears as $\ell n Q^2/m_q^2$, and the question is how to regulate the logarithm if $m_q = 0$. (In QCD m_q is transformed into Λ). In the $\gamma^* \gamma$ centre of mass frame, finite structure functions can be defined by excluding the kinematic regions where the outgoing q and $\bar{q}$ are within some angle δ of the γ^*, γ directions. Then $\ell n Q^2/m_q^2$ turns into $\ell n \delta$. However applying a similar condition in the e^+e^- centre of mass frame is rather different: the $q, \bar{q}$ jets are no longer back to back, and restricting q's momentum to lie outside an angle δ from the e^+e^- axis does not generally restrict the $\bar{q}$'s to lie outside too. If only one jet satisfies the constraint the singular configuration shown in fig. 2 is allowed: here one jet disappears down the beam pipe parallel to the real photon. Generally both jets must obey angular constraints in order to define a finite structure function.

GVDM

The Generalised Vector Dominance Model is like a pizza. You mould the dough to whatever shape you like, add some flavourings and cook it. The ingredients are infinite sequences of vector mesons V_α (usually assumed to have constant splitting in m^2), couplings to the photon em_α^2/f_α (independent of Q^2, and usually $f_\alpha \sim m_\alpha$ for scaling in e^+e^- annihilation), and the cross sections for the V_α to scatter on a target (independent of Q^2).

Applied to $\gamma^* \gamma$ we have in the diagonal approximation

$$\sigma(\gamma^* \gamma) = e^4 \sum_{\alpha, \beta} \frac{1}{f_\alpha^2 f_\beta^2} \left(\frac{m_\alpha^2}{m_\alpha^2 + Q^2} \right)^2 \sigma(V_\alpha V_\beta) . \tag{1}$$

The guesses made by Close et al.[3] gave the solid line in fig. 3.

Masso[4] described the dotted line, and Gotsman[5] the dashed (diffractive and Regge) and dashed-dotted (diffractive only) lines.

A ℓnQ^2 behaviour can be generated[4] by assuming $\sigma(V_\alpha V_\beta) \sim 1/(m_\alpha^2 + m_\beta^2)$ for large α, β. Through eq.(1) this gives $\sigma(\gamma^*\gamma) \sim \ell nQ^2/Q^2$.

Parametrisations of the real γ-real γ cross section should be affected by the infinite towers of vector mesons and Gotsman[5] proposed

$$\sigma(\gamma\gamma) = (295 + 460/W) \text{ nb} \qquad (2)$$

with W in GeV. He also emphasized the hadronic aspects of the virtual photon cross sections. See fig. 3 for an estimate of the energy dependence: at high W the Regge term disappears.

As $W \to 0$ $\sigma(\gamma\gamma)$ must turn over so that $\sigma(\gamma\gamma) = 0$ at $W = 0$. If $\sigma(\gamma\gamma)$ has its maximum at $W = W_0$, then in GVDM $F_2(x,Q^2)$ has[5] a corresponding maximum at $x_0 = Q^2/(W_0^2 + Q^2)$ and falls to zero as $x \to 1$. The calculations in ref.5 are in excellent agreement with published data on F_2, see e.g. fig. 4. It will be interesting to see if the dependence of the shape of F_2 on x and W^2 can be disentangled.

The question was raised as to whether GVDM was capable of explaining structure of final states. What cannot be done is rule out GVDM on this basis. Eg. (i) to explain final state structure in e^+e^- annihilation at $E_{beam} = 19$ GeV the decay characteristics of the 771st radial recurrence of the ρ are required, and you can guess them to be whatever you want, and (ii) the final state structure of $\gamma\gamma \to$ jet + jet at high p_T can probably be achieved by infinite sums of $\gamma\gamma \to R \to$ jet + jet where R is a pseudoscalar or tensor.

<u>Exclusive wide angle scattering</u>

The theory for exclusive wide angle scattering cross sections has matured from dimensional counting rules, giving power law dependence at fixed angle, to QCD calculations for the complete cross section via the Brodsky-Lepage[7] formalism using the constituent interchange mechanism CIM (not to be confused with pre-QCD CIM calculations), or Landshoff multiple scattering[8].

Calculations of $\gamma\gamma \to$ MM (M = meson) were done some time ago, and data have been presented at this workshop. The calculations used CIM. Grayson presented a multiple scattering calculation[10] of $\gamma\gamma \to \psi\psi$, $\psi\phi$. This appears at higher order in α_s than CIM, but is larger asymptotically for scattering angles less than about $45°$. A

sample diagram is shown in fig. 5. There are no $\ell ns/m^2$ terms, where
m is the mass of the quark, and so m = 0 may be taken. The
singularities appearing in individual diagrams all cancel in the sum.
To achieve this cancellation simply in a Monte Carlo integration, the
ε from the $i\varepsilon$ prescription was made finite, and the smooth limit of
successively smaller ε taken numerically. In fig. 6 we show the
asymptotic behaviour of $s^4 d\sigma(\gamma\gamma \to \psi\psi)/dt$ for the CIM and multiple
scattering mechanisms (α_s = 0.32 and the normalisation is given by
f_ψ = 0.26 GeV). An indication of the energy dependence in CIM[11] is
also shown. This has not yet been calculated for multiple scattering,
but is under investigation[10], along with another mechanism involving
a light quark loop. Finally we should point out that the cross section
for $\gamma\gamma \to \psi\psi$ is very small. Even at LEP energies it is only a few
$\times 10^{-38}$ cm^2 , and $B^2(\psi \to e^+e^-)$ = .005 .

Damgaard presented his calculation[12] of $\gamma\gamma \to p\bar{p}$ in the Brodsky-
Lepage formalism. The hard scattering amplitude $\gamma\gamma \to (qqq) + (\bar{q}\bar{q}\bar{q})$
contains over 200 diagrams, and the result was normalised to
$\psi \to p\bar{p}$ [13] . The cross section is shown in fig. 7 (the band indicates
the range given by a choice of wave functions). Also shown is the
asymptotic cross section from[14] pre-QCD CIM (also obtained from
$\gamma p \to \gamma X$ by Bloom-Gilman duality(!))

$$\frac{d\sigma}{dt} (\gamma\gamma \to p\bar{p}) = \frac{2\pi\alpha^2}{s^2} \left(\frac{11}{27}\right) \frac{t^2 + u^2}{tu} \ G_M^2(s) \ , \tag{3}$$

where G_M is the proton's form factor.

Data from TASSO[15] (and also data presented at this Workshop) have
been claimed to be in agreement with Damgaard's calculation. The
data have W = 2 - 3 GeV, barely above threshold whereas the calculation
is for asymptotic energies, so it is not at all clear how we should
view this agreement. Maybe the asymptotic theory is saved by
normalising to a measurement at W = 3 GeV , $\psi \to p\bar{p}$.

There are also data[16] for $\gamma p \to \gamma p$ (which in fact lend a little
support to the crossed version of eq. (3)), but singularities in some
of the hard scattering diagrams prevent a straightforward crossing.
The calculation is underway.

Decay

Singer[17] has a Veneziano-like dual amplitude for VP $\to$ V'P' with
V,V' vectors and P,P' pseudoscalars. The amplitude displays
crossing symmetry, Regge asymptotic behaviour, and the correct
behaviour at low and high energies. It is gauge invariant as

$m_V, m_{V'} \to 0$, so may be used for photon amplitudes (though a small cloud remains over isospin properties). Using the amplitude, couplings to leading resonances in different channels can be related. Eg. the $s \leftrightarrow t$ crossing shown in fig. 9 yields

$$4 \, g_{f\pi\pi} \, g_{f\gamma\gamma} = g^2_{\omega\pi\gamma} + g^2_{\rho\pi\gamma} \qquad (4)$$

Note that the $f_{\gamma\gamma}$ coupling is given in terms of measured quantities. There are corresponding relations for A_2, f^1 . The results for the $\gamma\gamma$ widths of f, A_2, f^1 are in excellent agreement with experiment. Note:

(i) the measured SU(3) breaking in the radiative decays of the vectors goes over into the $\gamma\gamma$ decays of the tensors.

(ii) VMD does not work for $A_2 \to \gamma\gamma$, this does.

(iii) in this scheme f^1 is consistent with being pure $s\bar{s}$.

A different set of decays was described by Kühn[18]. An example, $\psi \to \gamma f$ is shown in fig. 10. There are predictions[18] for many decays with 0^{-+}, 0^{++}, 1^{++} and 2^{++} states accompanying the photon, including helicity structure. Using non-relativistic vertices for both the ψ and f , $\Gamma(\psi \to \gamma f)$ can be calculated from $f \to \gamma\gamma$ via the derivative at the origin of the f's wave function. There is excellent agreement with experiment (though there may be problems for the detailed predictions of helicity structure). No gluon component of f, f^1 is needed.

The non-relativistic approximation is not appropriate for light states, and may be relaxed for $\psi \to \gamma +$ pseudoscalar[18] (the argument also shows why this talk appeared at a $\gamma\gamma$ meeting). What is required is the matrix element of two gluon fields with the pseudoscalar. The configuration with the two gluons far off-shell is dominant (in contrast to the case when the pseudoscalar has valence gluon components, for then the gluons will be on-shell). In this case the Brodsky-Lepage formalism could be used, but instead Kühn[18] chose to evaluate the matrix element $<0|JJ|PS>$ using the light cone algebra, in the same way as in calculations[19] of $\gamma^*\gamma^* \to$ pseudoscalar. The result can be expressed as

$$\frac{\Gamma(\psi \to \gamma PS)}{\Gamma(\psi \to e^+e^-)} \propto \frac{\alpha_s^4}{m_\psi^2} , \qquad (5)$$

which gives the complete dependence on the parent vector. This is in contrast to the mass dependence $1/m_\psi^6$ for a pseudoscalar with valence glue. In this context Kühn pointed out that in $\psi, \psi' \to \gamma\iota(1440)$, the

dependence on parent mass may be faster than inverse square, possibly
indicating the presence of valence glue.

Supersymmetry

While there is no experimental evidence for supersymmetry (SUSY),
the theory is very popular. It does not predict masses, but PETRA
data require that the scalar partners of the mundane quarks and leptons
weigh more than about 15 GeV. As usual, an industry has sprung up,
calculating familiar quantities in the new theory.

Reya[20] has calculated the structure function of the real photon
in SUSY-QCD. The box diagram for squark production (with seagull
attachments) gives

$$\tilde{F}_2 = \frac{3\alpha}{\pi} \sum_q e_q^4 x \{[1 - 8x(1-x) + \tau x(1-x)]v$$

$$+ [2x(1-x) + \tau x(3x-1) + \tfrac{1}{2}\tau^2 x^2] \ln(\frac{1+v}{1-v}) \} , \qquad (6)$$

$$\cdot \tau = 4m_{\tilde{q}}^2 / Q^2 , \qquad v = (1 - \tau x/(1-x))^{\frac{1}{2}} .$$

The sum runs over $q = u, d, \ldots$ (not antiquarks) and the 3 is for
colour. As well as providing a new contribution to F_2 , SUSY-QCD
modifies the old-fashioned QCD result by changing α_s and the
Altarelli-Parisi branching functions. The details of this of course
depend on masses, and SUSY-QCD evolution of squark distributions is
presumably academic. A sample result is shown in fig. 11: the dotted
line is ordinary leading log QCD; the dashed line shows how that gets
modified by SUSY-QCD α_s and P's ; the solid lines show the result
of adding the squark contribution $(n_f = 6 , \ln Q^2/\Lambda^2 = 10)$.

The squark contribution to F_L can also be calculated[21]. It is

$$\tilde{F}_L = \frac{6\alpha}{\pi} \sum_q e_q^4 x^2 \{ [1-x + \tfrac{1}{2}\tau x] \ln(\frac{1+v}{1-v}) - 3v(1-x) \} . \qquad (7)$$

In contrast to OF-QCD this exhibits $\ln Q^2$ behaviour.

Note that in (5) and (6) only squarks are considered. There
are also mechanisms involving sleptons, and photino exchange, which
may contribute at the same level.

Grifols[22] described a calculation of $e^+e^- \to$ SUSY particles via
$\gamma\gamma$, $\gamma\tilde{\gamma}$, $\tilde{\gamma}\tilde{\gamma}$ ($\tilde{\gamma}$ = photino). The cross sections for smuon production,

for various energies and masses are shown in fig. 12. Rates may not be negligible at LEP, but backgrounds are large, and signatures are uncertain and depend on details of masses.

It seems that sparticle production via $\gamma\gamma$ has no particular merit: it just exists. Any search for SUSY particles in e^+e^- interactions must consider also production from annihilation, and from t-channel $\tilde{\gamma}$ exchange.

QED, QCD corrections

Earlier results on QED corrections in $\gamma\gamma$ processes are given by Defrise, Ong, Silva and Carimalo[23], and on QCD corrections by Berends, Kunszt and Gastmans[24].

Vermaseren[25] presented preliminary results on radiative corrections to $e^+e^- \rightarrow e^+e^-$ pseudoscalar. This is the first stage of a programme of calculating corrections to $\gamma\gamma$ processes. The numerical integration scheme is very stable to changes in the cut-offs required to carry out real-virtual cancellations. A class of diagrams still remains to be calculated, e.g. the 5-point function, where a photon connects the incoming electron and positron. In fig. 13 we show the correction in % as a function of the mass of the pseudoscalar at PETRA and LEP energies. This is rather larger than the correction found by Defrise et al. The difference is because Defrise et al. only took electron loops in the vacuum polarisation graphs, while fig. 13 has electrons, muons and quarks. As well as total cross sections, differential cross sections in Q^2 can be calculated.

Daverveldt[26] reported on a Monte Carlo calculation of radiative corrections to $e^+e^- \rightarrow e^+e^-\mu^+\mu^-$, including an event generator. Corrections to the total cross section are < 1% as expected, but may be much bigger in certain restricted kinematic regions. The Monte Carlo can be modified to give radiative corrections to $e^+e^- \rightarrow e^+e^-q\bar{q}$.

As well as considering QED corrections, Missonier[27] calculated $0(\alpha_s)$ QCD corrections to the high p_T jet cross section in $\gamma\gamma$ collisions, at 90° in the $\gamma\gamma$ centre of mass frame. His definition of a jet depended on an angle δ : if two partons are within an angle δ of each other, then the two jets cannot be separated, and the resultant single jet momentum is just the sum of the momenta of the two partons. In fig. 14 we show $(\sigma^0 + \sigma^1)/\sigma^0$ as a function of p_T for several values of δ (E_{beam} = 15 GeV , and m_q = 0.3 GeV is taken to regulate collinear divergences). Also shown is the corrected cross section calculated by Berends et al.[24] for a _different_ jet definition: $\sigma(p_T^{jet} > p_T)$ where $p_T^{jet} = Max[p_T^q, p_T^{\bar{q}}, p_T^g]$. It would seem that, as in the case of QCD corrections to the thrust distribution in e^+e^- annihilation[28], the definition of a jet appropriate to experiment should be agreed on.

Direct photon-pair production

This topic has been under discussion for some time[29], Field's talk at this workshop[30] was devoted to it, and should be consulted for details. I would however like to mention that comparisons of $\gamma\gamma X$ to γX rates in πN collisions are troubled by trigger bias effects in the single γ cross section (unless the $\gamma\gamma$ sample comes from a single γ trigger). The 'bare' QCD calculation[31] is well below data[32] on $\pi^+ N \to \gamma X$ as shown in fig. 15. Much of the discrepancy is probably due to smearing a rapidly falling spectrum, and about a factor 2 is probably due to a K-factor (a similar effect is observed in $\pi N \to \mu^+\mu^- X$).

Kessler[33] has estimated $\gamma\gamma$ production through high mass resonances (e.g. $\chi_{b\bar{b}}$) and finds that resonance contributions may be comparable to those from $gg \to \gamma\gamma$ via a quark box, though this of course depends on experimental resolution. He also estimated backgrounds to $\gamma\gamma$, including a Brodsky-Lepage calculation of $gg \to \pi^0\pi^0$. Though presumably in the end the backgrounds will be an experimental, not theoretical, question.

Integer charge quark gauge model

The Pati-Salam gauge model[34] has integer charge quarks. The $SU(3)_c \times U(1)$ left over after $SU(3)_c \times SU(2) \times U(1)$ breaking subsequently breaks down into $U(1)_{em}$. As well as the W and Z there are 8 new massive gauge particles, the gluons. Of these $\tilde{U}$ is neutral and can mix with the photon. It turns out that for gluon current mass m_g and squared momentum transfer Q^2 , the part of the interaction which differs from standard QCD behaves as $m_g^2/(Q^2 - m_g^2)$ because of $\gamma, \tilde{U}$ interference. Hence in processes involving high Q^2 the results are identical to those in QCD. Clearly $\gamma\gamma \to$ jet + jet is an interesting place to look for differences from QCD, as the Q^2 in each photon can be very small, giving $0(1)$ $\gamma - \tilde{U}$ interference effects. This has been studied by Janah[35]. In fig. 16 is shown

$$R_{\gamma\gamma} = (\gamma\gamma \to \text{jet} + \text{jet}) / (\gamma\gamma \to \mu^+\mu^-) \qquad (8)$$

in the integer charge quark model, with $m_g = 10$ MeV , and in QCD (this is a preliminary result[35]). It would be interesting to see how the integer charge quark model compares with data over the wide range of processes usually described by fractional quarks and QCD.

Closing talk

Brodsky spoke on several topics, in particular

(i) From calculations of the pion form factor F_π for $\gamma^* \to \pi\pi$, and the photon transition form factor $F_{\pi\gamma}(Q^2)$ for $\gamma^*\gamma \to \pi$ one finds

$$\alpha_s(Q^2) = \frac{F_\pi(Q^2)}{4\pi Q^2 F_{\pi\gamma}^2(Q^2)} \ (1 + 0(\alpha_s)) \ , \tag{9}$$

providing a direct measurement of α_s from exclusive processes.

(ii) A gluon correction to $\gamma\gamma \to 4$ jets is shown in fig. 17. Factorisation holds, in that the cross section can be calculated from the usual parton model formula, but with QCD scaling violations in the quark density in a photon. However, the extra gluon exchange provides extra transverse momentum fluctuations in the final state.

(iii) Bagger and Gunion[36] have calculated the higher twist process $\gamma\gamma \to \pi X$ at high p_T . Here the π is isolated, and is formed from the $q\bar{q}\pi$ direct coupling. The fragmentation function $D_{q\to\pi}(z)$, where z is fractional momentum, has a calculable piece which falls off as $1/Q^2$, but is constant as $z \to 1$, i.e. $D_{q\to\pi}(z) \sim C/Q^2$ as $z \to 1$. This can provide an extra contribution to the cross section for isolated π's in $\gamma\gamma$ collisions.

Brodsky's final inspirational message was that we are working on the right subject. He said that $\gamma\gamma$ is the simplest initial state for hard hadronic scattering, we can understand complete reactions, and we can test QCD.

Acknowledgements

Peter Landshoff deserves thanks for organising this parallel session. Thanks go to him and James Stirling for reading the manuscript, Ch. Berger and colleagues for an enjoyable meeting, and the SERC of Great Britain for financial support.

References

1. G. Köpp, T. Walsh, P. Zerwas, I. Schmitt, in preparation.
2. M. Gorn, Munich preprint (1983).
3. F.E. Close, D.M. Scott and D. Sivers, Nucl. Phys. B117 (1976) 134.
4. E. Etim and E. Masso, TH 3260-CERN; E. Etim, E. Masso and and L. Schülke, TH 3423-CERN.
5. U. Maor and E. Gotsman, Tel Aviv preprint TAUP 1080-82.
6. PLUTO Collaboration, Ch. Berger et al., Phys. Lett. 107B (1981) 168.
7. S.J. Brodsky and G.P. Lepage, Phys. Rev. D22 (1980) 2157. See also W.J. Stirling, these proceedings.
8. P.V. Landshoff, Phys. Rev. D10 (1974) 1024.
9. S.J. Brodsky and G.P. Lepage, Phys. Rev. D24 (1981) 1808.
10. S.L. Grayson, R.R. Horgan and P.V. Landshoff, preprint DAMTP 83/8.
11. R.E. Ecclestone and D.M. Scott, DAMTP 82/31, Zeit. für Physik, to appear.
12. P.H. Damgaard, Nucl. Phys. B211 (1983) 435.
13. S.J. Brodsky, Tao Huang and G.P. Lepage, SLAC-PUB-2540.
14. D.M. Scott, Phys. Rev. D10 (1974) 3117.
15. A. Bäcker, Siegen preprint Si-82-05 (1982).
16. M.A. Shupe et al., Phys. Rev. D19 (1979) 1921.
17. N. Levy and P. Singer, Phys. Rev. D3 (1971) 1028, Phys. Rev. D4 (1971) 2177; N. Levy, P. Singer and S. Toaff, Phys. Rev. D13 (1976) 2662; P. Singer, TECHNION-PHYS-83-2.
18. J.G. Körner, J.H. Kühn, M. Krammer and H. Schneider, DESY 82-089; J.H. Kühn, Aachen preprint (1983).
19. G. Köpp, T.F. Walsh and P. Zerwas, Nucl. Phys. B70 (1974) 461.
20. E. Reya, TH 3504-CERN.
21. D.M. Scott and W.J. Stirling, DAMTP 83/13; E. Reya, unpublished.
22. J.A. Grifols and J. Sola, Barcelona preprint.
23. M. Defrise, S. Ong, J. Silva and C. Carimalo, Phys. Rev. D23 (1981) 663.
24. F.A. Berends, Z. Kunszt and R. Gastmans, Nucl. Phys. B182 (1981) 397.
25. W.L. van Neerven and J.A.M. Vermaseren, in preparation.
26. F.A. Berends, P.H. Daverveldt and R. Kleiss, Leiden preprint.
27. G. Missonnier, College de France preprint.
28. See A.J. Buras, proceedings of International Symposium on Lepton and Photon Interactions at High Energies, Bonn, 1981.
29. B. Combridge, Nucl. Phys. B174 (1980) 243; C. Carimalo, M. Crozon, P. Kessler and J. Parisi, Phys. Lett. 98B (1981) 105.
30. R. Field, these proceedings.
31. F. Halzen and D.M. Scott, Phys. Rev. D24 (1981) 2433.
32. J. Biel et al., preprint MSU-HEP 82/003.
33. C. Carimalo, M. Crozon, P. Kessler and J. Parisi, College de France preprint LPC 83-01, and in preparation.
34. J.C. Pati and Abdus Salam, Phys. Rev. D8 (1973) 1240; Phys. Rev. D10 (1974) 275.
35. A. Janah, Irvine preprint.
36. J. Bagger and J. Gunion, Davis preprint UCD 83/1.

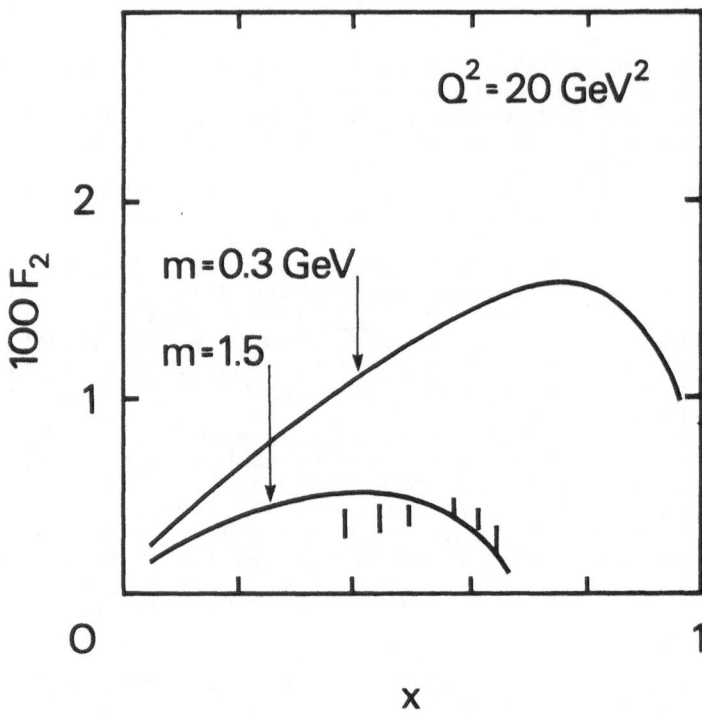

Fig.1   Contributions to F_2^γ . Solid lines form box diagram, vertical bars include $0(\alpha_s)$ corrections.

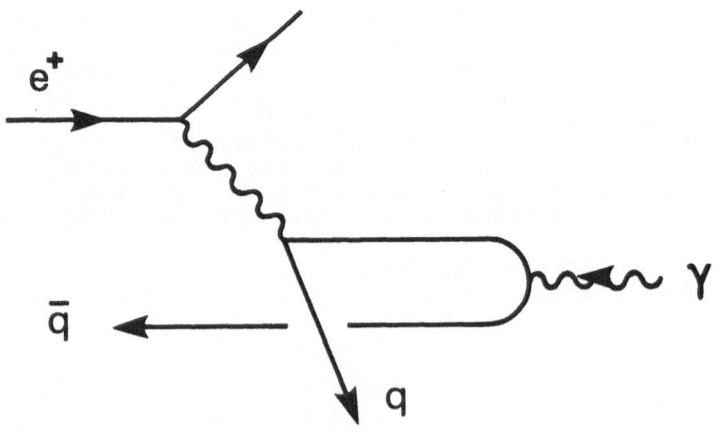

Fig.2   Singular configuration in deep inelastic scattering on a real photon.

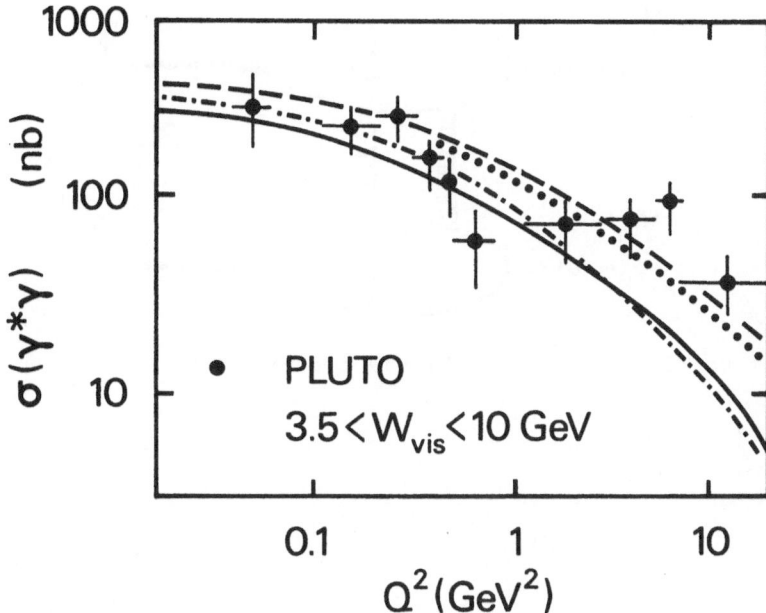

Fig.3   GVDM[3,4,5] versus PLUTO[6] .

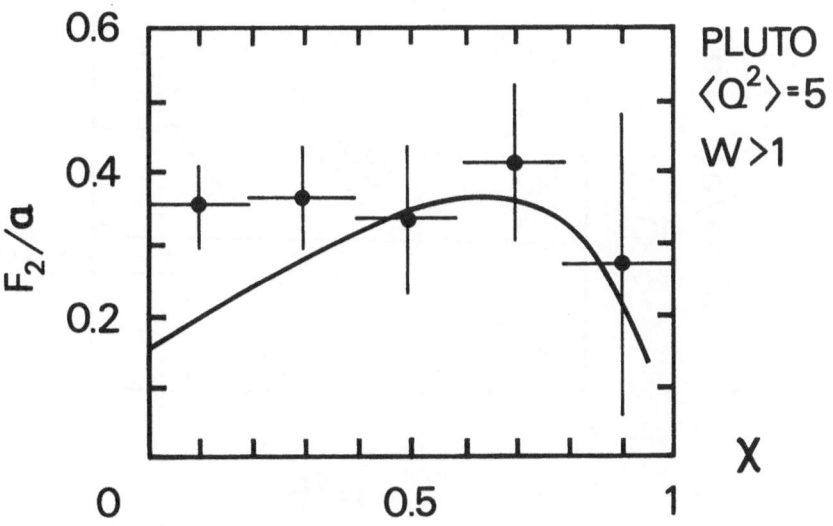

Fig.4   Gotsman and Maor[5] versus PLUTO[6] .

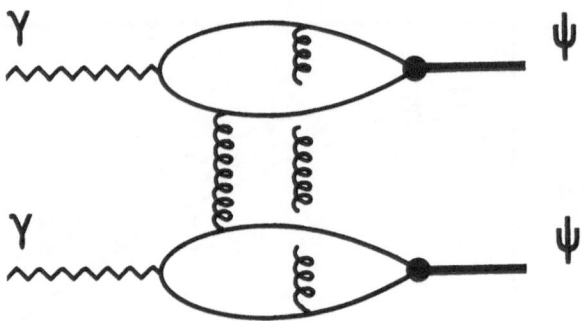

Fig.5   Double scattering diagram for $\gamma\gamma \rightarrow \psi\psi$.

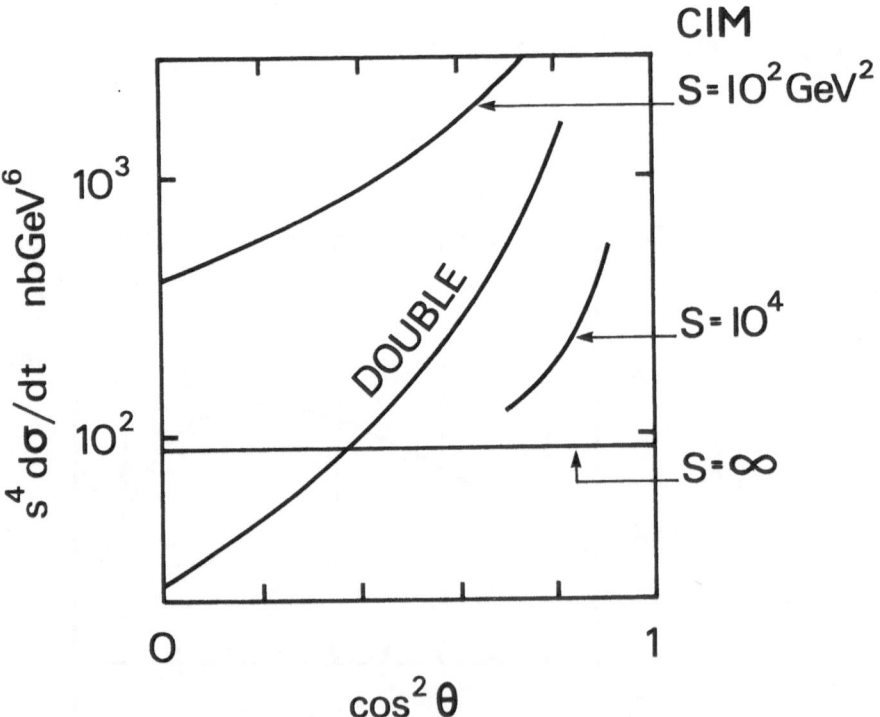

Fig.6   Calculations of $\gamma\gamma \rightarrow \psi\psi$.

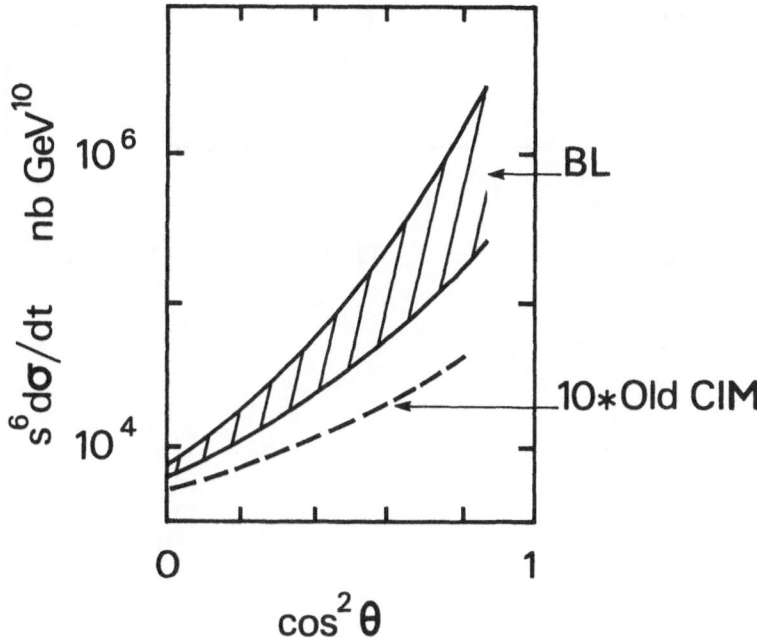

Fig.7   Calculations of $\gamma\gamma \rightarrow p\bar{p}$.

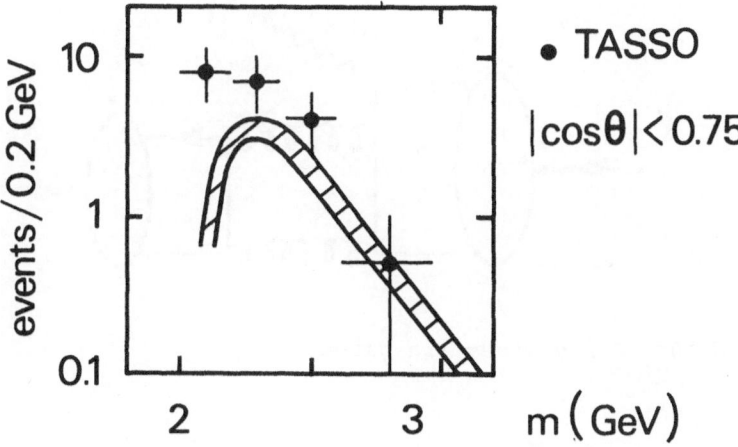

Fig.8   Damgaard[12] versus TASSO[15], in $\gamma\gamma \rightarrow \bar{p}p$.

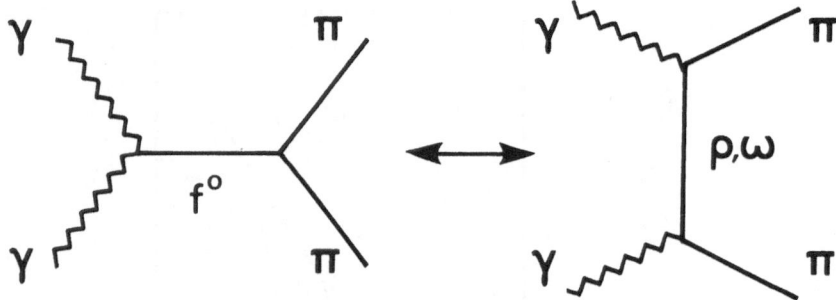

Fig.9

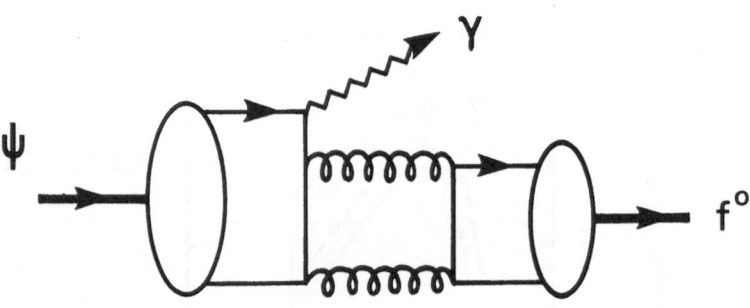

Fig.10   Curly lines are gluons.

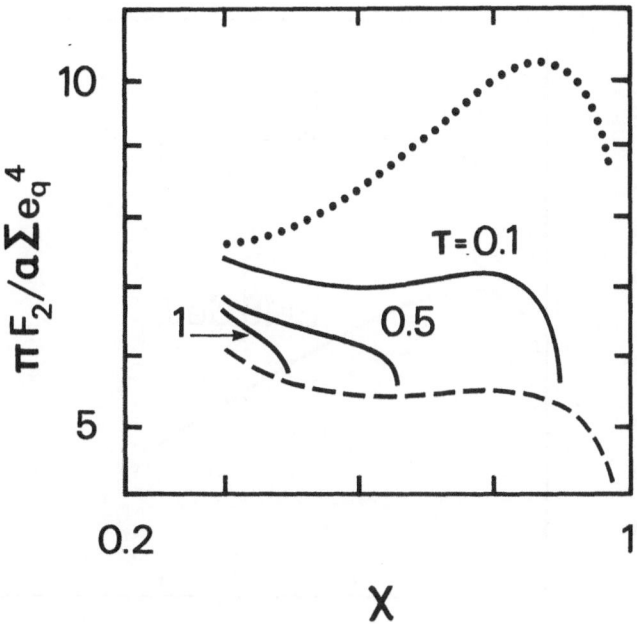

Fig.11 F_2^γ from SUSY-QCD[26].

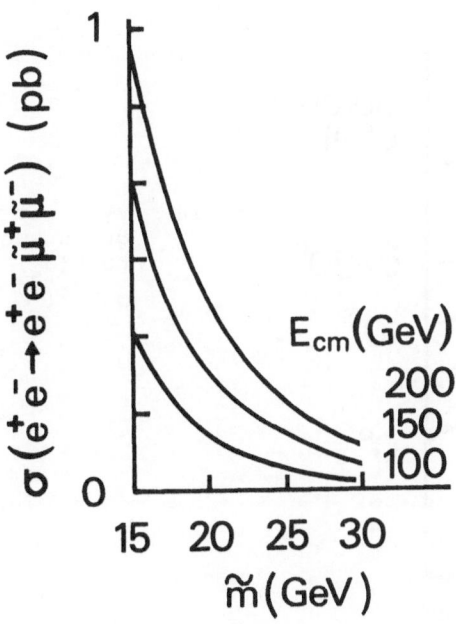

Fig.12 Smuon production cross sections[22].

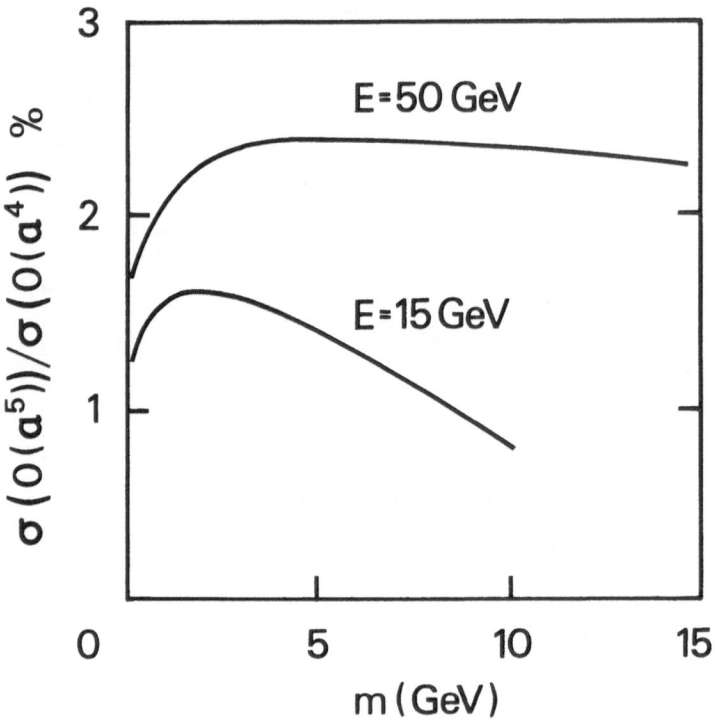

Fig.13   Corrections to $e^+e^- \to e^+e^-$ pseudoscalar (of mass m).

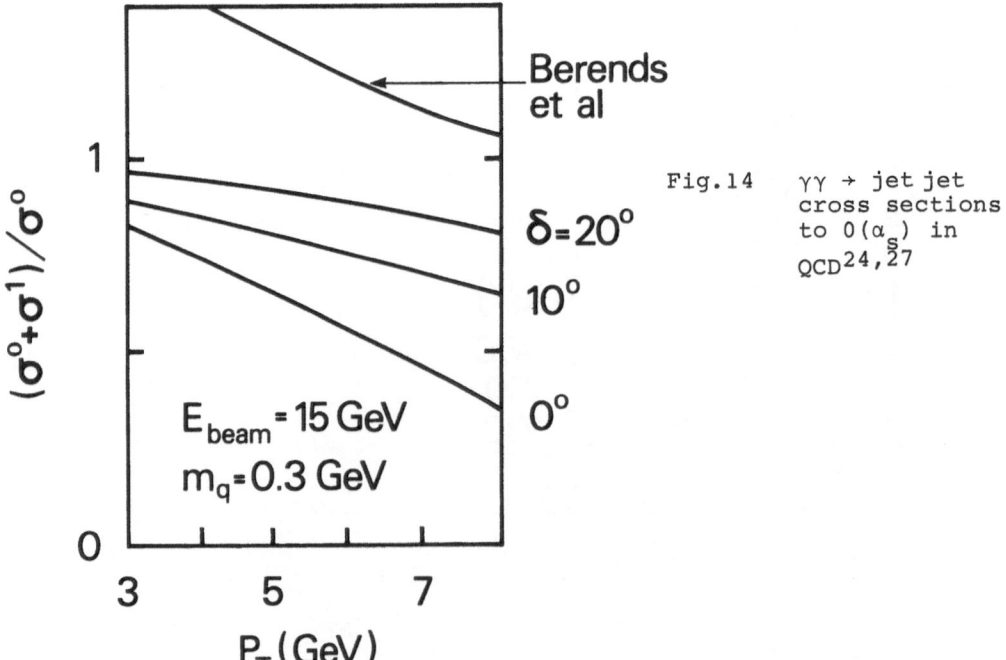

Fig.14 $\gamma\gamma \to$ jet jet cross sections to $0(\alpha_s)$ in QCD[24,27]

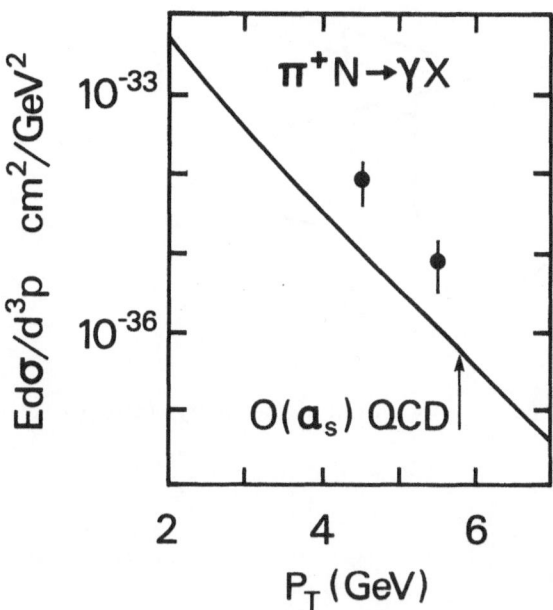

Fig.15 QCD calculation of $\pi^+N \rightarrow \gamma\chi$ versus data[32]. p_L = 200 GeV, $\theta = 90^\circ$.

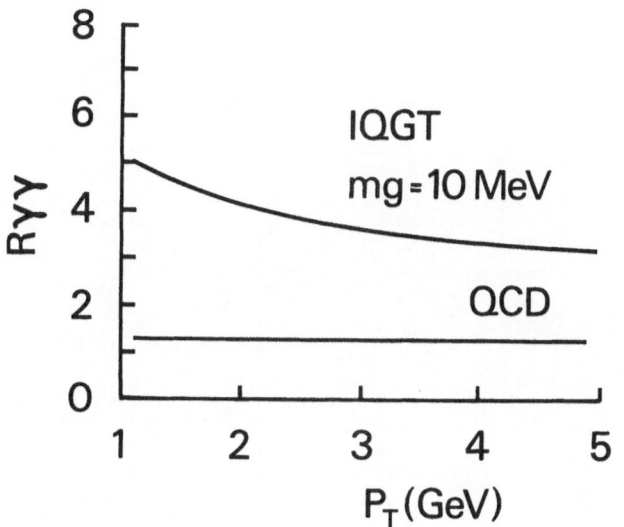

Fig.16 $R_{\gamma\gamma}$ from eq.(8) from QCD and integer charge quark model[35].

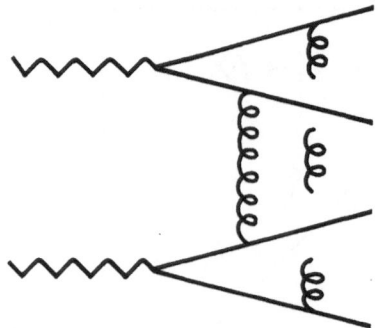

Fig.17   0(α_s) correction to $\gamma\gamma \rightarrow$ 4 jets. The wavy (curly) lines are
photons (gluons).

TWO-PHOTON PHYSICS, 1983: SUMMARY

William R. Frazer
University of California, San Diego
La Jolla, California 92093

The kinematics of the reaction which we call "two-photon physics" is shown in Fig. 1. At least one of the photons is very nearly real ($P^2 \approx 0$) in all the data

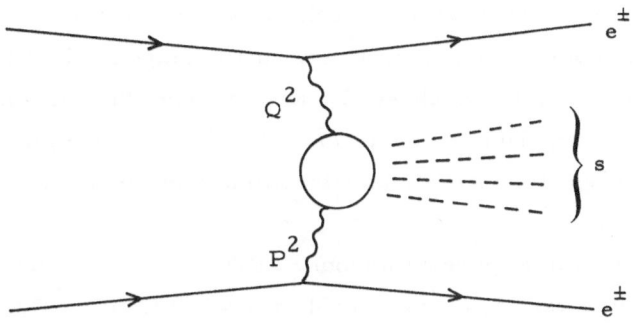

Fig. 1. Kinematics of two-photon reactions.

data presently available, except for some very interesting observations of "double-tagged" events by the PLUTO detector, which I shall discuss later. The organization of this paper is as follows:

Q^2	Final State	Theory
	leptons	QED
	hadrons:	
≈ 0	soft exclusive: resonances	QCD non-perturbative; resonance phenomenology
	hard exclusive	QCD perturbative (Brodsky-Lepage)
	jets: high p_T	QCD perturbative
large	inclusive: photon structure	QCD perturbative

Most of the current theoretical interest in two-photon physics is as a testing-ground of perturbative quantum chromodynamics (QCD) — a uniquely fruitful testing ground, where so far QCD is passing all the tests. But before we turn to QCD tests, we should make sure that QED tests are still in order. In a sense we are calibrating our "apparatus"; that is, we are checking to see that the leptonic content of the photon is as expected before proceeding to test the hadronic (quark) content. [1]

Tests of QED in $e^+e^- \rightarrow e^+e^-\ell^+\ell^-$

So far $\ell = e, \mu$; the τ rate is too low for measurement in the tests carried out to date. A very brief summary of all the results presented is that so far there have been no surprises. I will present only a very small sample. The first one, the PLUTO "single-tag" data, I have chosen because it shows the only example of a disagreement between the data and the theory — but as we shall see, this is no cause for alarm because the theoretical calculation is incomplete.

In Fig. 2 one sees preliminary results from PLUTO in which one observes $e^\pm\mu^+\mu^-$ in the final state. The variable x is defined as $x = Q^2/(Q^2 + s)$. One sees a very high point at $x \rightarrow 1$. But in fact the theoretical calculation includes only the diagram:

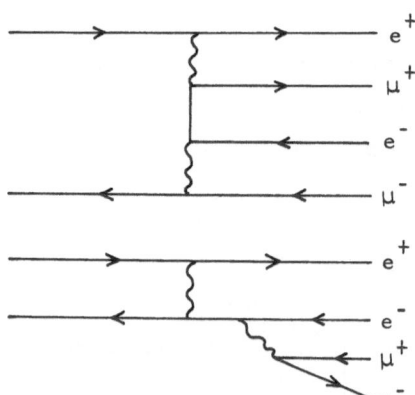

whereas the bremsstrahlung diagram has not been included:

Another interesting example is shown in Fig. 3: the muon contribution to the photon structure, which has been measured by CELLO and also in new data from PEP-9. The solid curve is the theoretical expectation at the value of Q^2 appropriate to the CELLO data. The lower Q^2 data from PEP-9 clearly exhibit that the structure function rises with Q^2 as expected.

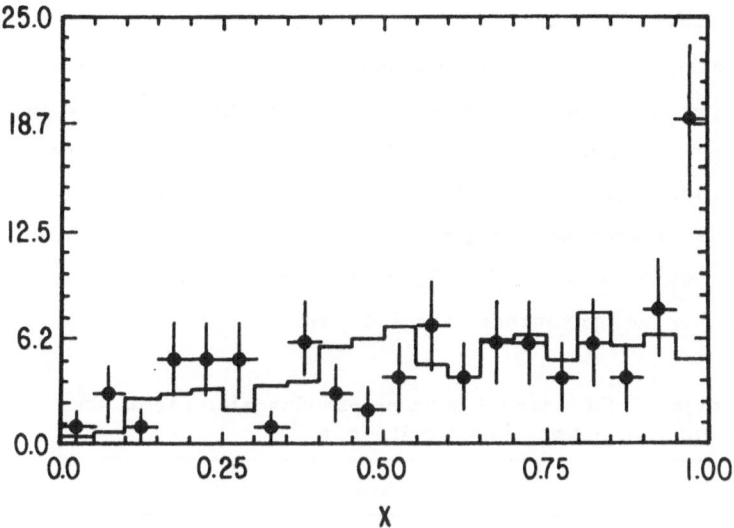

Fig. 2. Preliminary PLUTO data compared with QED prediction (bremsstrahlung diagram shown above not yet included). [1]

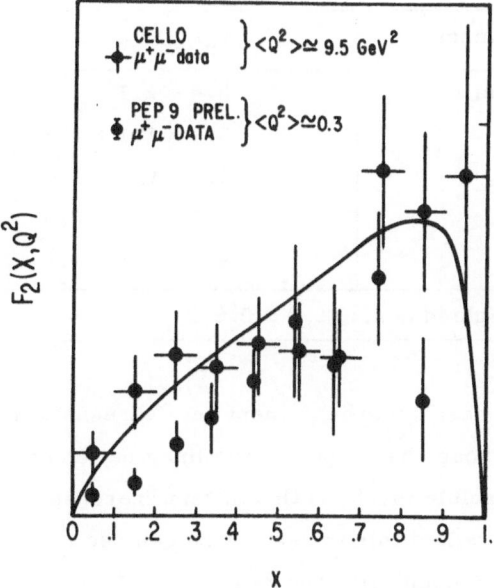

Fig. 3. Muon contribution to the photon structure ($e^+e^- \rightarrow e^+e^-\mu^+\mu^-$) as measured by CELLO at $\langle Q^2 \rangle \approx 9.5$ GeV^2 and by PEP-9 (preliminary) at $\langle Q^2 \rangle \approx 0.3$ GeV^2. The QED prediction is calculated at CELLO value. [1]

Resonance formation in $\gamma\gamma$ collisions [2]

The past two years have seen the emergence of a rich resonance spectroscopy in the reaction $\gamma\gamma \to$ resonance $\to$ hadrons. The resonances observed, with their $\gamma\gamma$ decay widths measured, now include η, η', f, f', and A_2. Searches have been carried out for $\iota(1440)$, $\theta(1640)$, and $\eta_c(2981)$, and upper limits have been given for the $\gamma\gamma$ mode. Also observed in the TASSO detector is an enhancement in $\gamma\gamma \to \pi^+\pi^-\pi^+\pi^-$ at $M = 2.1 \pm 0.01$ GeV, with a width comparable to the 60 MeV solution. The observed resonances are summarized in Tables 1 and 2.

Table 1. Summary of data on pseudoscalar resonance formation in two-photon collisions. [2]

	$\gamma\gamma \to \eta \to \gamma\gamma$	
	Experiment	$\Gamma_{\eta\gamma\gamma} \pm$ stat $\pm$ sys keV
η	Cornell (Primakoff)	0.324 ± 0.046
	Crystal Ball	$0.56 \pm 0.12 \pm 0.09$

	$\gamma\gamma \to \eta \to \pi^+\pi^-\gamma$	
	Experiment	$\Gamma_{\eta'\gamma\gamma} \pm$ stat $\pm$ sys keV
	Binnie et al.	5.4 ± 2.1
	Mark II	$5.8 \pm 1.1 \pm 1.2$
η'	CELLO	$6.2 \pm 1.1 \pm 0.8$
	JADE	$5.0 \pm 0.5 \pm 0.9$
	TASSO	$4.1 \pm 0.4 \pm 1.5$

Weighted avg.: 5.3 ± 0.6

One final note on resonances: an enhancement near threshold in $\gamma\gamma \to \rho^0\rho^0$, observed by the TASSO group, has captured the imagination of many theorists who have speculated about possible identification of this "resonance." New data from JADE on $\gamma\gamma \to \rho^+\rho^-$ have laid this to rest. One can see from Fig. 4 that no single resonance of $I = 0, 1$ or 2 could fit these data.

Table 2. Summary of data on 2^+ resonance formation in two-photon collisions. [2]

	Experiment	$\Gamma_{f\gamma\gamma} \pm$ stat $\pm$ sys keV		Decay mode obs.
f	PLUTO	$2.3 \pm 0.5 \pm 0.35$		$\pi^+\pi^-$
	Mark II	$3.6 \pm 0.3 \pm 0.5$		$\pi^+\pi^-$
	TASSO	$3.2 \pm 0.2 \pm 0.6$		$\pi^+\pi^-$
	CELLO	$2.7 \pm 0.2 \pm 0.2$		$\pi^+\pi^-$
	Crystal Ball	$2.7 \pm 0.2 \pm 0.6$	$\lambda = 2$	$\pi^0\pi^0$
	Crystal Ball	$2.9 \pm 0.6 \pm 0.6$ / 0.4	λ fit	$\pi^0\pi^0$
	JADE	$2.3 \pm 0.2 \pm 0.5$	prel.	$\pi^0\pi^0$

Weighted avg.: 2.8 ± 0.2 keV

	Experiment	$\Gamma_{A_2\gamma\gamma} \pm$ stat $\pm$ sys keV		Decay mode obs.
A_2	Crystal Ball	$0.77 \pm 0.18 \pm 0.27$		$\eta\pi^0$
	CELLO	$0.81 \pm 0.19 \pm 0.27$		$\rho^\pm\pi^\mp$
	JADE	$0.84 \pm 0,07 \pm 0.15$	prel.	$\rho^\pm\pi^\mp$

Weighted avg.: 0.82 ± 0.13 keV

f′   TASSO $\Gamma_{f'\gamma\gamma} \cdot BR(f' \to K\overline{K}) = 0.11 \pm 0.02 \pm 0.04$ keV

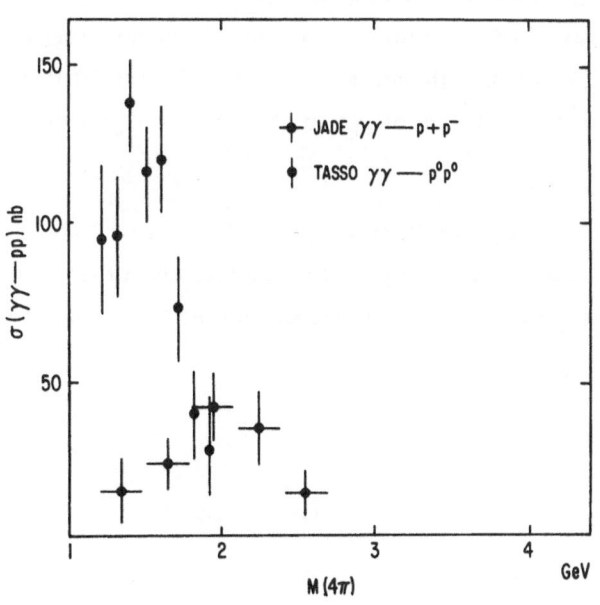

Fig. 4. TASSO data on $\gamma\gamma \to \rho^0\rho^0$, compared to new JADE data on $\gamma\gamma \to \rho^+\rho^-$.

Hard exclusive final states - Brodsky-Lepage analysis

In 1978 Brodsky and Lepage initiated a highly original development in QCD perturbation theory: the application of that theory to exclusive reactions. [3, 4] They observed that in a "hard" exclusive process, such as $\gamma\gamma \to \pi\pi$ at high energy and large angle, one can analyze the amplitude into a hard scattering subprocess folded into wave functions. The wave functions, which describe the "hadronization" of the quarks produced in the hard process, are not calculable in perturbative QCD, but can often be normalized to an observed decay. This situation is quite analogous to the classical application of QCD to deep inelastic hadron scattering, in which one factor in the result is an unknown hadronic matrix element. For example, $\gamma\gamma \to \pi\pi$ is described as follows:

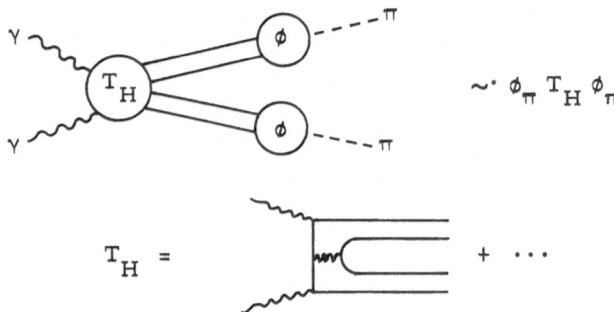

At this conference we have seen the first detailed confrontation of the Brodsky-Lepage analysis with data. [5] Although the data are in such a low energy range that the authors of the theory might not have had the temerity to apply it, nevertheless the results are very encouraging. The first of these, shown in Fig. 5, are PLUTO data on $\gamma\gamma \to h\bar{h}$, where the prediction is the sum of $\gamma\gamma \to \pi\pi$, $K\bar{K}$, and $p\bar{p}$ predictions.

The second example, shown in Fig. 6, compares the theory, worked out by Damgaard, [6] with the experimental data on $\gamma\gamma \to p\bar{p}$. The uncertainty in the theoretical curve represents various choices for the distribution function of three quarks in a proton.

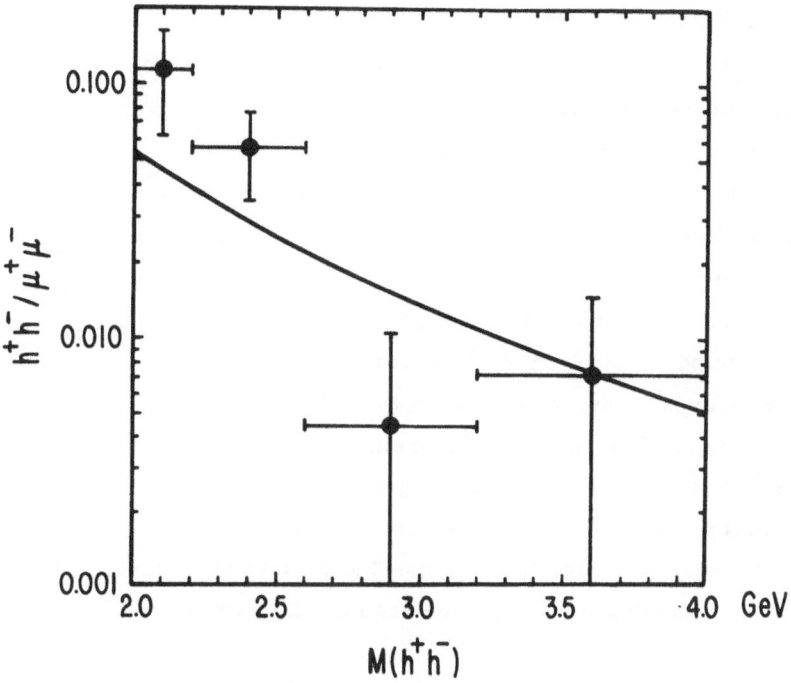

Fig. 5. Comparison of preliminary PLUTO data on $\gamma\gamma \to h\bar{h}$ with Brodsky-Lepage QCD prediction. [5)]

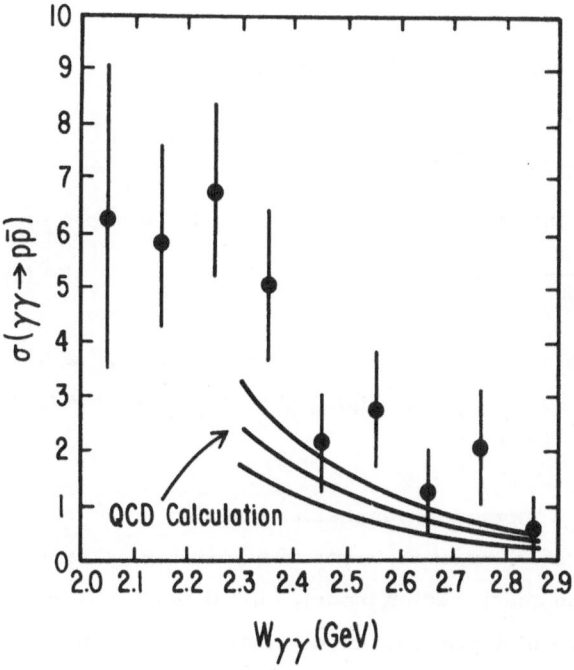

Fig. 6. Comparison of cross section for $\gamma\gamma \to p\bar{p}$ with QCD prediction.

High p_T jets in two-photon reactions

One of the first predictions of QCD to show
the unique character of two-photon physics was
the prediction of events in which only two large-
angle jets are produced, with nothing going forward or backward. The very simple production
mechanism is shown in Fig. 7. Evidence for
this process, at the predicted rate, is now
very convincing.[7] See Fig. 8, in which the
characteristic p_T^{-4} tail is clearly shown in
the TASSO data. The newer, still-prelimi-
nary PLUTO data show remarkable agree-
ment with the $\gamma\gamma \to q\bar{q}$ prediction correspond-

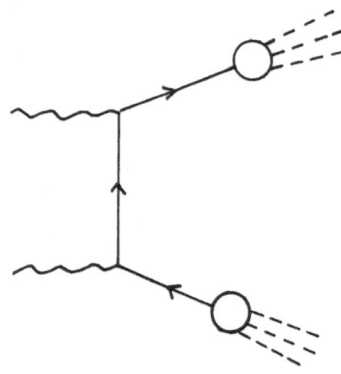

Fig. 7. $\gamma\gamma \to$ 2 jets.

ing to the diagram of Fig. 7. Note that the contribution of Fig. 7 relative to background has been suppressed by choosing high-Q^2 events.

The magnitude of the two-jet process is given simply by

$$R_{\gamma\gamma} = \frac{d\sigma/dt\,(\gamma\gamma \to q\bar{q} \to 2\text{ jets})}{d\sigma/dt\,(\gamma\gamma \to \mu\bar{\mu})}$$

$$= \Sigma\, q_i^4 \tag{1}$$

The PLUTO group were able to measure a different quantity,

$$\tilde{R}_{\gamma\gamma} = \frac{d\sigma/dt\,(\gamma\gamma \to \text{jet} + x)}{d\sigma/dt\,(\gamma\gamma \to \mu\bar{\mu})} \tag{2}$$

At large $x_T = 2\,p_T/\sqrt{s}$ one expects the two-jet process to dominate, so

$$\tilde{R}_{\gamma\gamma} \xrightarrow[x_T \to 1]{} R_{\gamma\gamma}\,. \tag{3}$$

In Fig. 9 the PLUTO results are shown.[7] Again, the selection of high-Q^2 suppresses background and leads to good agreement at high p_T of $R_{\gamma\gamma}$ with the QCD
prediction.

Before moving on, it is worth taking a moment to emphasize that the $\gamma\gamma \to$ jet prediction is a completely clean test of QCD, which it seems to be passing. One might
object that this prediction is "only the quark-parton model," but without QCD one
would not have the systematic theory which predicts the dominance of the one

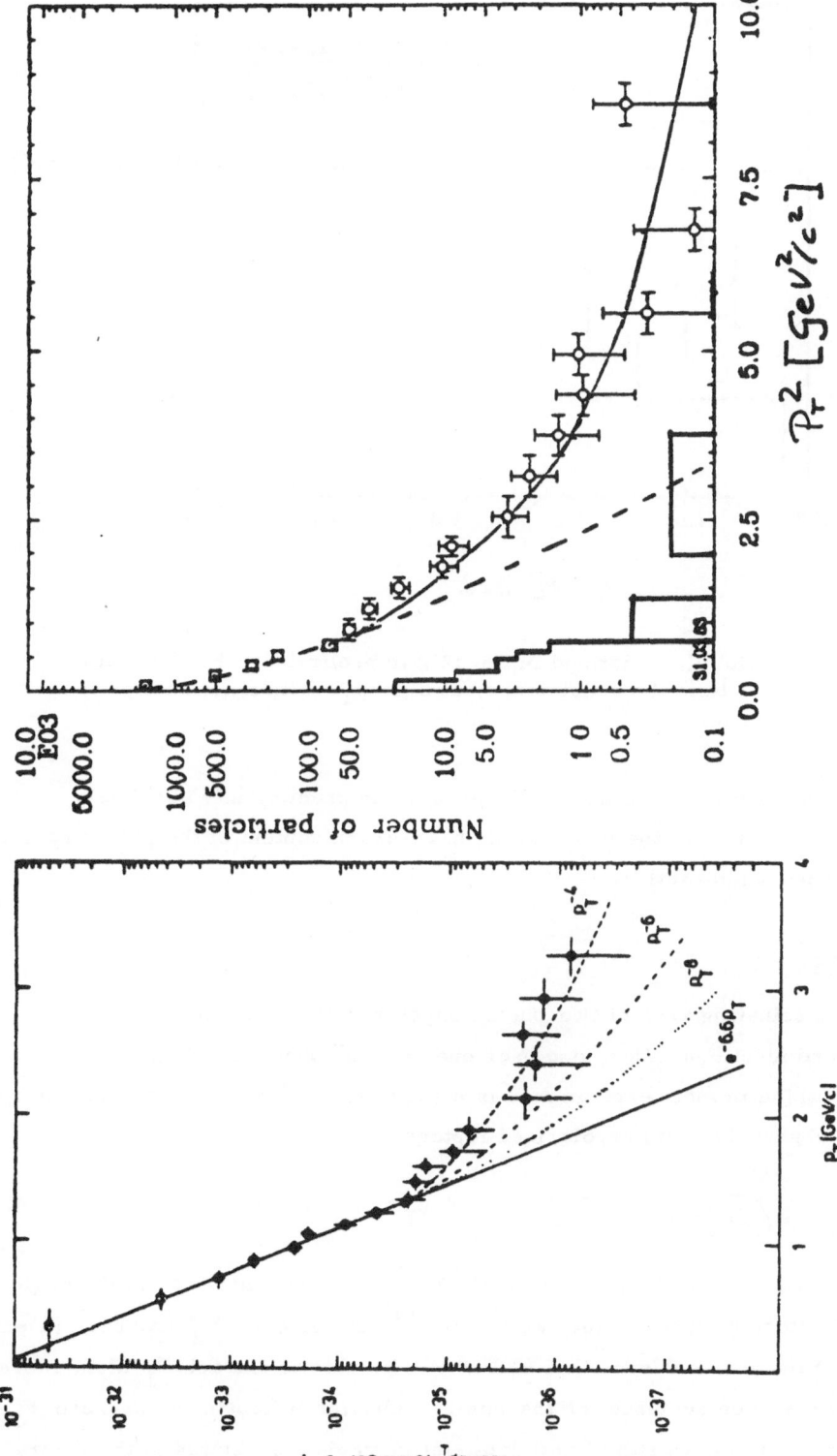

Fig. 8. Data on $e^+e^- \to e^+e^- + h^\pm + x$, showing evidence for the two-jet process diagrammed in Fig. 7. a) TASSO data show p_T^{-4} tail, and b) preliminary PLUTO data at high Q^2 show good agreement with calculation based on diagram of Fig. 7.7)

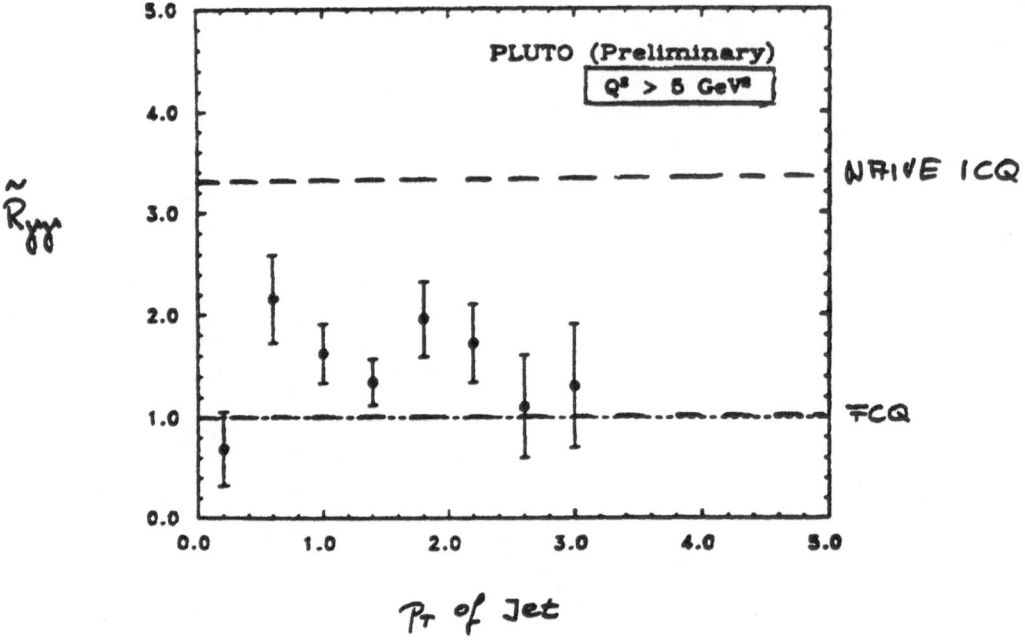

Fig. 9. Ratio $\tilde{R}_{\gamma\gamma}$, defined in Eq. (2), in preliminary PLUTO data at high Q^2. Note the consistency, at high p_T, with fractionally charged quarks.

diagram we have used to fit the data. Moreover, the prediction is absolutely normalized, a reflection of the fact that the hard-quark content of the photon results from a simple point interaction.

Photon structure

Another very fascinating area of two-photon physics is the determination of the photon structure functions. Here one uses one high-Q^2 photon as a probe of the other photon, which in most experiments is nearly real $(P^2 \approx 0)$. Three structure functions are observable with unpolarized leptons,

$$\frac{d\sigma}{dE_1'\, dE_2'\, d\Omega} = K(F_1 + \epsilon_2 F_L + \epsilon_1 \epsilon_2 F_3 \cos 2\phi) \qquad (4)$$

where K, ϵ_1, and ϵ_2 are known kinematical factors, and where ϕ is the angle between the scattering planes of the two leptons.[8] So far only F_2 has been investigated, but we have seen evidence that PEP-9 will be able to extract F_L also. Moreover, PLUTO has seen evidence for the $\cos 2\phi$ behavior indicating a non-zero F_3. But before looking at the wealth of new data, let us review the status of the theory in light of some new developments.

The calculation of F_2^γ by perturbative QCD yields a result for the moments of the structure function which is of the form[9]

$$F_2^\gamma(n, Q^2) = \frac{1}{\alpha_s(Q^2)} a_n^{(0)} + a_n^{(1)} + \sum_{m=2}^{\infty} \alpha_s^{m-1} a_n^{(m)}$$

$$+ \sum_i \sum_{\ell=0}^{\infty} h_{i,\ell}(n) \alpha_s^{d_i^n + \ell} \qquad (5)$$

The first three terms are the point-like part, which is calculable (in principle) in perturbation theory. The first two have, in fact, been calculated, by Witten[10] and by Bardeen and Buras.[11] The last term, the hadron-like part, is characterized by coefficients $h_{i,\ell}(n)$ which are not calculable in perturbative QCD. Since the d_i^n are non-negative, the hadron-like term can be neglected at sufficiently high Q^2 and an absolute prediction for F_2^γ results.

At currently-available values of Q^2 the hadron-like part is probably not negligible, at least at small x. The magnitude of this term has traditionally been estimated by a vector dominance model. The validity of this estimate — in fact, the validity of the separation of point-like and hadron-like terms at finite Q^2 — has been called into question by recent work by G. Rossi.[12]

The difficulty concerns singularities at $x = 0$. These were already present in the Witten result,[10]

$$a^{(0)}(x) \sim x^{-1.60} \qquad (6)$$

which gives rise to a small spike near $x = 0$. In the next order one finds

$$a^{(1)}(x) \sim x^{-2} \qquad (7)$$

Duke and Owens observed that the singularity results in a negative structure function at small x.[13] Nevertheless, it was possible not to be too concerned about this problem, since we knew that the VDM contribution already makes the result uncertain at small x. Rossi, however, has pointed out that the singularity becomes increasingly worse the higher the order,

$$a^{(2)}(x) \sim x^{-5.33} \qquad (8)$$

$$a_n^{(m)} \sim \frac{1}{d^n + 1 - m}. \qquad (9)$$

It is hard to ignore a singularity as severe as $x^{-5.33}$ by hoping that its effect will be confined to small x!

Now for some good news: there really is no singularity in F_2^γ after all! The reader must have suspected that we were leaving out some term which would cancel the singularity. In fact, there must be singular terms in the hadron-like part which cancel the singularities we have found in the point-like part. This can be seen explicitly for the case of a massive target photon, $P^2 \gg \Lambda^2$, which has been investigated by Uematsu and Walsh[14] and by Rossi.[12] The cancellation is of the form

$$\frac{1}{D_n} \left[1 - \left(\frac{\alpha_s(Q^2)}{\alpha_s(P^2)} \right)^{D_n} \right] \quad , \tag{10}$$

which in the limit $D_n \to 0$ goes harmlessly to $\ell n[\alpha_s(P^2)/\alpha_s(Q^2)]$. Note that the cancellation involves the point-like term 1 and the hadron-like term $[\alpha_s(Q^2)/\alpha_s(P^2)]^{D_n}$. We call this "hadron-like" because it has $\ell n\, Q^2$ dependence characterized by an anomalous dimension, just like hadronic structure functions. So the singularity of the point-like term is completely cancelled, and there is no problem — for a massive target photon.

On the other hand, at $P^2 \approx 0$ we know that a similar cancellation must occur, but it involves the incalculable hadronic term. If we subtract off the singularity we are left with a finite ambiguity. The region over which the ambiguity could be sizeable can be estimated by noting the region over which the original singularity was sizeable. In the absence of higher-order calculations we can only speculate, but doubt has been cast on the reliability of our predictions even at sizeable values of x.

What has happened, of course, is not a breakdown of the QCD perturbation expansion. In fact, the singularity has been seen to be spurious. What has happened, however, is that the point-like and hadron-like parts have been inextricably mixed, whereas we relied on separating them if we were to have an absolute prediction. Are we now in the position of retracting our earlier optimistic statements about the unique calculability of F_2^γ and admitting that it is no more calculable than a hadronic structure function?

Before turning to that question, let me first point out that we should not have found these singularities surprising. After all, we learn in beginning field theory that a typical diagram in the ladder which makes up the Witten result,

in frared singular. To cancel this singularity one must add the diagrams

But the first of these is of the type we associate with vector meson dominance,

Therefore we should have realized that the VMD terms must contain singularities needed to cancel those of the point-like terms. The old model of F_2^γ = point-like + VMD, where VMD is non-singular, is nonsense! This does not mean that the VMD contribution is not present, but that we must try to construct more sophisticated models with the right singularity structure, if they are to be useful. I don't know whether this will be possible, but it would be interesting to try.

Now to return to the question of how seriously to worry about this development. Let me organize the argument in a dialectical form:

Optimist: We have shown that the $x = 0$ singularities are absent. We now have no reason to believe the series will not converge well, and the residual uncertainties will be small.

Pessimist: The effect of a singularity like $x^{-5.33}$ is not confined to small x! The residual ambiguity after subtraction adds an arbitrary function of x, and F_2^γ is no more calculable than F_2^h !

Optimist: But you must admit that we have no reason to doubt the QCD perturbation expansion of F_2^γ. For Q^2 sufficiently large, F_2^γ is given by our calculable point-like terms.

Pessimist: Yes, for some Q^2 — but at today's Q^2?

Experimentalist: Enough! Let's measure it! You theorists may be confused, but by careful measurement of $F_2^\gamma(x, Q^2)$ we can establish whether the point-like term dominates and we can then go ahead to determine Λ.

In fact, a wealth of new data have become available at this conference and in the two years since the previous two-photon conference.[15] Figures 10-13 give a sample. Note that although present data are of the same shape and magnitude as the QCD prediction, they are not yet sufficiently precise to distinguish easily between the parton model, low-order QCD, or higher-order QCD.

In Fig. 14 is shown the first evidence of detection of the third structure function F_3, which is predicted to be given in leading order by the simple box diagram.

In Fig. 15 is shown the first measurement with a massive target photon, with an average value of $\langle P^2 \rangle = 0.4$ GeV. Although this is a bit low to be comfortable with the Uematsu-Walsh theory, its prediction is shown and is in good agreement! This will be another impressive test of QCD, although again it is dominated by the simple box diagram with no gluons.

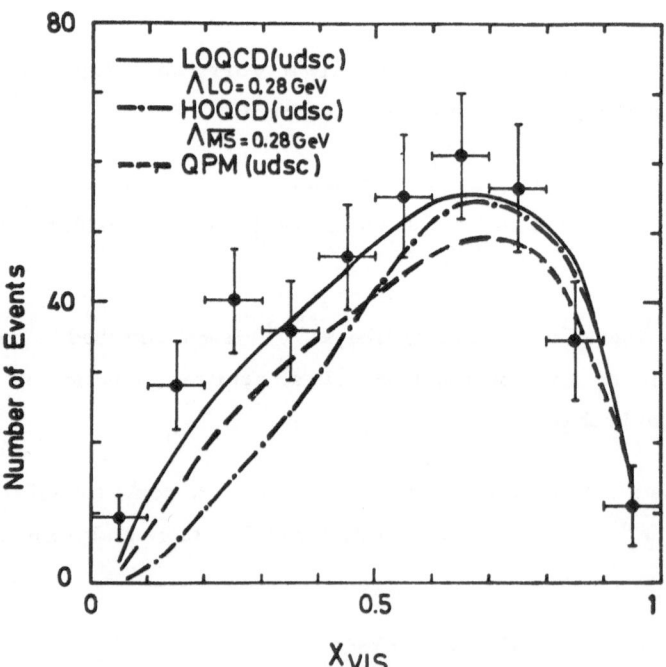

Figure 10. Preliminary JADE data on the photon structure at $\langle Q^2 \rangle =$ 24 GeV^2; compared to lowest-order QCD, QCD including first two orders, and the parton model.

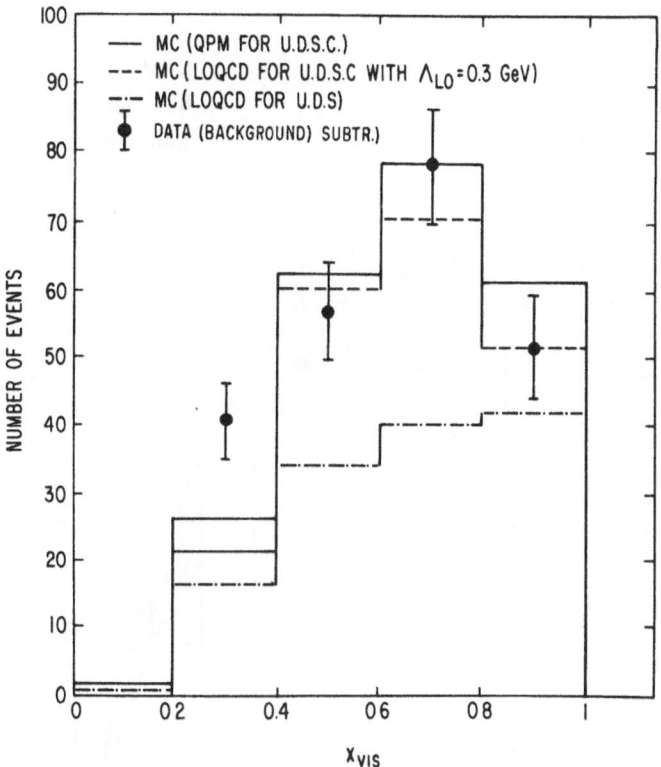

Fig. 11. Preliminary TASSO data on the photon structure at $\langle Q^2 \rangle =$ 23 GeV^2, compared to the parton model, and to lowest-order QCD with and without charmed quarks included.

Finally, in Fig. 16 we see a very interesting summary in which all the structure function results are integrated over x (caution - different ranges) to show the rise of $\int dx\, F_2(Q^2, x)$ with Q^2.

This is a good place to conclude, with an emphasis on the wonderful accumulation of new experimental results in the past two years. Two-photon physics is living up to expectations as the definitive testing ground of QCD!

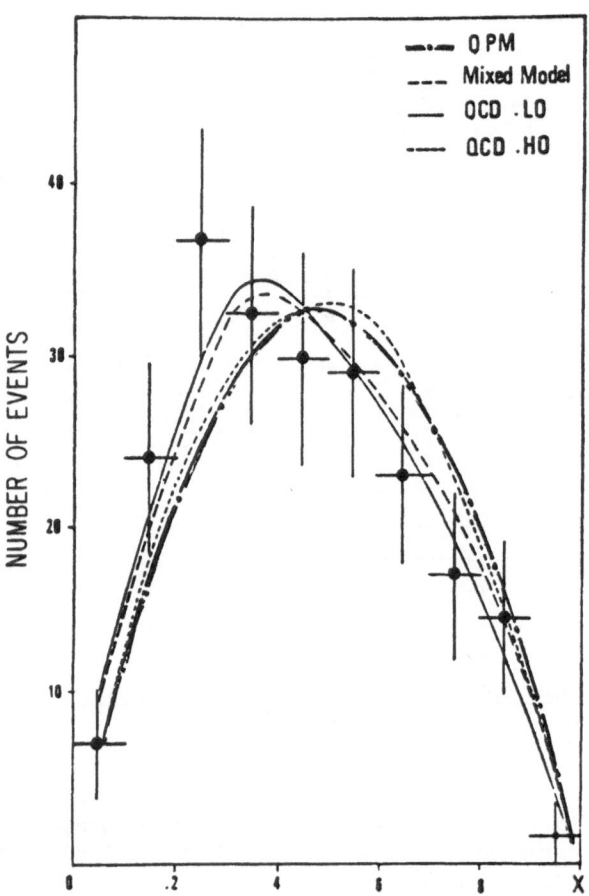

Fig. 12. CELLO data on photon structure at $\langle Q^2 \rangle = 9 \, \mathrm{GeV}^2$, compared to parton model, low- and higher-order QCD.

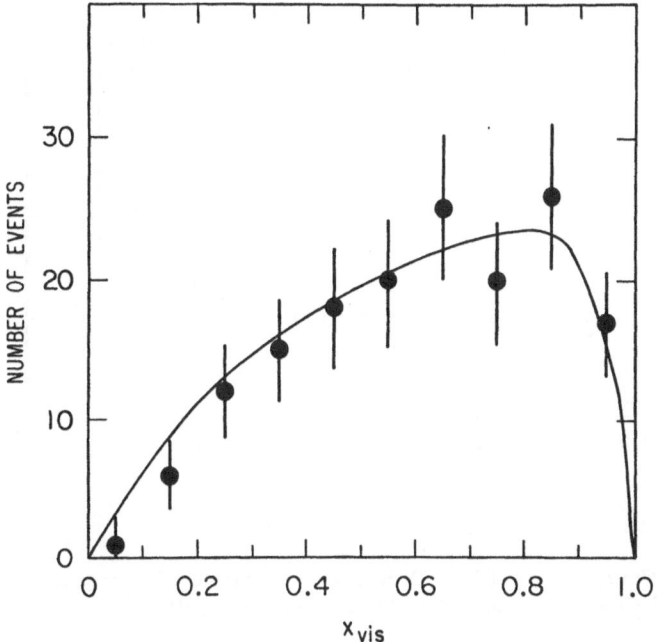

Fig. 13. Preliminary MAC data on photon structure at $\langle Q^2 \rangle$ = 32 GeV^2 , compared to parton model prediction.

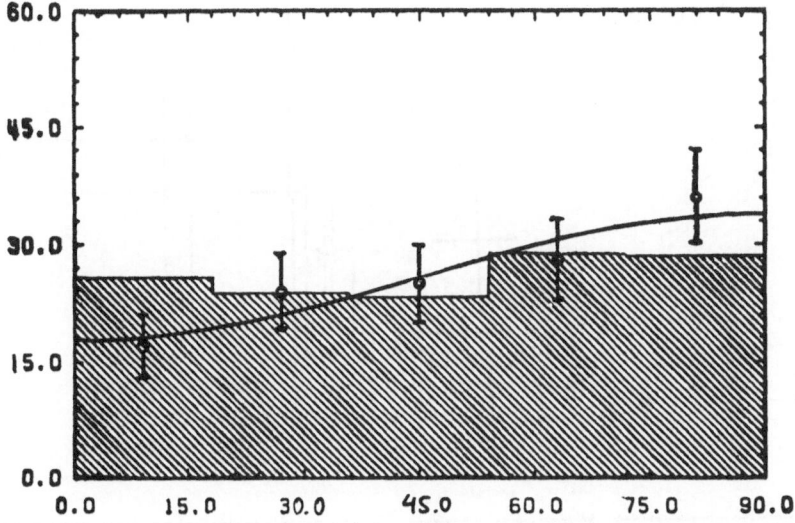

Fig. 14. Preliminary PLUTO data[15] on the cos 2ϕ dependence (see Eq. (4)) in $e^+e^- \rightarrow e^+e^- + x$. The shaded area is the normalized QCD prediction. Result shows $F_3 < 0$ at the 3σ level.

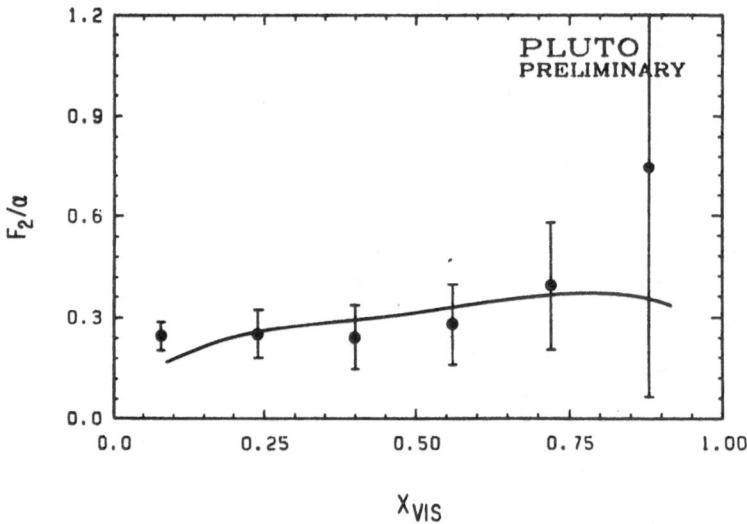

Fig. 15. Preliminary PLUTO data on structure of a massive photon $\langle P^2 \rangle = 0.4\,\text{GeV}^2$, compared to QCD prediction.[16]

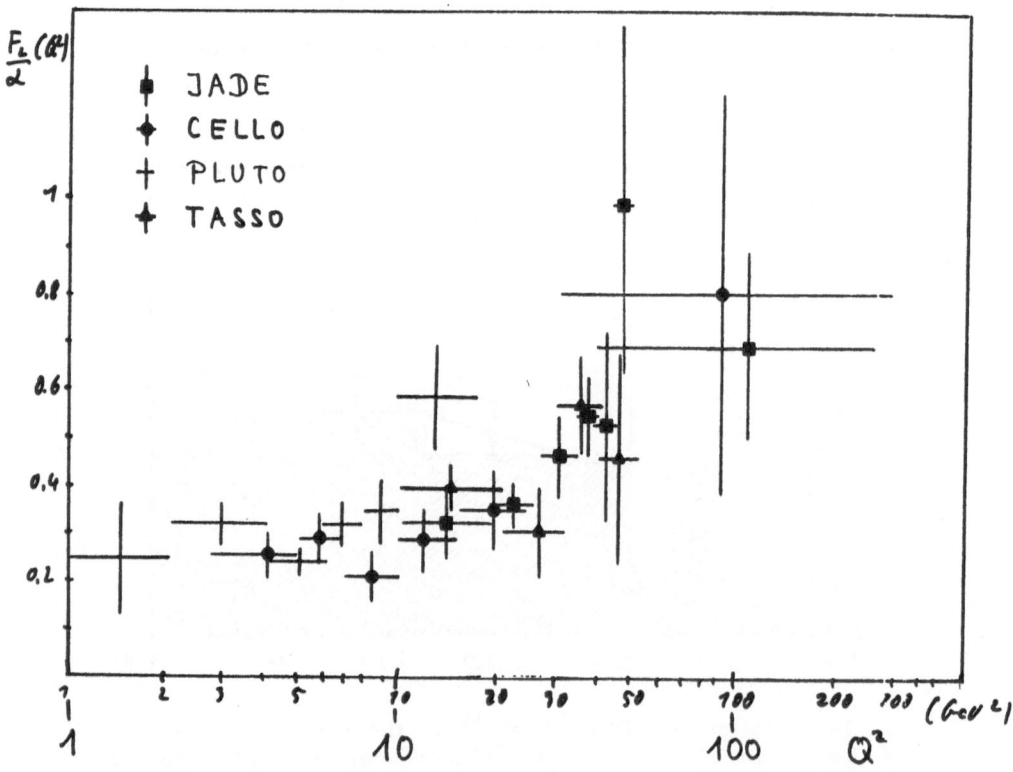

Fig. 16. Summary of all data on $F_2(Q^2)$, integrated over x.

References

1) For a more detailed discussion and a list of references, see the talk by M. Pohl in these proceedings.

2) For more detailed discussion and a list of references, see the talk by J. Olsson in these proceedings.

3) G. P. Lepage and S. J. Brodsky, Phys. Rev. $\underline{D22}$, 2157 (1980), and Phys. Lett. $\underline{87B}$, 359 (1979); S. J. Brodsky and G. P. Lepage, SLAC-PUB-2294, published in "Quantum Chromodynamics," Proceedings of the 1978 La Jolla Institute Summer Workshop, Wm. Frazer and F. Henyey (eds.), AIP (1979), Phys. Rev. Lett. 43, 545 (1979), Erratum, ibid. $\underline{43}$, 1625 (1979), and SLAC-133, Proceedings of the SLAC Summer Institute, 1979; S. J. Brodsky, Y. Frishman, G. P. Lepage and C. Sachrajda, Phys. Lett. $\underline{91B}$, 239 (1980).

4) W. Stirling, these proceedings.

5) H. Kolanowski, these proceedings.

6) P. H. Damgaard, Nucl. Phys. B211, 435-46 (1983).

7) W. Wermes, these proceedings.

8) R. P. Worden, Phys. Lett. $\underline{51B}$, 57 (1974); W. Frazer and G. Rossi, Phys. Rev. $\underline{D21}$, 2710 (1980).

9) For a more detailed discussion and references, see D. Duke, these proceedings.

10) E. Witten, Nucl. Phys. $\underline{B120}$, 189 (1977).

11) W. A. Bardeen and A. J. Buras, Phys. Rev. $\underline{D20}$, 166 (1979).

12) Giuseppe Rossi, Lawrence Berkeley Laboratory preprint LBL-15912; University of California, San Diego, preprint UCSD-10P10-227.

13) D. W. Duke and J. F. Owens, Phys. Rev. $\underline{D22}$, 2280 (1980).

14) T. Uematsu and T. F. Walsh, Nucl. Phys. $\underline{B199}$, 83 (1982).

15) W. Wagner, these proceedings.

16) The "QCD prediction" shown in Fig. 15 is preliminary, since the mass of the charmed quark has not been taken into account.[15]

DISCUSSION SESSION

S. Brodsky, SLAC

As chairman of this concluding session, I also want to voice my thanks to Ch. Berger and his colleagues at Aachen for an outstanding and unusually productive workshop. I think all of us are very impressed with the remarkable experimental achievements in two-photon physics during the past two years. Photon-photon collisions, especially at PETRA, has now developed into a sophisticated and essential element of hadronic dynamics. From the point of view of QCD, the two photon exclusive amplitudes allow an accessible and unique insight into coherent, perturbative and non-perturbative phenomena over a natural kinematic range important for testing QCD. Similarly, the $\gamma\gamma \to$ jet and photon structure function inclusive measurements are providing severe tests of basic QCD predictions in the simplest possible channels. It is clear that the detailed experimental work is now just beginning, which suggests to me that new experiments, designed specifically to maximize two-photon physics would certainly be justified.

1. EXCLUSIVE PROCESSES (except resonances)

1.1 Analysis of $e^+e^- \rightarrow e^+e^- \mu^+\mu^-$. First results from PEP-9

C.WILLIAMS(PEP-9)

Data on muon pair production in photon-photon collisions were taken with the detector PEP-9 using also' the Berkeley-TPC (PEP-4). A preliminary analysis was carried out in terms of structure functions F_2 and xF_1, which were determined seperately.

1.2 Two-Gamma Exclusive Production of Muon and Hadron Pairs

R.KELLOGG (PLUTO)

We have measured the exclusive production of muon and charged hadron pairs in two-photon interactions at large angles ($|\cos\Theta_{LAB}| < 0.6$) and at large invariant mass ($W > 2.0$ GeV). The muon signal is found to be in agreement with QED, while the hadron signal is only a small fraction of what would be expected for pointlike particles. The exclusive hadron pair production has been calculated according to a QCD model in the large t limit. We compare our results to these predictions.

1.3 Exclusive Proton-Antiproton Production in Two Photon Collisions

H. KUECK (TASSO)

The reaction $\gamma\gamma \to p\bar{p}$ has been measured with the TASSO detector at the e^+e^- storage
ring PETRA. Data have been analyzed for an integrated luminosity of 74 pb^{-1} at
beam energies of about 17 GeV. Proton-antiproton pairs were identified using the
information of the time of flight counters of the TASSO inner detector. 72 events
of the type $e^+e^- \to e^+e^- + p\bar{p}$ with a background of less than 6 events have been
found. These events cover an invariant mass range between 2.0 and 2.9 GeV and a
c.m. angular range $\cos(\Theta*) \lesssim 0.7$. Within this kinematical region the differential cross section shows a flat angular behaviour (see H.Kolanoski, this conference).
The data points are roughly one order of magnitude smaller than expected from a
QED Born term calculation. The extrapolation of a recent QCD calculation (see
J.Stirling, this conference) to the low mass range of our data yields differential
cross sections of roughly the same order of magnitude as our data. Assuming a flat
angular behaviour the total cross section is about 6 nb above 2.0 GeV and drops
to about 1 nb at 2.9 GeV.

1.4 Analytic Distributions of Various Parameters in ee $\to$ eex^+x^- under Standard
Experimental Conditions

A.COURAU (ORSAY)

We consider various distributions of physical parameters for relativistic-pair
production in $\gamma\gamma$ collisions (ee$\to$ eex^+x^-) expected in no-tag or anti-tag experiments,
taking into account experimental constraints. Invariant mass, visible energy and
transverse momentum distribution are derived analytically, with use of the double
equivalent-photon approximation, in an easily computable integral form, at least
whenever the Q^2-dependence of the $\int$ cross section can be neglected.

1.5 A Detailed Analysis of the Reaction $\gamma\gamma \to \pi^+\pi^-\pi^+\pi^-$

M.WOLLSTADT (TASSO)

Below 2 GeV invariant mass the reaction $\gamma\gamma \to \pi^+\pi^-\pi^+\pi^-$ is dominated by the production of two neutral ρ^0-mesons. Results of a detailed partial wave analysis of
this process are already published (Z.Phys. 16 (1982) 13). With the now available

integrated luminosity (80 pb^{-1} instead of 41 pb^{-1}) the analysis was redone. The results of the publication were confirmed: dominant contributions from the negative parity states $J^P = 0^-$ and $J^P = 2^-$ to the reaction $\gamma\gamma \to \rho^0\rho^0$ are excluded. The data indicate sizeable contributions from $J^P = 0^+$ for $M_{4\pi} < 1.7$ GeV and from $J^P = 2^+$ for $M_{4\pi} > 1.7$ GeV. The data are also well described by a model with isotropic and uncorrelated isotropic decay of the ρ's. Cross sections for the process $\gamma\gamma \to \rho^0\rho^0$ are determined.

1.6 A Simple Method to Detect Interferences in $W \to x^0(\to \pi^+\pi^-) + x^0(\to\pi^+\pi^-)$ near Threshold and its Application to $\gamma\gamma \to \rho^0\rho^0$

A.SHAPIRA (TASSO)

Interesting information on the $\gamma\gamma \to \rho^0\rho^0$ reaction can be inferred from the measurement of simple invariant quantities like $\mu_+ = (m_{++}^2 + m_{--}^2)/(\Sigma m_{ij}^2)$ and similar ones, where m_{ij} is the invariant mass of particles i and j and the sum is over the six possible combinations (m_{++}, m_{--} and four m_{+-} combinations) for the $\pi^+\pi^-\pi^+\pi^-$ final state.The average values of these quantities as a function of the four-pion mass are sensitive to interferences and will behave differently for the cases: (a) phase space, (b) resonance production with free and uncorrelated decays or (c) $\rho^0\rho^0$ production with a Bose-symmetrized amplitude. Particular low μ_+ values are expected in case (c) when the $\rho^0\rho^0$ are produced around and below threshold. The results are insensitive to possible ρ^0 decay distributions. Application to the TASSO $\gamma\gamma \to \rho^0\rho^0$ data indicate significant deviations from free decay which are roughly consistent, but probably larger than expected for a simple symmetrized Breit-Wigner amplitude.

1.7 The Reaction $\gamma\gamma \to \pi^+\pi^-\pi^+\pi^-$ Measured with the CELLO-Detector

H.BEHREND (CELLO)

The reaction $\gamma\gamma \to \pi^+\pi^-\pi^+\pi^-$ was measured with the CELLO-detector at PETRA. Results on a model free calculation of the 4π-cross section are given. Fits of the two-dimensional distributions of the two pion invariant masses to contributions from $\rho\rho,\rho$ 2π and 4π show that only 40% of the cross section are due to the $\rho\rho$ process. The angular distributions of the 2π-rest system, are given. For forward produced 2π-system the angular distribution is compatible with helicity conservation as is expected from production of a $\rho\rho$-pair.

2. RESONANCES

2.1 Observation of the Two-Photon Production of η Mesons by the CRYSTAL BALL at SPEAR

A.J.WEINSTEIN (CRYSTAL BALL)

Using 2700 nb^{-1} of data taken at $\sqrt{s}$ from 5.00 to 7.25 GeV with a new trigger sensitive to decays of lower mass particles produced in two-photon collisions we have observed 56 $\pm$ 12 η events consistent with production in the reaction $e^+e^- \rightarrow \gamma\gamma e^+e^- \rightarrow \eta\ e^+e^-$. Background has been subtracted using separated beam data. We obtain $\Gamma_{\gamma\gamma}$ (η) = 0.56 $\pm$ 0.12 $\pm$ 0.09 and Θ_{PS} = -17.6° $\pm$ 3.0° which agree within 2 standard deviations with 1974 Cornell results using the Primakoff effect.

2.2 Production of η'(958) in Photon-Photon Scattering. Preliminary TASSO Results

R.MIR (TASSO)

The reaction $\gamma\gamma \rightarrow \eta' \rightarrow \rho^0\gamma$ was measured in the final state $\pi^+\pi^-\gamma$ with the TASSO detector at PETRA. Two different measurements are reported: One in which the photon is detected in the liquid argon barrel calorimeter (LABC), and the second in which the photon is detected in the lead-scintillator shower counter (HASH). The integrated luminosity is 65.8 pb^{-1} (LABC) and 75.6 pb^{-1} (HASH). The integrated luminosity is 65.8 pb^{-1} (LABC) and 75.6 pb^{-1} (HASH), mainly at beam energies around 17 GeV. The cuts used were: 1) acoplanarity $(\pi^+\pi^-)$ > 10°; 2) 1γ only: E(γ) > 0.1 (0.16) LABC (HASH) GeV; 3) sum (p transverse of $\pi^+\pi^-\gamma$) < 0.1 GeV; 4) ABS (COS (angle between the π^+ (in the $\pi^+\pi^-$ rest frame) and the $\pi^+\pi^-$ direction)) < 0.85. The η'(958) is observed, and its radiative width is measured (preliminary values): $\Gamma_{\gamma\gamma}$(η') = 4.1 $\pm$ 0.5 (stat.) $\pm$ 1.5 (syst.) keV (LABC) and $\Gamma_{\gamma\gamma}$ (η') = 4.2 $\pm$ 0.7 (stat.) $\pm$ 1.5 (syst.) keV (HASH).

2.3 Production of the f_0(1270) Meson in γγ-Collisions

F.KOVACS (CELLO)

f_0 production in two photon collisions has been observed in the CELLO detector at PETRA. The $\pi^+\pi^-$ final state was observed. The f_0 peak was found to lie on a dipion continuum and to be shifted downwards in mass by $\approx 60 \text{MeV/c}^2$. The entire $\pi\pi$ mass spectrum from 0.8 to 1.5 GeV/c^2 was found to be well fitted by the model of Mennessier using only a unitarized born amplitude and an f_0 amplitude. The previously observed mass shift and distortion of the f_0 amplitudes is explained by strong interference between the Born and f_0 amplitudes. The only free parameter in the fit to the model is the radiative width $\Gamma_{\gamma\gamma}(f_0)$. It was found that: $\Gamma_{\gamma\gamma}(f_0) = 2.9 \pm 0.3 \pm 0.2$ keV where the first (second) quoted errors are statistical (systematic).

2.4 Single Resonance Production in Two-Photon Interactions

M.ZACHARA (PLUTO)

Preliminary results on $\eta'(958)$ and $f^0(1270)$ production in two-photon interactions are presented. A production of the η' meson in untagged $\gamma\gamma$ interactions has been studied. A clean signal is seen in the $\pi^+\pi^-\gamma$ mass distribution. A first estimate of $\Gamma_{\gamma\gamma}(\eta')$ agrees well with published results. A precise measurement is expected soon, upon completion of an acceptance study. The Q^2 dependence of $f^0(1270)$ production has been investigated using singly tagged events. The f^0 signal observed in the $\pi^+\pi^-$ final state, at an average Q^2 of 0.4 GeV^2, is found to be suppressed by at least a factor of 0.6 when compared to the untagged sample. A more exact determination of the Q^2 dependence of the radiative width of f^0, as well as those of η' and $A_2(1320)$, is under study.

2.5 Excitation of the Tensor Meson f'(1520) in Photon-Photon Collisions

L.KÜPKE (TASSO)

The exclusive production of charged and neutral kaon pairs in photon-photon collisions was observed with the detector TASSO. The data show a signal at the photon-photon center-of-mass energy of 1520 MeV, which is interpreted as production of the resonance f'(1520). The measurement of the decay width $\Gamma_{\gamma\gamma}(f')$ gives information on the quark content of the mesons in the $J^{PC} = 2^{++}$ nonet. Assuming the f' to be produced in a helicity 2 state, we determine $\Gamma_{\gamma\gamma}(f') \cdot B_{K\overline{K}}(f') = 0.11 \pm 0.02 \pm 0.04$ keV.

2.6 Study of the reactions $\gamma\gamma \rightarrow \pi^+\pi^-\pi^0$ and $\gamma\gamma \rightarrow \pi^+\pi^-2\pi^0$

J.OLSSON (JADE)

Exclusive final states with 2 charged pions and 2 or 4 photons were reconstructed in the JADE detector at PETRA. The $\pi^+\pi^-\pi^0$ final state is dominated by the reaction $\gamma\gamma \to A_2(1310) \to \rho^{\pm}\pi^{\mp}$. The radiative width $\Gamma_{\gamma\gamma}(A_2)$ was measured to be $\Gamma_{\gamma\gamma}(A_2) = 0.84 \pm 0.07 \pm 0.15$ keV (prel.). In the final state $\pi^+\pi^-2\pi^0$ the reaction $\gamma\gamma \to \rho^+\rho^-$ was searched for and upper limits for the cross section $\sigma(\gamma\gamma \to \rho^+\rho^-)$ were obtained. These limits exclude a number of models proposed for the low mass enhancement in the reaction $\gamma\gamma \to \rho^0\rho^0$. In the 4-pion final state upper limits were also obtained for the branching ratio $BR(f(1270) \to \pi^+\pi^-2\pi^0)$ and for the product $\Gamma_{\gamma\gamma}(\eta_c)$ $BR(\eta_c \to \pi^+\pi^-2\pi^0$).

2.7 Search for Unusual and Heavy States in $\gamma\gamma$-Collisions

U.KARSHON (TASSO)

Upper limits of the product of the $\gamma\gamma$ partial width times the branching ratio to various final states are given for the glueball candidates (1440) and $\Theta(1690)$. Assuming they are isoscalars with spin J = 0, we get with 95% confidence level (C.L.): $\Gamma_{\gamma\gamma}(\quad) \cdot B(\quad \to K\bar{K}) < 7.0$ keV (preliminary), $\Gamma_{\gamma\gamma}(\quad) \cdot B(\quad \to \rho\rho) < 3.0$ keV, $\Gamma_{\gamma\gamma}(\Theta) \cdot B(\Theta \to K\bar{K}) < 0.3$ keV, and $\Gamma_{\gamma\gamma}(\Theta) \cdot B(\Theta \to \rho\rho) < 3.6$ keV. A narrow structure ($\Gamma = 30 \pm 34$ MeV) is seen at 2.1 GeV in the 4π spectrum of $\gamma\gamma \to \pi^+\pi^-\pi^+\pi^-$ with a significance of 4.3 standard deviations. If interpreted as a resonance, we get $\Gamma_{\gamma\gamma}("2.1") \cdot B(4\pi) \cdot (2J + 1) = 1.25 \pm 0.5$ (stat.) ± 0.5 (syst.) keV. No internal structure is seen in the 4π system of the "2.1", and we obtain the 95% C.L. ratios $B(\rho^0\rho^0)/B(4\pi^{\pm}) < 0.6$ and $B(K_s^0 K_s^0)/B(4\pi^{\pm}) < 0.06$. Finally, 95% C.L. limits are given for various decay modes of the $\eta_c(2980)$, assumed to be an isoscalar with J = 0: $\Gamma_{\gamma\gamma}(\eta_c) \cdot B(\eta_c \to p\bar{p}) < 0.4$ keV, $\Gamma_{\gamma\gamma}(\eta_c)$ $B(4\quad) < 0.7$ keV, and $\Gamma_{\gamma\gamma}(\eta_c) \cdot B(K\bar{K}\pi) < 27$ keV (preliminary). Using the known η_c branching ratios, our best limit for $\Gamma_{\gamma\gamma}(\eta_c)$ is extracted from the $4\pi^{\pm}$ decay mode:$\Gamma_{\gamma\gamma}(\eta_c) < 36$ keV (95% C.L.).

EXPERIMENTAL SESSION II, Thursday 14.4.1983

3. TWO-PHOTON PHYSICS AT LEP

3.1 Plans for Small Angle Tagging at LEP

D.MILLER (LONDON)

Available information on LEP tagging detectors is reviewed. Some physics possibilities will be discussed.

4. TOTAL CROSS SECTION

4.1 Difficulties and Inherent Problems in the Determination of $\sigma_{\gamma\gamma}(W)$

N.WERMES (TASSO, now at SLAC)

A detailed investigation of the systematic uncertainties and the model dependence of a measurement of $\sigma_{tot}(\gamma\gamma \to$ hadrons) is presented. The analysis has been carried out by the TASSO Collaboration. Particular emphasis is laid on the attempt to determine the explicit $W_{\gamma\gamma}$-dependence. The study was induced by the discrepancies observed between the first PLUTO and TASSO analyses.

5. HIGH P_T PHENOMENA

5.1 Hard Processes in $\gamma\gamma$-Collisions

E.DUCHOVNI (TASSO)

The point-like structure of the photon is expected to be seen by the presence of two high p_t jets in $\gamma\gamma$-collisons as well as other processes characterized by high p_t particles in the final state. Recently experimental reports (JADE, PLUTO, TASSO) dealing with such final states were published. In these papers the single tag mode was used in order to keep the data as clean as possible by reducing the background of the e^-e^+ annihilation process. This leads to very limited statistics. We report here on a similar study that was done by the TASSO Collaboration at DESY using the no-tag mode. Although the background problems are more intricate, the statistics are much higher (roughly 50 times higher than the publsihed one) and after a thorough study of the possible backgrounds we believe that using the right cuts, reliable results can be obtained. At the present moment the data cannot be fully explained by the simplest $\gamma\gamma \rightarrow q\bar{q}$ diagram (nor in the single tag mode), but the presence of a large high p_t tail in $\gamma\gamma$ collision is clearly proven. In this tail the cross section goes down as p_t^{-4} and the jet structure is clearly exhibited by the sphericity distribution.

5.2 High p_T Jets in the PLUTO Detector

SUSAN CARTWRIGHT (PLUTO)

We have studied high p_T hadron production using single tagged events with $0.1 < Q^2 < 15$ GeV^2. We observe in both the inclusive and the jet p_T2 distribution a high p_T tail which is not described by the VDM model. The number of events in this tail approaches the QPM prediction from above as p_T increases at fixed Q^2, and also as Q^2 increases at fixed p_T. Selecting events where the jet p_T in the hadronic c.m.s. exceeds 2 GeV, we find that the thrust distribution of such events is consistent with QPM Monte Carlo predictions for $Q^2 > 5$ GeV^2, but disagrees with expectation for $Q^2 < 1$ GeV^2.

5.3 Search for High p_T Jets in $\gamma\gamma$-Collisions without Photon Tag

H.LIERL (CELLO)

We studied the production of multihadronic events in photon-photon scattering without a photon tag in the CELLO detector at the e^+e^- storage ring PETRA. Events with a clear 2-jet structure are observed. Their production rate is higher than expected from the process gamma-gamma into quark-antiquark. P_T-distributions of hadrons and jets are shown and high p_T tails are discussed.

6. PHOTON STRUCTURE FUNCTION

6.1 Preliminary Results on Double Tag Events

St.J.MAXFIELD (PLUTO)

A preliminary investigation of the process e^+e^- e^+e^- + hadrons where both scattered leptons are detected ("double-tagged" events) has been made. About 250 such events have been observed with the PLUTO detector. The gross features of the events are accounted for by a sum of quark-parton and generalised vector dominance model expectations. An asymmetry in the distribution of the angle between the lepton scattering planes is seen (with low statistical significance) and is consistent with QPM expectations. A structure function, F_2 for an average target photon mass of 0.4 GeV has been extracted and is consistent with next-to-leading-log box diagram predictions.

6.2 A Method to Unfold the Photon Structure Function $F_2(x)$

V.BLOBEL (PLUTO)

The analysis of structure function data is generally complicated by the limited
acceptance and finite resolution of the detectors. The calculation of the acceptance and resolution usually requires the Monte Carlo simulation of the
full process in the detector. A general unfolding method based on statistical
principles, which includes a regulation to suppress insignificant local oszil-
lations of the result, es explained. The application to the unfolding of the photon
structure function $F_2(x)$ is described.

6.3 Experimental Study of the Photon Structure Function $F_2(X,Q^2)$ in the High Q^2 Resion

F.J.KIRSCHFINK (TASSO)

Data on deep inelastic electron-photon-scattering were taken with the TASSO detector
at beam-energies of 17.0 - 17.4 GeV. Only such events have been considered where on
of the scattered electrons is detected in the liquid argon endcap shower
counters. After background subtraction 216 events were observed with an average Q^2
of 23 GeV^2. The selected data are investigated in terms of the photon structure
function, F_2, and compared to different model calculations. All event distributions
are in good agreement with QPM- and Lowest-Order QCD predictions, including
c-quark contributions.

6.4 Experimental Study of the Photon Structure Function with the CELLO Detector at PETRA

G.CARNESECHI (CELLO)

A measurement of the processes: $e^+e^- \rightarrow e^+ e^- + L^+ L^-$ (L = electron or muon)
and $e^+ e^- \rightarrow e^+ e^- +$ hadrons has been carried out in the single tag condition,
using the CELLO detector at PETRA, at a mean beam energy of 17 GeV. Data is interpreted in terms of deep inelastic scattering of electrons on quasi-real photons:
$e + \gamma \rightarrow e + X$ (X = L^+L^- or hadrons). The off-mass shell effects of the target photon are taken into account as well as the radiative corrections to the scattered
electron. 130 electron pairs, 110 muon pairs, and 215 hadronic events are observed at an average Q^2 of ~ 9 $(GeV/c)^2$ and at an average W^2 of ~ 16 $(GeV/c^2)^2$.

Leptonic events agree with the exact Feynman graph calculation. Also the structure function F_2 of the photon when it fragments into $\mu^+\mu^-$ is compared with data and a good agreement is obtained. Hadronic events are compared to the calculation based on the estimation of the photon structure function F_2 in the quark parton model and in QCD, at the leading and next to leading order. Values of the QCD scale parameter Λ range between 0.07 and 0.28 GeV/c^2, depending on specific features of the model used.

6.5 Experimental Study of the photon Structure Function F_2 In the Q^2 Range from 10 to 200 GeV^2

T.NOZAKI (JADE)

An analysis of the process $e^+e^- \rightarrow e^+e^- +$ hadrons, where one of the scattered electrons is detected at large angles, was made with the JADE detector at PETRA. The analysis has been extended in statistics and Q^2 range from the previous one. The scattered electron was detected either by the endcap or by the barrel lead glass counters. 379 endcap events and 24 barrel events were observed at $\langle Q^2 \rangle$ = 24 GeV^2 and 110 GeV^2, respectively. The results were analysed in terms of the photon structure function F_2 and compared with QCD and QPM predictions. Leading order QCD describes the x_{vis} distributions well in the whole x_{vis} region. Higher order QCD and QPM describes the x_{vis} distributions well in the range $x_{vis} > 0.4$. The QCD parameter was determined to $\Lambda_{\overline{MS}} = 0.15$ (+0.07, -0.05) GeV. The Q^2 dependence of the function F_2 is described well in the Q^2 range from 10 to 200 GeV^2 by leading order QCD if the contribution from c quark production is included.

6.6. Measurement of the Photon Structure Function F_2 in the Intermediate Q^2 Range (1 - 16 GeV^2)

B.KING (PLUTO)

From the data taken with the PLUTO detector at PETRA during 1981/1982, events from the process $e^+e^- \rightarrow e^+e^- +$ hadrons have been observed, in which one of the scattered electrons is detected within the large angle tagger, whilst the other is scattered at sufficiently small angle to escape detection. The mean measured Q^2 for such events is 5.3 GeV^2. Such events can be considered in terms of the deep inelastic scattering of an electron off a real photon. The selected data have been used to investigate the photon structure function F_2 with particular emphasis on the effects of fragmentation on the measurement of F_2 from the visible x to the true x distribution.

6.7 Deep Inelastic Electron-Photon Scattering at Q^2 from 10 to 130 GeV^2

G.KNIES (PLUTO)

The PLUTO data analysis has been extended to include events in which the tagging electron has scattering angles from 21 to 39 degrees. They allow to compare deep inelastic e γ scattering to QED (77 events of the type ee → ee $\mu\bar{\mu}$), and to QCD (122 events of the type ee → ee + hadrons), via predictions for the photon hadronic structure function F_2 (x,Q^2). The $\mu\bar{\mu}$ data agree with the QED prediction. The hadronic data which have $\langle Q^2 \rangle$ = 45 GeV^2, are inconsistent with $F_2(x)$ constant or falling with x, but require a clearly rising behavior in x. This latter property is a signature of a running coupling constant α_s.

6.8 A new Determination of the Photon-Structure Function F_2

Frank A. Raupach

Due to the kinematical situation in photon-photon scattering efficiency losses are unavoidable. A consequence of this is that in many cases the reconstruction of the final state x is incomplete and so during the determination of the photon-structure function $F_2(x)$ large corrections are necessary (unfolding procedure). Searching for conditions which separate only $\gamma\gamma$-events which are fully reconstructed ("fully contained") in the detector one avoids these large corrections. A Monte-Carlo study for the Pluto detector shows that for example only charged balanced final hadronic states with more than three charged tracks give a sample of events which is a good approximation to total measured events.

It is interesting to compare the resulting photon-structure function $F_2(x)$ with that one getting from the complete data sample. Within the errorbars the results agree over the whole x-region but the corrections depending on unfolding decrease rapidly for fully measured events.

With the data sample of fully contained events one can calculate the charm contributions in the hadronic final state which are used to determine F_2. We require an additional cut in the invariant mass $W_{\gamma\gamma}$ of the $\gamma\gamma$-system of 1.5<$W_{\gamma\gamma}$<3.5 GeV. It turns out that the charm contributions are negligible ($\langle Q^2 \rangle$= 5 GeV^2).

1. Direct-Photon Pair Production

C. Carimalo, M. Crozon, P. Kessler and J. Parisi
(Laboratoire de Physique Corpusculaire, Collège de France, Paris, France)

Reactions of the type A B → γγX are studied. It is shown that those reactions should give about the same yields (under similar conditions) as the Drell-Yan process. The relative contributions of the $q\bar{q}$ and the gg mechanism (the latter proceeding via a quark box) are considered, and it is shown that, at very high energy and moderate invariant mass of the photon pair, the latter may become as important as the former. It is also shown that massive resonant structures coupled to two photons should be visible in the γγ spectrum, provided Γ (Res. → γγ) > 1 keV. Finally, the background problem, due to indirect photons mostly originating from π^0's, is studied; it is shown that, measuring two photons emitted at large angle, with large, opposite and equal transverse momentum, without any accompanying hadrons, it might be possible - at collider energies and moderate photon-pair invariant masses - to reduce that background to the level of a few percent.

2. Signatures For Supersymmetry In Two-Photon Processes

E. Reya (Institut für Physik, Universität Dortmund, 4600 Dortmund 50, West-Germany)

The effects of supersymmetry (SUSY) are studied in 2γ processes, which are the dominant reactions at LEP energies, in particular for the photon structure function $F_2^\gamma(x,Q^2)$. Associated squark ($\tilde{q}$) production in the threshold region ($s_{\gamma\gamma}, Q^2 >> 4m_q^2$) amounts to about 20% as compared to the dominant quark contribution which also will be modified by SUSY QCD to become softer, i.e. more depleted at large x (≥ 0.4) than in standard QCD. These combined effects result in a flatter $F_2^\gamma(x,Q^2)$ in the large-x region than in standard QCD and in an increase of F_2^γ for decreasing values of x. Such effects should be observable at LEP provided squark masses are not larger than of the order of $m_{W^\pm}/2$; the absence of these effects puts lower limits on squark (slepton) masses.

3. γγ → Squarks and Sleptons

J.A. Grifols and J. Solã(Department de Física Teórica Universitat Autònoma de Barcelona BELLATERRA (Barcelona) Spain

High energy γγ-type reactions are investigated as a means for producing pairs of scalar superpartners of leptons and quarks. The calculated rates are within detectability limits. A detailed discussion of both signatures and background is given. It turns out that for a small enough supersymmetry breaking scale √d̄ and for a sufficiently light photino, one should be able to see photons coming from photino decay(since $\tilde{f} \to f\tilde{\gamma}$). In models where √d̄ is large, one might rather expect the sfermion to be almost stable and thus be directly detected.

4. Radiative Correction to Two Photon Processes

F.A. Berends, P.H. Daverveldt, R. Kleiss (Instituut-Lorentz, Leiden, (Netherlands)

Two event generators have been constructed for the computer simulation of the two photon processes $e^+e^- \to e^+e^-\mu^+\mu^-$ and $e^+e^- \to e^+e^-\mu^+\mu^-(\gamma)$. Soft and hard radiation from the external electron and positron lines, vertex corrections on the electron and positron vertices, electron and positron self energy corrections and the vacuum polarization are taken into account.

The simulation programs can deal with any kind of experimental set-up. The second type of program gives a small radiative correction (1%) for the total cross section. In general the radiative correction can be sizeable depending on the applied experimental cuts.

The program can easily be changed to the simulation of $e^+e^- \to e^+e^-q\bar{q}(\gamma)$ which is useful for the study of high p_T jets in two photon physics.

5. Radiative Corrections to Two Photon Physics

W.L. van Neerven and J.A. M. Vermaseren (NIKHEF-H, Amsterdam)

We have calculated radiative corrections to the reactions $e^+e^- \to e^+e^-X$ in which X is a pointlike pseudoscalar particle of arbitrary mass M. Only the corrections to the eeγ vertices were taken into account. The in and out state interactions between the electron and the positron were ignored, due to their complexity.

The result are that the total cross-section increases by a little over 1% at PETRA and PEP energies and by about 2% at LEP energies when we take M between 0.5 and 5 GeV. The corrections to $d\sigma/dQ^2$ can be larger. Between 1 GeV² and 100 GeV² they are about: +3% to +5% at PETRA and PEP energies and +5% to +6% at LEP energies. This does not include detector effects that occur when the photon is not observed.

6. Radiative Quarkonium Decays and $\gamma^-\gamma^-$ Collisions

J. Kühn (Institut für Theoretische Physik, RWTH Aachen, Germany)

In the framework of perturbative QCD we calculate exclusive radiative quarkonium decays. The coupling of the $^3S_1$ state to the photon and the two gluons is given by the standard Ore Powell matrix element, the coupling of the two gluons to the light meson is related to its two photon amplitude. $\Gamma(J/\psi \rightarrow f$ and $f')$ is found in good agreement with experiment [1]. For decays into light pseudoscalars the nonrelativistic approximation can be avoided [2] and the absolute rate is predicted in terms of $<0/\delta^\mu A /PS >$. In lowest order α_s we find a different scaling behaviour of the ratio

$$\frac{\Gamma(^3S_1 \rightarrow PS+\gamma)}{\Gamma(^3S_1 \rightarrow e^+e^-)} \quad \frac{\Gamma(J/\psi \rightarrow PS+\gamma)}{\Gamma(J/\psi \rightarrow e^+e^-)} = (M_{J/\psi}/M_{^3S_1})^n$$

for $q\bar{q}$ components (n=2) and gluon components (n=6) of the pseudoscalar wave function [2]. The relatively low experimental bound for $\Psi' \rightarrow \gamma\iota$ suggests, that ι might be dominantly a glueball, in contrast to η' with a scaling behaviour consistent with n=2.

1) J.G. Körner, J.H. Kühn, H. Schneider, Phys. Letts. 120B,444(83)
2) J.H. Kühn, Phys. Letts., in press.

7. Exclusive Processes in QCD: Two-Photon Physics Involving Baryons

P.H. Damgaard (Laboratory of Nuclear Studies, Cornell University, Ithaca, New York, USA)

Exclusive two-photon processes involving baryons have been calculated within the framework of perturbative QCD. In particular, the angular dependence of the process $\gamma\gamma \rightarrow p\bar{p}$ is predicted as a function of the universal baryonic distribution amplitudes $\phi(X_i)$. The overall normalization is then found by a comparison with the process $\psi \rightarrow p\bar{p}$, so that we can present a unique, absolutely normalized QCD prediction for the two-photon processes. We argue that these predictions, based on the $\psi \to p\bar{p}$ decay rate, should be more reliable than those based on baryonic form factors. Presently not enough data exist to give a meaningful differential cross section for the process $\gamma\gamma \to p\bar{p}$. Nevertheless, a total cross section has been obtained. Excluding the region close to threshold, quite good agreement between our predictions and experimental data is found. The analysis in this paper is trivially generalized to other baryons as well.

8. Exclusive Production of Heavy Mesons in Photon-Photon Collisions: The Double-Scattering Mechanism

S.L. Grayson, R.R. Horgan and P.V. Landshoff (Department of Applied Mathematics and Theoretical Physics, University of Cambridge, Silver Street, Cambridge CB3 9EW U.K.

We present two mechanism which can compete with constituent interchange in the exclusive production at wide angle of heavy mesons in photon-photon collisions. We show that they have the same energy behaviour as constituent interchange, and that their amplitudes are free from large logarithms. We exploit this to calculate the double-scattering diagrams numerically. We find that while constiuent interchange dominates at 90°, double scattering becomes dominant for angles less than about 45°.

9. Photon-Photon Reactions Wihtin The Context Of The Generalized Vector Dominance Model

E. Gotsman and U. Maor (Department of Physics and Astronomy, Tel Aviv University, Tel Aviv

We show that the General Vector Dominance Model (GVDM) provides a reasonable fit to the presently available data, both for photon-photon cross section $\sigma_{\gamma\gamma}$ and for the photon structure function F_2^γ for values of $Q^2 < 25$ GeV^2. We emphasize that the maxima that appears in F_2^γ for large x, is a kinetic effect i.e. a consequence of the threshold enhancement of $\sigma_{\gamma\gamma}$ and not an intrinsic prediction of QCD. The GVDM approach allows us to circumvent the problem of double counting concerning the hadron and point-like pieces of the photon, and yields a value of $b_2 = -0.04$, for the undermined coefficient of the second moment of F_2^γ.

10. A Hadronic Point of View for γγ-Interactions

E. Etim and E. Massō ((Cern), Genf and L. Schülke (Siegen), Germany)

One may describe photo-induced processes in two alternative (dual) ways: coupling the photon to $q\bar{q}$-pairs (continuum way) or to an infinite set of vector mesons (discrete way). We have shown the validity of this idea (Q^2-duality) for deep inelastic scattering on photons. Indeed, in the framework of an extended VMD, the photon structure function, F_2 (x, Q^2), is simply related to the vacuum polarization amplitude, and the scaling analysis of F_2 (x, Q^2) leads to its scale log-breaking for large values of Q^2. On the other hand, the extrapolation to low values of Q^2 may be analitically done, giving a prediction for all values of Q^2, in satisfactory agreement with the PETRA data.

11. Photon Structure Function for Massless Quarks

M. Gorn (Universität München, Germany)

The possibility of defining a photon structure function F_2 for massless quarks is investigated. The mass singularity being created by the kinematicel configuration of a $q\bar{q}$ pair parallel to the photon - photon axis, the log Q^2/m_q^2 can be regularized for $m_q = 0$ by means of a polar angle cut-off for very narrow jets if both of them are observed. In case only one jet is observed ($x_{vis} = 1$) there is however a kinematical region where the cut-off is irrelevant and the singularity remains. In both cases the obtained F_2 differs substantially from the massive parton result.

12. Determining the Quark Charges by a Two Photon Process

A. Janah (University of California, Irvine, USA)

Two photon processes such as $e^+e^- \rightarrow e^+e^-$ + 2 jets distinguish between the gauge theories of integer and fractionally charged quarks (icq and fcq) even below the threshold energy for the production of color non-singulet states. To obtain the icq predictions for this process, one must take into account (a) the momentum dependent color suppression effect and (b) the added contribution from pair production of charged gluons. This is done, and it is observed that: (i) for icq, the ratio R(γγ, 2 jet) is not simply a number given by the quark charges; it depends on the gluon mass, on kinematics and on the particular differential cross-section considered; (ii) the deviation of icq cross-section from the fcq values depends crucially on whether "untagged"

events are included; if this is done, the deviation is large; the
charged gluon contribution is mainly responsible for this deviation,
the quark contribution being smaller than naively expected. Finally,
these predictions are compared with experiment.

13. Heavy Quark Contributions to Photon Structure Functions

I. Schmitt (Universität Wuppertal), G. Köpp (RWTH Aachen), T.F. Walsh
(DESY, Hamburg) and P. Zerwas (RWTH Aachen)

We investigate the properties of the transverse and longitudinal photon
structure functions. Varying Q^2 from low to large values at fixed
Bjorken x, the charm threshold is crossed, and so we expect a substantial increase of the structure functions. In the quark parton model
which describes the gross features of the structure functions, we find
that for x between 0.4 and 0.8 significant contributions from charmed
quarks are already obtained at Q^2 = 20 GeV². To check the stability
of these results against QCD corrections we calculate gluon corrections
to the parton model for heavy quarks in the one loop approximation. We
find that QCD corrections are small in the energy region where multiple
gluon bremsstrahlung is negligible.

14. Testing Quantum Chromodynamics in Two-Photon Reactions

Stanley J. Brodsky (Stanford Linear Accelerator Center, Stanford, CA,USA)

Because of the simplicity of the initial state and the predicted slow
fall-off at large-momentum transfer, two-photon hadronic production
reactions provide important tests of coherent effects perturbative
QCD, e.g. the determination of mesonic and baryonic distribution amplitudes [1], and the determination of quark mass effects [2].

The following subjects are briefly discussed (1). An interpolating
formula[1] is given for the $\gamma^*\gamma \to \pi^\circ$ amplitude which connects the current
algebra Q^2 = 0 and asymptotic QCD large momentum transfer predictions
(2). The QCD coupling constant α_s (Q^2) can be measured at large Q^2 independently of the form of the pion distribution amplitude by combining
data for the $F_{\gamma\pi^\circ}$ (Q^2) transition form factor and the pion form factor[1]
(3). The large momentum transfer QCD predictions for $\gamma\gamma \to$ M$\bar{\text{M}}$ can be used
as a model for resonance background studies, as an alternative to the
usual point-like meson Born term models. This opens up the possibility
of constructing analytically correct theoretical models for $\gamma\gamma \to$ M$\bar{\text{M}}$
amplitudes from threshold to large invariant pair mass, including
the photon-mass and polarization dependence (4). An explicit demon-

stration of the presence of a J = 0 fixed Regge singularity in the
γM→γM QCD Compton amplitude for transversely polarized vector mesons
is given [1] (5). A simple QCD model is discussed which can be used
to study the relationship of vector-meson-dominated and point-like
contributions to the photon structure function.

1.) S.J. Brodsky and G.P. Lepage, Phys. Rev. $\underline{D24}$, 1808 (1981)
2.) Tao Huang and Zhui-jian Shi, preprint BIHEP-TH-82-17 (1982)

15. Radiative Decays of Tensor Mesons and SU(3)$_f$ Symmetry Breaking

Paul Singer (Dept. of Physics, TECHNION-Israel, Institute of Technology,
Haifa, Israel

A Veneziano-type dual amplitude for the scattering of pseudoscalar
mesons on vector mesons is constructed. The amplitude displays crossing
symmetry and Regge asymptotic behaviour determined by leading trajectories in all channels and is gauge invariant for massless vector mesons.
Exploiting the duality property we derive relations between the couplings of tensor mesons dominating the s-channel and those of vector
mesons contributing in the crossed channel.

The f°, A_2^o, f'→γγ decays are thus expressed [1] in terms of experimentally measured one-photon transitions of vector mesons to pseudoscalar
mesons, the calculation giving $\Gamma(f°→γγ)$ = 2.66±0.45 KeV; $\Gamma(A_2^o→\dot{γ}γ)$ =
0.90±0.36 KeV; $\Gamma(f'→γγ)$. $B(f'→K\bar{K})$ = 141±39 eV, in excellent agreement
with experiment. The model requires the SU(3)-flavor symmetry breaking,
as observed in vector meson radiative transitions, to determine the
pattern of symmetry breaking in radiative decays of tensor mesons.
Using the same procedure, we present also predictions [2] for the one-
photon decays T→Vγ of f°, A_2^o, K** and D** tensor mesons.

1.) P. Singer, Phys. Lett. $\underline{124B}$, 531 (1983)
2.) P. Singer, Phys. Rev. $\underline{D27}$, 2223 (1983)

List of Participants

Badier, J.	Palaiseau, France
Bardeen, W.	Fermilab, USA
Berends, F.	Leiden, Netherlands
Berger, Ch.	Aachen, Germany
Blobel, V.	Hamburg, Germany
Braunschweig, W.	Aachen, Germany
Brodsky, S.J.	SLAC, USA
Burkart, D.	Hamburg, Germany
Bussey, P.	Glasgow, England
Carnesecchi, G.	Saclay, France
Cartwright, S.	Glasgow, England
Colas, P.M.	Orsay, France
Cooper, S.	DESY, Germany
Coupland, M.	London, England
Courau, A.	Orsay, France
Dainton, J.	Glasgow, England
Damgaard, P.	NORDITA, Denmark
Daverveldt, P.-H.	Leiden, Netherlands
Deuter, A.	DESY, Germany
Donnachie, A.	Manchester, England
Duchovni, E.	Weizmann, Israel
Duke, D.	Florida, USA
Field, J.	Paris, France
Field, R.D.	Gainsville, USA
Frazer, W.R.	Berkeley, USA
Genzel, H.	Aachen, Germany
Gorn, M.	München, Germany
Gotsmann, E.	Tel Aviv, Israel
Grayson, St.	Cambridge, England
Grifols, J.A.	Barcelona, Spain
Grupen, C.	Siegen, Germany
Haissinski, J.	Orsay, France
Hilger, E.	Bonn, Germany
Janah, A.	Irvine, USA
Jarlskog, C.	Bergen, Norway
Kapusta, F.	Paris, France
Karshon, U.	Weizmann, Israel
Kastrup, H.	Aachen, Germany
Kellogg, R.	Maryland, USA

Kessler, P.	Paris, France
King, B.	Glasgow, England
Kirschfink, F.-J.	Aachen, Germany
Kleiss, R.	Leiden, Netherlands
Knies, G.	DESY, Germany
Ko, W.	Davis, USA
Köpke, L.	Bonn, Germany
Kolanoski, H.	Bonn, Germany
Kovacs, F.	Paris, France
Kück, H.	Bonn, Germany
Kühn, J.	Aachen, Germany
Landshoff, P.	Cambridge, England
Le Diberder, F.	Paris, France
Leu, P.	Hamburg, Germany
Lewendel, B.	DESY, Germany
Lierl, H.	München, Germany
Lillestol, E.	Bergen, Norway
Lübelsmeyer, K.	Aachen, Germany
Luit, E.-J.	NIKHEF, Netherlands
Maor, U.	Tel Aviv, Israel
Martyn, H.-U.	Aachen, Germany
Masso, E.	CERN, Switzerland
Menessier, G.	Montpellier, France
Meyer, J.	DESY, Germany
Miller, D.	London, England
Mir, R.	DESY, Germany
Missennier, G.	Paris, France
Nozaki, T.	Heidelberg, Germany
Olsson, J.	DESY, Germany
Orenstein, St.	Palaiseau, France
Parisi, J.	Paris, France
Petronzio, R.D.	CERN, Switzerland
Pfeil, W.	Bonn, Germany
Poggioli, L.	Paris, France
Pohl, M.	Aachen, Germany
Raupach, F.A.	Aachen, Germany
Renard, F.	SLAC, USA
Reya, E.	Dortmund, Germany
Rippich, Ch.	Pennsylvania, USA
Schmitt, I.	Wuppertal, Germany
Schmitz, D.	Aachen, Germany
Schülke, L.	Siegen, Germany

Scott, J.	Cambridge, England
Shapira, A.	Weizmann, Israel
Singer, P.	Haifa, Israel
Skillicorn, I.	Glasgow, England
Smith, J.	New York, USA
Söding, P.	DESY, Germany
Soergel, V.	DESY, Germany
Spitzer, H.	Hamburg, Germany
Stella, B.	Roma, Italy
Stirling, J.	Cambridge, England
Strauch, K.	Harvard, USA
Timm, U.	DESY, Germany
Tylka, A.	Maryland, USA
Vermaseren, J.	NIKHEF, Netherlands
Wacker, K.	SLAC, USA
Wagner, W.	Aachen, Germany
Wedemeyer, R.	Bonn, Germany
Weinstein, A.	SLAC, USA
Wermes, N.	SLAC, USA
Wollstadt, M.	Bonn, Germany
Xiao, Ch.-G.	DESY, Germany
Zachara, M.	DESY, Germany
Zerwas, P.M.	Aachen, Germany

Zeitschrift
für Physik C **Particles
and Fields**

 Europhysics Journal

Editors in Chief:
G. Kramer, II. Institut für Theoretische Physik,
Luruper Chaussee 149, 2000 Hamburg 52;
H. Satz, Fakultät für Physik, Universität Bielefeld,
Postfach 8640, 4800 Bielefeld 1

Aims abnd Scope:
Zeitschrift für Physik C, **Particles and Fields** is devoted to
the experimental and theoretical investigation of elementary
particles. In view of the steadily growing interplay of theory
and experiment in this field, particular emphasis is given to
a clear and complete presentation of research.

The topics covered include:

- Experimental and theoretical particle physics
- Structure of elementary particles
- High energy processes
- Strong, electromagnetic and weak interactions
- Symmetry principles
- Unification schemes
- S-matrix theory
- Quantum field theory
- Lattice field theory

Special features: Rapid publication, no page charge.
Language of publications: English.

Zeitschrift für Physik appears in three parts: A: Atoms and
Nuclei; B: Condensed Matter; C: Particles and Fields
Each part may be ordered separately.

Subscription information and/or **sample** copies are available
from your bookseller or directly from Springer-Verlag,
Journal Promotion Dept., P. O. Box 105280,
D-6900 Heidelberg, FRG

Orders from North America should be addressed to:
Springer Verlag Inc., Journal Sales Department,
175 Fifth Avenue, New York, NY 10010, USA

Springer-Verlag
Berlin
Heidelberg
New York
Tokyo

Lecture Notes in Physics

Selected Issues from
Lecture Notes in Mathematics